BIOLOGICAL STRUCTURE AND FUNCTION 8

THE MAMMALIAN SKULL

BIOLOGICAL STRUCTURE AND FUNCTION

---

1. *Biology of Bone*: N. M. HANCOX
2. *The Macrophage*: B. VERNON-ROBERTS
3. *The Integument*: R. I. C. SPEARMAN
4. *The Pituitary Gland*: R. L. HOLMES AND J. N. BALL
5. *The Mammalian Kidney*: D. B. MOFFAT
6. *Blood Vessels*: W. J. CLIFF
7. *Biology of Cartilage Cells*: R. A. STOCKWELL
8. *The Mammalian Skull*: W. J. MOORE

THE MAMMALIAN SKULL

W. J. MOORE

Professor in Anatomy, University of Leeds

CAMBRIDGE UNIVERSITY PRESS

CAMBRIDGE
LONDON NEW YORK NEW ROCHELLE
MELBOURNE SYDNEY

CAMBRIDGE UNIVERSITY PRESS
Cambridge, New York, Melbourne, Madrid, Cape Town, Singapore, São Paulo, Delhi

Cambridge University Press
The Edinburgh Building, Cambridge CB2 8RU, UK

Published in the United States of America by Cambridge University Press, New York

www.cambridge.org
Information on this title: www.cambridge.org/9780521113328

First published 1981
This digitally printed version 2009

A catalogue record for this publication is available from the British Library

ISBN 978-0-521-23318-7 hardback
ISBN 978-0-521-11332-8 paperback

CONTENTS

PREFACE

It was with a feeling akin to impudence that I took upon myself the task of attempting to give a unified account of the structure, evolution and functional adaptations of the mammalian skull. The principal reason for my misgivings was the range of topics that would have to be included – a range so vast that clearly no single person, and certainly not myself, could claim to have direct personal knowledge of more than a small part. I was encouraged to proceed, however, by the conviction that such an account is now needed to bring together the magnificent work of the early comparative anatomists and the findings of more recent, and no less impressive, researches in the fields of embryology, palaeontology and functional anatomy, as well as by the reflection that whoever had undertaken this task would have found him- or herself labouring under deficiencies similar to, although probably less severe, than my own. The only way out of this type of difficulty is by means of multiple authorship but this, to my mind, introduces more problems than it solves.

This work is, therefore, highly derivative, which is both a strength and a weakness – a strength in the breadth of attack that it has allowed; a weakness in the room for error that it has provided. It will probably be helpful for the reader to know, at the outset, that my own research interest is the development and growth of the skull and it is here that I write with most (though far from complete) authority. When dealing with other aspects of my subject I have had to rely, perforce, almost entirely on the published literature. I have striven for accuracy but doubtless errors remain. I ask the indulgence of those authors who find that I have misinterpreted their views.

I have many colleagues to thank for the direct or indirect help that they have given me in the preparation of this book. I owe a tremendous debt of gratitude to Professor Lord Zuckerman, OM, KCB, DSc, FRS, who first introduced me to the fascination of comparative anatomy and who taught me the little skill I have in critically synthesising scientific information. A debt of almost equal magnitude is owed to the team of workers that Lord Zuckerman gathered around him in the Birmingham Anatomy Department and who helped provide the exciting milieu in which I began my academic career. I should like especially to pick out for thanks Professor Eric Ashton, my colleague now for more than two decades. Over the years, I have derived much enjoyment and stimulation from conversation and correspondence with Dr P. Atkinson, Professor B. K. Hall, Dr R. Jakeways, Dr D. R. Johnson, Professor F. P. Lisowski, Professor M. L. Moss, Professor C. E. Oxnard and Mr T. F. Spence. I am particularly grateful to Dr Robert Presley for his help and the unstinting way in which he has allowed me access to unpublished material. My thanks are also due to Professor Robert Holmes and the staff of the Leeds Anatomy Department for having borne my two-year obsession with the skull without too obvious irritation. Particularly valuable help has been provided by the staff of the Medical and Dental Library of the University of Leeds by the diligence with which they have dealt with innumerable requests for inter-library loans, many of which were for old and now obscure works. I am also indebted to Mr H. Grayshon Lumby for preparing Figs. 27–29. Finally, but by no means least, I should like to thank Mrs Celia Peters for typing the manuscript and for her help in translating some of the more ancient German papers.

It just remains for me to add that the responsibility for any errors, misinterpretations or omissions is entirely my own.

September 1979 W.J.M.

ABBREVIATIONS

CRANIAL NERVES (OR THEIR FORAMINA)

I	= olfactory	VII H	= hyomandibular branch of facial
II	= optic	VII P	= palatine branch of facial
III	= oculomotor	VIII	= acoustic
IV	= trochlear	IX	= glossopharyngeal
V1	= profundus (trigeminal)	X	= vagus
V2	= maxillary (trigeminal)	XI	= accessory
V3	= mandibular (trigeminal)	XII	= hypoglossal
VI	= abducens		
VII	= facial		

OTHER ABBREVIATIONS

A	alisphenoid
AB	auditory bulla
AC	acrochordal cartilage
Add F	adductor fossa
AF	acoustic foramen
Ang	angular
Ang RL	reflected lamina of angular
Art	articular
Art DP	dorsal process of articular
AT	ala temporalis
AT Al	alar process of ala temporalis
AT Asc	lamina ascendens of ala temporalis
Bb	basibranchial
Bc C Ant	anterior basicapsular commissure
Bc F Ant	anterior basicapsular fenestra
Bcran F	basicranial fenestra
Bo	basioccipital
BP	basal plate
Bpt P	basipterygoid process
Bs	basisphenoid
Btr P	basitrabecular process
Bul	ethmoidal bulla
C2	secondary cartilage
CA	columella auris
CAD	dorsal process of columella auris
CA Exs	extrastapedial part of columella auris
Car C	carotid canal
Cb	ceratobranchial
Cb P	cribriform plate
CC	cranial cavity
CE	cavum epiptericum
Ch	ceratohyal
Ch L	ceratohyal ligament
Co	coronoid
C On	cavum orbitonasale
Cr G	crista galli
Cr P	crista parotica
C Sc	cavum supracochleare
D	dentary
DS	dorsum sellae
E	epipterygoid (intermediate part of palatoquadrate cartilage or ossification)

EAM	external acoustic meatus
Eb	epibranchial
Ect	ectopterygoid
EN	external naris
Ent	entotympanic
Eo	exoccipital
Es	extrascapular
F	frontal
FC	fenestra cochleae
FF1	primary facial foramen
FF2	secondary facial foramen
FS	frontal sinus
F So P	supraorbital process of frontal
FV	fenestra vestibuli
H	hyomandibula
Hb M	hypobranchial muscle
HH	hyoid process of hyomandibula
H Op	opercular process of hyomandibula
HQ	quadrate process of hyomandibula
HS	hiatus semilunaris
I	incus
ICA	internal carotid artery
IF	incisive foramen
IN	internal naris
Inf Con	inferior concha
Ip	interparietal
IS	interorbital septum
It	intertemporal
J	jugal (zygomatic)
L	lacrimal
L On	lamina orbitonasalis
M	malleus
Man	mandible
Mas	mastoid
Mas P	mastoid process
MC	Meckel's cartilage
Me	mesethmoid
MF	metotic fissure
Mid Con	middle concha
MS	maxillary sinus
Mt	maxilloturbinal
Mx	maxilla
N	nasal
NB	nasolacrimal bulla
N Cap	nasal capsule
N Cav	nasal cavity
Not	notochord
Nt	nasoturbinal
O	occipital
OA	occipital arch
OFA	olfactory foramen adhevens
OFE	olfactory foramen evehens
On F	orbitonasal fissure
Oo	opisthotic
Op	opercula
Orb	orbit
Orb C	orbital cartilage
Os	orbitosphenoid
Ot C	otic capsule
P	parietal
PA	pila antotica
Pal	palatine
Par P	paroccipital process
Pars	parasphenoid
Pb A	palatobasal articulation
Per	periotic
Pet	petrous
PF	parietal foramen
Phb	pharyngobranchial
Ph T	pharyngotympanic tube
Pit F	pituitary foramen
PM	pila metoptica
Pmx	premaxilla
Postf	postfrontal
Posto	postorbital
Postp	postparietal
Postt F	posttemporal fenestra
PP	pila prooptica
Pq	palatoquadrate cartilage (or ossified palatoquadrate complex)

Pq Asc ascending process of palatoquadrate
Pq Bas basal process of palatoquadrate
Pq Ot otic process of palatoquadrate
Pq Pal palatine process of palatoquadrate
Pq Q quadrate ramus of palatoquadrate
Prart prearticular
Prf prefrontal
Prfac C prefacial commissure
Prop preopercular
Pres presphenoid
Pro prootic
Ps planum supraseptale
Pt pterygoid
Q quadrate (or its precursor the quadrate part of the palatoquadrate cartilage)
Qj quadratojugal
QL quadrate ligament
RP retroarticular process
S stapes
Sbop subopercular
Sbt F subtemporal fossa
Sc R supracribrous recess
SDP dorsal process of stapes
Se sphenethmoid
Se C sphenethmoidal commissure
Sf C suprafacial commissure
Sm F stylomastoid foramen
Smx septomaxilla
So supraoccipital
Sp splenial
Sq squamous
SS sphenoidal sinus
St supratemporal
Sth stylohyal
Sup Con superior concha
Sur surangular
T tabular
TL transverse lamina
TM taenia marginalis
Tr trabecula
TT tegmen tympani
Ty tympanic
Tyh tympanohyal
UP uncinate process
V vomer

INTRODUCTION

AIMS

The purpose of this book is to present an account of the structural components, evolution and adaptive functional potential of the eutherian skull. Section I consists of a description of the basic elements that contribute to the structure of the skull and, in general terms, of the major steps in their phylogenetic and ontogenetic histories. Since this section inevitably includes a substantial amount of detailed and at times complicated morphology, I have repeated descriptions of certain essential facts and concepts in the hope that this will facilitate reading by eliminating the need for extensive cross-reference. Section II deals with the changes by which the skull of the mammal-like reptiles was progressively modified to give rise to the characteristic mammalian structure. In Section III a description is given of the wide range of functional modifications encountered in modern eutherian mammals in three regions of the skull (the masticatory apparatus, the middle ear structures and the nasal cavity) that are generally regarded as having been critical for the realisation of the full adaptive potential of this group. Section IV deals with the postnatal growth of the skull.

It is perhaps as well to state quite clearly what the book does not aim to do. It is not an anatomy text for the mammalian skull. Any attempt to include the vast amount of material necessary for such a purpose would have been prevented by considerations of space alone as well as running counter to the main theme of this work, which is to give a unified and coherent account of the structure and function of the skull uncluttered by morphological minutiae. For similar reasons, detailed accounts are not given of cranial ontogeny in individual mammalian species. In any case, this was done, in unsurpassable fashion, by de Beer (1937) in his classical work *The development of the vertebrate skull* which requires, as yet, no sequel. Although the masticatory apparatus figures large, descriptions of the dentition are limited to the bare few necessary for understanding the functional aspects of jaw usage since there are already available several excellent modern works (notably *Dental morphology and evolution*, edited by A. A. Dahlberg, 1971, Chicago, University of Chicago Press, and *Development, function and evolution of teeth*, edited by P. M. Butler and K. A. Joysey, 1978, New York and London, Academic Press) which give far more authoritative and comprehensive accounts of the dental structures than would be possible in the present work.

There are two omissions which I find more regrettable. The first is the lack of any attempt to trace the evolution of the skull within the mammalian orders. This was contemplated but rejected because of the very unequal nature of the information available. Thus the early phylogenetic history of the mammalian skull is poorly known whereas the later evolutionary changes in the skull of a few mammalian orders have been worked out in great detail. I thought it impossible, with knowledge in this state, to present a general and coherent account of later mammalian cranial evolution of the type appropriate to this book. Secondly, I regret the exclusion of the marsupials and monotremes which was necessitated entirely by limitations of space. To do justice to these fascinating creatures would require one, or possibly, two separate volumes.

Finally, in this list of disclaimers, I should point out that I have not tried to be inclusive – a clearly impossible aim – but have selected topics on the basis of what seems to me to be of general significance or particular interest.

TERMINOLOGY IN RELATION TO THE DEVELOPMENT OF THE SKELETON

Much confusion exists in the usage of terms such as dermal bone, membrane bone, cartilage bone and endochondral bone which are employed extensively in descriptions of the phylogeny and ontogeny of skeletal elements. This has arisen principally because of the widely disparate viewpoints and backgrounds of workers in these fields.

Patterson (1977) has recently put forward an admirably simple way of avoiding this confusion. His recommendations, with slight modification, have been used here and can be summarised as follows. It is first necessary to distinguish between the skeletal system, the skeletal organs (i.e. individual bones) and the skeletal tissues. So far as skeletal organs are concerned, the terms are used with the following meanings. A *cartilage replacing bone* is one preformed in cartilage and which first begins to ossify, during embryonic development, beneath the perichondrium and, somewhat later, also ossifies endochondrally (i.e. within the cartilage). Subsequently, a large part of a cartilage replacing bone may be formed by the activity of osteoblasts differentiating in the osteogenic membranes (periosteum and endosteum) covering its surfaces. Endochrondrally formed bone thus constitutes only a part of a cartilage replacing bone and clearly the terms endochondral bone and cartilage replacing bone cannot be used as though they are synonymous. A *membrane bone* is one which ossifies in membrane with no ontogenetic or phylogenetic connection with the ectoderm. Bone of this type may be found in the skull in sites where it is homologous with a cartilage replacing bone in more primitive forms but in which the cartilaginous stage has been secondarily lost. Such membrane bones are not preformed in cartilage in ontogeny but were so in phylogeny. Patterson also includes neoformations, such as intramembranously ossifying outgrowths from cartilage bones, and sesamoid bones under the heading membrane bone. *Dermal bones* are not preformed in cartilage, either in ontogeny or phylogeny, and are directly related to the ectodermal basement membrane by their surface coverings or are homologous with such bones in more primitive forms.

Cartilage replacing bones and membrane bones make up the endoskeleton; the dermal bones constitute the dermal skeleton (or exoskeleton).

Unfortunately, as Patterson points out, it is less easy to clear up the confusion in terminology relating to the skeletal tissues. The difficulty here is that the terms cartilage, membrane and dermal bone were coined to describe whole bones. To take the femur as an example – this is clearly a cartilage replacing bone but it is often described in texts of microanatomy as being composed, in the adult, entirely of membrane bone because during its growth the endochondrally ossifying bone is resorbed and replaced by bone formed periosteally or endosteally. On the few occasions in Sections I and II when discussion involves skeletal tissues or the processes of ossification, an attempt will be made to avoid confusion by using the localising terms perichondral, endochondral, periosteal, endosteal and intramembranous as precisely as possible.

HOMOLOGY

Section I of this work is to a considerable extent based on the assumption that it is possible to trace the basic elements of skull structure throughout, or at least for long periods of, vertebrate evolution. This assumption raises the general biological problem of homology which indeed underlies the whole science of comparative anatomy. It would be inappropriate to enter here into a long discussion of this rather controversial topic but briefly I have used the definition proposed by de Beer (1971) that 'homology between organs is based on their correspondence with representatives in a common ancestor of the organisms being compared, from which they were descended in evolution'. This, like all definitions of homology which are sufficiently explicit to have any value, contains an element of tautology in some of its applications in comparative studies but nevertheless seems to work reasonably well in the practical situation.

ANATOMICAL AND ZOOLOGICAL TERMINOLOGY

So far as possible the anatomical terminology used is that of modern human anatomy (specifically *Nomina Anatomica*, 3rd edn). In some instances, however, this terminology is inapplicable or inappropriate in the comparative context and here the terms in common use by workers in the particular field under consideration have been adopted.

In general, common names have been used for species where these are sufficiently accurate and likely to be known to the majority of readers. Elsewhere standard zoological nomenclature has been employed.

REFERENCES

In referring to original papers, I have striven to provide ready access to rather than an exhaustive bibliography of what is, without exaggeration, a vast literature. Thus, where an author has published extensively on a particular topic, I have referred either to his later papers (from which his earlier work can be traced) or to review articles. Similarly, I have not thought it necessary to refer readers in detail to much of the pre-1940 literature but instead have indicated sources which provide summaries and extensive bibliographies of this older work. Believing that most of my readers will be English speaking, I have chosen source references in that language wherever possible.

SECTION I

THE CRANIOGENIC ELEMENTS

1

HEAD MESODERM

SEGMENTATION IN THE HEAD REGION

HEAD SOMITES

Despite certain specialisations, such as its burrowing habit, the modern marine animal *Branchiostoma lanceolatum* (amphioxus) shows many of the morphological features believed to have characterised the primitive level of chordate organisation (Fig. 1). The adult form of this animal has an elongated body some 5 cm long, flattened from side to side, and stiffened by a median notochordal rod composed of cuticularised cells. Arranged along the body on each side of the notochord is a series of discrete blocks of muscle, the myomeres. The myomeres are bilaterally paired, each myomere being derived from the myotomic division of a somite and separated from its anterior and posterior neighbours by bands of connective tissue, the myocommas. The myomeres contract in an anteroposterior sequence with the two sides out of phase, the effect being to produce an S-shaped bend which travels along the body in a posterior direction, so imparting a backward momentum to the suspending medium and hence propelling the animal forwards.

Associated with this metameric segmentation of the musculature is a corresponding segmentation of the peripheral nervous system. Emerging from the dorsal nerve cord, opposite each segment, are paired motor and sensory nerves. The motor nerve, consisting of several roots, lies immediately opposite the myomere which it supplies. The sensory nerve emerges from a point which is slightly posterior and dorsal to the motor nerve, running out between the myomere of its own segment and that of the segment behind, and carries motor fibres for non-myotomal muscles as well as sensory fibres to the segment. Although the two nerves do not fuse to form a mixed spinal nerve, they are generally regarded as equivalent to the ventral and dorsal roots of the spinal nerves of vertebrates.

In amphioxus, metameric segmentation is apparent throughout the length of the body. Anteriorly, a further segmentation, having no obvious correlation with the metameric segmentation, is produced by the presence of numerous gill pouches which open internally into the pharynx. In the tissue between the gill pouches are rods of a cartilage-like material. Apart from this visceral arch material and the anterior extension of the notochord, amphioxus has no cranial skeleton.

In craniate vertebrates, apparent somitic segmentation ceases at the caudal

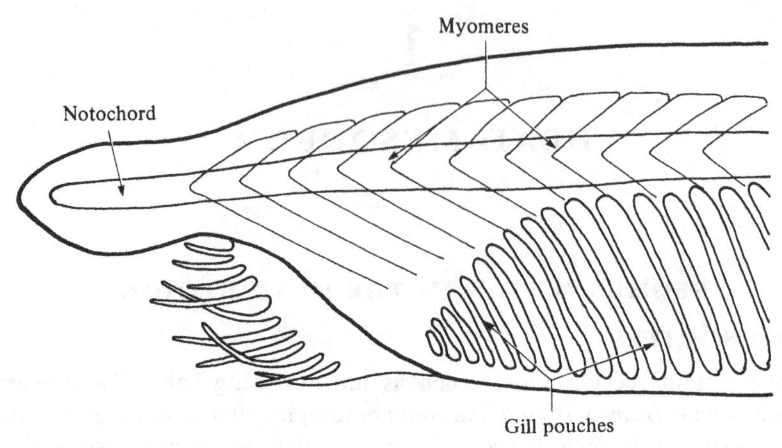

Fig. 1. Anterior end of amphioxus to show metameric and branchiomeric segmentation. (After Young, 1950.)

limit of the head, the only manifest segmentation in the head region being that produced by the gill pouches. But in view of the arrangement seen in amphioxus the question arises as to whether the head is a region that was never metamerically segmented, or whether primitively it was segmented like the remainder of the body and, if so, whether any trace of this segmentation remains in modern craniates. On superficial inspection the latter possibilities seem unlikely. In adult living forms the head muscles show no obvious segmental arrangement and the cranial nerves lack the orderly disposition seen in the spinal region. However, the many detailed analyses of the structure of the head region in developing vertebrates carried out in the nineteenth and early twentieth centuries (see de Beer (1937) for a summary and full bibliography of this work) produced evidence suggesting that an underlying metameric segmentation might indeed be present. Such analyses seemed at first to support the idea, based on Goethe's well-known 'Theory of the skull', that the cranium is composed throughout its length of fused vertebrae (see, for example, Owen, 1866) but compelling reasons for rejecting this 'Theory' as an oversimplification were put forward by Huxley (1859), who demonstrated that a major portion of the craniate skull is made up of the continuous chondrocranium and its replacing bones, which show no trace of a vertebral structure during their ontogeny. Huxley also provided evidence that the skull is at least as old, phylogenetically, as the vertebral column and that while both started from the same primitive elements they began at once to diverge.

Despite the rejection of the vertebral theory of skull structure, the fundamental basis on which it rests – that the head is segmental in origin – has persisted and has been shown to be at least partly correct. The basis for the modern view was provided by Balfour (1876–78), who demonstrated that the dorsal head mesoderm in *Scyllium* becomes segmented during embryonic

TABLE 1. *The derivatives and nerve supply of the head somites and visceral arches in* Squalus *according to the classical scheme of head segmentation* (*based on de Beer* (*1937*) *and incorporating Balfour's* (*1876–78*) *main conclusions derived from the study of* Scyllium)

Segment	Somite	Derivatives of myotome		Ventral nerve	Visceral arch	Branchial nerve	
1	1 (First prootic)	Inferior oblique; superior, inferior and medial recti muscles		Oculomotor (III)	Premandibular (?)	Profundus (V_1) (?)	
2	2 (Second prootic)	Superior oblique muscle		Trochlear (IV)	Mandibular	Maxillary and mandibular of trigeminal (V_2 and V_3)	
3	3 (Third prootic)	Lateral rectus muscle		Abducens (VI)	Hyoid	Facial (VII)	
4	4 (First metotic)	Disappears		—	First branchial	Glossopharyngeal (IX)	
5	5 (Second metotic)	Disappears (may develop a temporary myomere)		—	Second branchial	First branchial branch	
6	6 (Third metotic)	First permanent myomere			Third branchial	Second branchial branch + rudimentary dorsal ganglion	
7	7 (Fourth metotic)	Second permanent myomere	Hypoglossal (tongue) muscles	Occipital (hypoglossal) nerves	Fourth branchial	Third branchial branch + rudimentary dorsal ganglion	Vagus (X)
8	8 (Fifth metotic)	Third permanent myomere			Fifth branchial	Fourth branchial branch + rudimentary ganglion	
9	9 (Sixth metotic)	Fourth permanent myomere			—	Rudimentary dorsal ganglion	
10	10 (Seventh metotic)	Fifth permanent myomere		First mixed spinal nerve	—	First mixed spinal nerve	

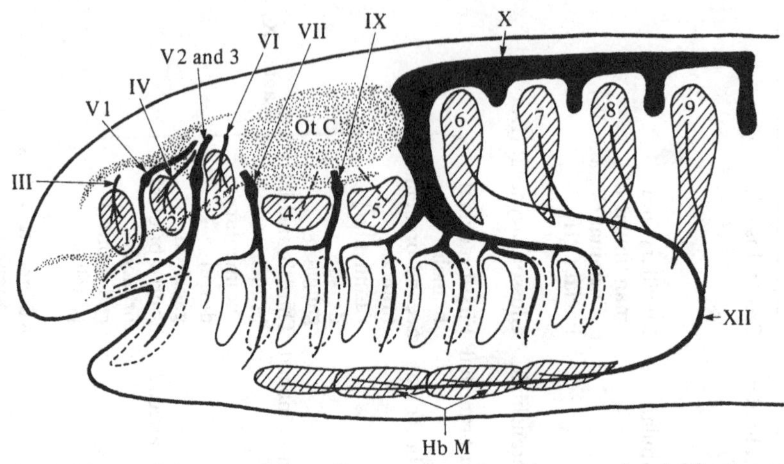

Fig. 2. Schematic representation of innervation of gnathostome head. Myotomes (and derived muscle) hatched; branchial arches (between gill pouches) indicated by dotted lines. Myotomes numbered (arabic numerals) according to the classical scheme. For labelling see list of abbreviations. (After Goodrich, 1930.)

development right up to its anterior end. This finding has been confirmed by later workers (e.g. Goodrich, 1918; de Beer, 1922, 1924) in the embryos of a variety of cartilaginous fishes. In more advanced vertebrates the head somites often appear in an irregular sequence, making interpretation difficult. Nonetheless, studies in those forms where development is favourable for analysis (e.g. *Amia* – de Beer, 1924; axolotl – Goodrich, 1911; duck – de Beer & Barrington, 1934; turtle (*Chelydra serpentina*) – Johnson, 1913; mouse – Dawes, 1930) have indicated that in these groups too the paraxial head mesoderm shows segmentation during development.

From this body of work a scheme of metameric segmentation in the gnathostome head emerged which was developed principally by Goodrich (1930) and de Beer (1937). The scheme can be summarised as follows (see also Table 1). The paraxial mesoderm becomes segmented to form three prootic and a series of metotic somites, all of which are serially homologous with the trunk somites. The myotomes derived from these somites are supplied by segmental ventral motor nerves (Fig. 2). The sclerotomic components of the somites contribute to the chondrocranium, the apparent lack of segmentation of which is, therefore, misleading. The lateral plate mesoderm, lying ventral to the paraxial mesoderm, is pierced by gill pouches with one such pouch associated with each paraxial segment. A branchial nerve, which is held to be the dorsal nerve companion to the myotomal ventral nerve, runs behind the myotome to the gill pouch. The branchial nerve contains a motor component for the muscles derived from the visceral arch posterior to the gill pouch. In the scheme of head segmentation, therefore, this arch is described as corresponding to the somite by whose segmental dorsal nerve it is supplied.

Although no arch is apparent corresponding to the first (first prootic) somite in modern gnathostomes, it is assumed that such an arch (termed premandibular) was originally present but has been lost as a separately identifiable structure. In consequence, the first segment is frequently termed the premandibular somite. The second (mandibular) and third (hyoid) somites correspond to the mandibular and hyoid arches respectively. The fourth (first metotic) somite is believed to have been crushed by enlargement of the otic capsule and its myotome fails completely to develop while the myotome of the fifth somite is, at most, a transitory structure. As will become apparent, several of the main propositions contained in this scheme have recently been challenged.

Marshall (1881) traced the origin of the extraocular muscles in cartilaginous fishes from the first three (prootic) somites. These muscles have, therefore, been generally taken by subsequent authors to be equivalent to myotomal muscles of the trunk. The superior, medial and inferior recti and the inferior oblique muscles are believed to be derived from the myotome of the first somite and the lateral rectus muscle from the myotome of the third somite. Hence the oculomotor and abducens are generally assumed to be the ventral nerves of the first and third segments respectively. Several workers (e.g. Johnson, 1913) have traced the superior oblique muscle from the second myotome, which suggests that its nerve, the trochlear, is the ventral nerve of the second segment, but this interpretation has been challenged (see below).

Although the origin of the eye musculature can be followed from the prootic segments in cartilaginous fishes and some lower vertebrates, this is usually not possible in mammals. Here, the musculature differentiates characteristically *in situ* from a sheet of mesenchyme which stretches forwards from the better developed somites of the metotic region. A further factor complicating analysis is the presumed loss in gnathostomes of the fourth and fifth myotomes. Nevertheless, the nerve supply of the mammalian extraocular muscles follows the same pattern as in lower vertebrates and it seems reasonable to assume that the mesoderm from which these muscles are derived is homologous with the prootic myotomes of the lower forms.

The myotomes of the more posterior (metotic) segments of the head region are supplied by ventral nerves which fuse to form the hypoglossal nerve. The muscles derived from these myotomes migrate into the floor of the mouth where they form the hypobranchial muscles of fishes and the musculature of the tongue in tetrapods. In higher vertebrates these muscles may appear to differentiate *in situ* as do the extraocular muscles.

The fate of the sclerotomic components of the head somites is even more difficult to trace. Theoretically they would be expected to contribute to the chondrocranium and there is evidence that they do so, at least to its posterior part which develops on either side of the rostral end of the notochord. The methods which have been used to study this aspect of skull development and the findings to which they have led are described in Chapter 2. It is sufficient here to mention that the number of segments contributing to the chondrocranium appears to be quite variable, a finding which is generally taken to indicate that the posterior limit of the skull has been transposed up and down the metameric segments during the course of evolutionary history. A similar transposition is seen in many other segmental structures, notably, and significantly in the present context, in the vertebral column where the boundary between the different types of vertebrae can be quite variable, even within a single species.

VISCERAL ARCHES

The possession of gill pouches at some stage in life is a vertebrate characteristic. The number of pouches varies widely, being generally greater in primitive than in more advanced vertebrates. According to the classical scheme of head segmentation, as expounded by Goodrich (1930) and de Beer (1937), the early agnathans are believed to have possessed a terminal mouth behind which was the series of bilaterally paired gill pouches piercing the lateral plate mesoderm intersegmentally and being innervated by the dorsal nerve of the segment in front. Within the lateral plate mesoderm located between the pouches the skeletal structures of the visceral arches developed to act as supports for the gill apparatus. The original first and second gill pouches were located, it is held, ventral to the intersegmental regions between the first and second and second and third somites, respectively. The visceral (premandibular) arch,[1] presumed to have been situated between these two pouches, would have been innervated by the dorsal nerve (the profundus) of the first paraxial segment. In gnathostomes the greatly enlarged mouth is believed to have extended posteriorly, obliterating the first and second pouches.

As already described, the premandibular arch is not apparent in modern gnathostomes, but its existence has been inferred from the presence during early development of the first or premandibular somite. Its skeletal element may be contained in the anterior part of the braincase. In the classical scheme of head segmentation the premandibular somite and arch are taken together to represent the first segment of the primitive chordate body. Yet a further arch may have been present, however, rostral to the first pouch and it has been suggested that this arch was innervated by the nervus terminalis (see below). In view of the recent criticisms of the classical scheme, yet to be

[1] As a consequence of this complex and still incompletely known evolutionary history, there is considerable variation in the naming of gill pouches and visceral arches. The system most widely used in gnathostomes, and the one which will be adopted here, is to term the arch from which the skeleton of the jaws and related structures is derived the *mandibular* or *first* arch and the subsequent arch the *hyoid* or *second* arch despite the fact that in the traditional view of the evolution of the arches the mandibular and hyoid arches are actually the second and third, respectively, not first and second, of the complete series postulated for ancestral forms. The arches posterior to the hyoid are termed, first, second, third, etc. *branchial* arches (Fig. 2). The arch often assumed to have been originally present anterior to the mandibular is termed the *premandibular*.

described, such obsessive attempts to equate the original branchiomeric and metameric segments are probably unjustifiable.

The skeleton of the mandibular arch is postulated as having been originally of the typical branchial arch type but to have become modified at the beginning of gnathostome evolution to form the skeletal elements of the upper and lower jaws (palatoquadrate and Meckel's cartilage, respectively, and their replacing bones). First arch mesoderm also gives rise to numerous dermal bones which complete the skeleton of the jaws in higher vertebrates. The hyoid arch skeleton is thought also to have undergone profound structural modification to form, in this case, the jaw suspension. As a result of these modifications the opening between the mandibular and hyoid arches, presumed to have been originally a typical gill pouch, has become restricted to a dorsal opening termed the spiracle. The size of the spiracle varies greatly between different species of fish, being greatest in bottom dwellers where it functions as a channel for the inflow of clear water to the gills.

The number of branchial arches varies from one vertebrate group to another. It is greatest amongst living forms in the cyclostomes, being eight in *Petromyzon* and up to fourteen in *Bdellostoma*. In fishes the number is generally five, while in amphibians it has been reduced to four and in amniotes to three or four. Although the arches retain their respiratory function in fishes, they have become greatly modified in land vertebrates to form skeletal structures of the pharynx and larynx.

The structure of the cranial nerves merits further consideration at this point because of the crucial part it has played in the analysis of head segmentation and the interpretation of cranial ontogeny and phylogeny. At first sight the cranial nerves of gnathostomes appear to be arranged in a more haphazard fashion than the spinal series. Their apparent lack of regularity may be explained on several grounds. First, the dorsal (branchial) and ventral nerves of the cranial region remain separate whereas in the spinal region they fuse in each segment to produce a mixed spinal nerve. Secondly, because of the complicated structure of the head and the fact that certain adjacent ventral nerves fuse with each other, as do some adjacent branchial nerves, the segmental correspondence of the ventral and branchial nerves is not always readily apparent. Thirdly, the branches of the cranial nerves vary in their number, distribution and types of constituent neurons, even between closely related forms. Fourthly, the cranial nerves contain a greater variety of functional types of neuron than do the spinal nerves because of the presence of the visceral arches in the head region. Finally, while the dorsal root of a spinal nerve is composed largely, if not entirely, of sensory fibres, the assumed equivalent branchial nerves of the cranial region contain, in addition, visceral motor fibres. In the first and last of these respects the structure of the cranial nerves is, in fact, nearer to the primitive condition than that of the spinal nerves, being similar to the arrangement found along the entire length of the body in amphioxus and the Agnatha.

The types of neuron within the cranial nerves can be analysed on a functional basis. This was done in great detail by Strong (1895) for amphibians and subsequently for a wide variety of other vertebrate groups by Herrick (1899), Landacre (1914) and Norris (1925) amongst others. From this comprehensive body of work it has been established that six types of neuron occur in cranial nerves, five of these being found in the branchial nerves and the remaining one constituting the ventral nerves. The five components of a complete branchial nerve are: (1) somatic sensory (general cutaneous, proprioceptive, and special cutaneous also termed acousticolateral); (2) general visceral sensory (from the lining of the alimentary canal); (3) special visceral sensory (taste); (4) general visceral motor (parasympathetic to glands and smooth muscle); (5) special visceral motor or branchiomotor (to muscle which, though derived from the lateral plate mesoderm of the visceral arches, resembles myotomal muscle in being striated in appearance and voluntary in function). The cell bodies of the sensory neurons in categories (1) to (3) are found in centrally located ganglia of the branchial nerves which are thus equivalent to the spinal dorsal root ganglia. The general visceral motor fibres synapse in more peripherally located ganglia while the special visceral motor fibres are non-synapsing. The ventral cranial nerves contain the remaining somatic motor component (to myotomal muscles); in the spinal region the ventral roots also contain the visceral motor fibres.

The branches of the branchial nerves vary from one species to another in which of the possible five functional components they contain. It is helpful, therefore, to begin an account of the branchial nerves by outlining the structure of a theoretically complete branchial nerve. Such a nerve gives rise to dorsal and lateral branches soon after its escape from the skull. It continues towards its gill pouch, gives off a pharyngeal branch, and then splits into pretrematic and posttrematic divisions (Fig. 2). The posttrematic division enters the visceral arch behind the pouch and divides into one internal and two external branches. The pretrematic division passes in front of the gill pouch where it divides into external and internal branches. The general and special cutaneous fibres (both somatic sensory components) are distributed by the dorsal and lateral branches, respectively; visceral sensory fibres pass in the pharyngeal branch and pre- and posttrematic divisions; visceral motor fibres are distributed to the visceral arch behind the gill pouch through the posttrematic branch of the nerve (because the posttrematic branch is generally much larger than the pretrematic, it is convenient and conventional to refer to the nerve supplying the posttrematic branch to an arch as the 'nerve of that arch' – that is, the branchial nerve is allocated to the posttrematic arch rather than to the gill pouch). This theoretically complete arrangement is rarely found; more frequently some components are absent or travel by alternative pathways, and branches are missing or fused.

A major advance in interpreting the pattern of distribution of the cranial nerves was achieved by Balfour's (1876–78) study of the development of the skull in the dogfish (*Scyllium*), which formed the basis of the classical scheme of head segmentation developed by Goodrich (1930) and de Beer (1937) (see Table 1 and Fig. 2). According to this scheme the profundus nerve is described as innervating the original first (premandibular) arch which is not seen as a separate morphological entity in any modern gnathostome (an alternative possibility, already noted, is that the nervus terminalis is the most rostral branchial nerve, although the associated visceral arch has never been identified in any chordate; this nerve is a minute filament in most higher vertebrates which runs in close association with the olfactory nerve to the mucous membrane of the anterior part of the nasal septum and probably contains sensory and visceral motor components). The profundus nerve has two main branches, the nasociliary and frontal. In mammals it has become incorporated in the trigeminal nerve as its ophthalmic branch, but in lower vertebrates it may be present as an independent nerve with its own separate sensory ganglion. Although an important nerve supplying a large area of skin on the front of the head with somatic sensory fibres, the profundus is obviously reduced, in comparison to a theoretically complete branchial nerve, in the number of different types of functional components that it contains. The ventral nerve corresponding segmentally with the profundus in the classical scheme is the oculomotor.

The maxillary and mandibular division of the trigeminal nerve are allocated to the mandibular arch. Both divisions contain somatic sensory fibres. The mandibular division, which may be equivalent to the posttrematic branch of a more typical branchial nerve, contains also special visceral motor fibres to the muscles derived from the mandibular arch. The visceral sensory component of the trigeminal nerve has been lost.

Balfour believed that all the extraocular muscles were derived from the first somite, but subsequent analyses (Marshall, 1881; van Wijhe, 1882) indicated that only the muscles innervated by the oculomotor are so derived and that the superior oblique muscle arises from the second somite and the lateral rectus from the third somite. If this interpretation is correct the trochlear is the ventral nerve to the second somite and, therefore, corresponds segmentally with the maxillary and mandibular divisions of the trigeminal nerve. However, the fact that the fibres of the trochlear emerge from the dorsal aspect of the brainstem has led to the view that this nerve is a detached visceral motor component of the trigeminal. Edgeworth (1935), in an extensive study of the development of the cranial muscles in vertebrates, returned to Balfour's original point of view. He observed that all the extraocular muscles in mammals, as well as in Dipnoi and amphibians, appear to derive from the first somite and suggested that the trochlear and abducens nerves are not segmental but should be regarded as separated portions of the oculomotor nerve. The difficulties involved in interpreting the development of the extraocular muscles have already been described and are doubtless responsible for these differing interpretations.

The hyoid arch is innervated by the facial nerve, the corresponding ventral nerve, according to the classical scheme, being the abducens. In teleosts the facial nerve contains all the components of a typical branchial nerve, but in tetrapods the general cutaneous and special cutaneous components have been reduced, although part of the latter may be included in the acoustic nerve supplying the internal ear. The main posttrematic or hyomandibular branch supplies special visceral motor fibres to the muscles derived from the second arch. In fishes this branch contains also special cutaneous and visceral sensory fibres. Neurons of the latter type are conveyed also in the pharyngeal (or palatine) branch of the facial nerve to the roof of the mouth. The pharyngeal branch in mammals, where it is known as the greater superficial petrosal, contains, in addition, general visceral motor fibres to the lacrimal gland and nasal mucosa. General and special visceral sensory and general visceral motor fibres form the internal mandibular branch which leaves the hyomandibular trunk to supply the mucous membrane and glands of the lower jaw and floor of the oral cavity. In amniotes the internal mandibular branch is termed the chorda tympani and may have fused with it the original pretrematic branch (the homologies of the chorda tympani are especially difficult to interpret).

The acoustic nerve is a development of the acousticolateral component of the branchial nerves. It seems likely, from similarities in their embryological development, that the internal ear represents a specialised deeply sunk part of the lateral line system, with the ampullae of the semicircular canals and the maculae of the utricle and saccule being modifications of the neuromast sensory organs found in the lateral line system. It is possible that the acoustic nerve represents parts of the acousticolateral components of the facial, glossopharyngeal and vagus nerves.

The glossopharyngeal nerve innervates the first branchial arch. In fishes it closely approximates the theoretically complete branchial nerve described above, the only component missing being the general cutaneous and even this is present in the cartilaginous fishes. In amniotes also, the general cutaneous component is absent. In mammals the posttrematic branch contains special visceral motor fibres (to the muscle of the first branchial arch) as well as general and special visceral sensory fibres to the posterior part of the mouth and tongue and to the pharynx. General visceral motor fibres (to the parotid gland) leave in the pharyngeal (tympanic) branch. The lack of a ventral nerve corresponding to glossopharyngeal is attributed, in the classical scheme of head segmentation, to the loss of the myotome of the first metotic somite crushed out of existence by enlargement of the otic capsule.

The remaining branchial arches are innervated by the vagus which is the biggest and most variable of the branchial nerves. According to the scheme of head segmentation there has been a partial polymerisation of the four segmental branchial nerves corresponding to the second to fifth branchial arches in which the special somatic sensory and general and special visceral (sensory and motor) components have been gathered forward to form the vagus. The general cutaneous component has been less completely gathered forward and the portion that remains unincluded makes up the vestigial dorsal ganglia which have been described as appearing (usually in a transitory fashion in mammals) in each vagal segment posterior to the first or second (e.g. Goodrich, 1918). The general cutaneous component of the vagus is usually small, being limited, in tetrapods, to the auricular branch. The special visceral motor branches are conveyed to the musculature derived from the second to fifth arches by posttrematic branches. These are represented in mammals by the pharyngeal, superior laryngeal and recurrent laryngeal branches. General and special visceral sensory fibres are conveyed in the fish by pre- and posttrematic branches at each gill slit. In land vertebrates the general visceral component is represented, in the head and neck region, by the sensory fibres to the mucous membrane of pharynx and larynx conveyed in the pharyngeal and laryngeal branches; the special visceral component innervates the taste buds at the laryngeal inlet. The largest branch of the vagus is the intestinal supplying general visceral sensory and motor fibres to the majority of the trunk viscera.

A characteristic of mammals is the presence of a spinal accessory nerve. This arises as a series of rootlets from the lateral part of the cervical spinal cord and ascends through foramen magnum to join the vagus. After a short distance it parts company again, passing posteriorly to innervate the trapezius and sternomastoid muscles. The nature of the spinal accessory nerve and its associated musculature is uncertain. Probably the most widely held view is that the nerve represents a detached component of the vagus. Straus & Howell (1936), for example, described the trapezius and sternomastoid as being closely associated with the posterior visceral arches during embryonic development in a variety of vertebrate types. If these muscles really are of visceral arch origin it would imply that their motor nerve supply belongs to the branchial series. Straus & Howell suggested that this is, indeed, the case, the spinal accessory being, in their opinion, a detached special visceral motor component of the vagus and its nucleus of origin a part of the dorsal motor nucleus which has migrated caudoventrally. They further suggested that the spinal accessory was originally detached as a mixed nerve whose sensory fibres have been lost by migration into the dorsal roots of the adjacent cervical spinal nerves. These suggestions explain the double innervation of the trapezius and sternomastoid without the necessity of invoking a myotomic contribution to the muscles to account for the spinal nerve component. An alternative view is that the spinal accessory is the modified ventral root (or roots) of a spinal nerve (or nerves). This view has been advanced most strongly by Pearson (1938) on the basis that, in human embryos, the cells of origin of the spinal accessory are somatic in appearance and form a column which is part of the ventral horn of grey matter (a vestigial sensory component possibly being represented by ganglia occasionally found along the nerve trunk). In this case, the trapezius and sternomastoid would be of myotomic origin. The difficulty with this interpretation is accounting for the segments which provide the myotomal contribution.

The hypoglossal nerve is of special importance in the present context because of the numerous attempts that have been made to relate its marked structural variations within the vertebrates to the evolutionary history of the occipital region of the skull. The number of ventral nerves emerging through the occipital region varies with the number of segments that have become included in this part of the skull (Table 1 and Fig. 2). These occipital nerves arise from a column which represents the continuation into the medulla oblongata of the ventral column of the spinal cord, and supply muscles derived from occipital somites. In fishes the occipital myotomes migrate around the pharynx during development to form the hypobranchial musculature concerned with moving the gill arches. In land vertebrates this musculature has become modified to form the tongue muscles (although its migration may not be observable, the tongue muscles appearing to differentiate *in situ*) and the occipital nerves now comprise the last cranial or hypoglossal nerve. In amphibians a tongue is not always present, but when it is, as in frogs and toads, it is supplied by a hypoglossal nerve formed from the rostral rootlets of the first spinal nerve. This nerve emerges behind the caudal limit of the skull, reflecting the reduction of the cranial segments to five in these forms. In amniotes the number of cranial segments appears to have been stabilised at nine with a corresponding stabilisation in the number of occipital nerves compounded into the hypoglossal nerve.

Greater precision in defining the developmental history of the hypoglossal nerve is likely to be unobtainable in view of the probability that neurobiotaxic influences have brought about a forward migration of its cells of origin and roots (Ariëns Kappers, Huber & Crosby, 1936), thus disrupting their original relationships with other cranial structures. Because of this the exact relationships of the hypoglossal nerve to the vagus cannot be determined for all vertebrate groups. According to the scheme set out in Table 1 for *Squalus* the first three occipital nerves are the ventral roots corresponding to the dorsal roots represented by the second to fourth branchial branches of the vagus (although, since the vagus is only a partial polymerisation, a part of each of these dorsal roots is present in the rudimentary ganglia seen in the caudal vagal segments). The rudimentary ganglion found with the fourth occipital nerve presumably represents the whole of its corresponding dorsal root. However, even within the selachians the number of occipital nerves varies and the correspondence between them and vagal units is not always clear.

The attempts by early comparative anatomists to distinguish between occipital and occipitospinal nerves (the former being associated with the protometameric neurocranium and lacking dorsal roots; the latter emerging through the auximetameric neurocranium and possessing at least vestigial dorsal roots) in explaining the variations in the structure of the hypoglossal nerve is based on over-dogmatic assumptions about segmental homologies (see de Beer (1937) for a discussion of this old controversy). For mammals it is giving a spurious accuracy to say more than that probably the rostral segments of the hypoglossal represent the ventral nerve equivalents of the branchial elements comprising the vagal segments behind the first, while the caudal segment is formed from an originally separate spinal nerve. If the spinal accessory is accepted as representing one or more spinal nerves, rather than a detached part of the vagus, the segmental history of the occipital region of the skull and its associated nerves in mammals is rendered even more complex.

The classical scheme of segmentation just outlined has never received universal acceptance despite the fact that it appears to be consistent with, and indeed explains, some of the morphological features of the gnathostome head. One of the most cogently argued cases against the scheme was put forward by Kingsbury in 1926. His principal reasons for rejecting many of the assumptions on which it is based can be summarised as follows.

(1) There is no conclusive evidence that the prootic somites are in reality the serial homologues of the metotic somites. The latter develop in a cephalocaudal sequence, being undoubtedly part of a continuous series with the trunk somites, while the development of the former follows a sequence in the reverse direction and frequently does not begin until well after the metotic somites have appeared. Furthermore, the division of the paraxial mesoderm in the prootic region is usually very irregular so that it is often difficult to determine the exact boundaries and even the number of the resulting prootic segments.

(2) The number of gill pouches varies from one vertebrate group to another and the visceral arches seldom bear a clear numerical, topographical or, in their development, a chronological correspondence with the head somites. As we have seen, the lack of numerical correspondence is usually attributed to the loss of arches or somites. This type of explanation raises the general question of homology in serially repeated structures of the type encountered in branchiomeric and metameric segmentation. Kingsbury cited the great variation in the number of visceral arches within the vertebrates as evidence that the individual arches cannot be regarded as morphogenetic entities, individually inherited, and cannot, therefore, be homologised on a numerical basis between widely separate vertebrate groups. Rather, the branchial region should be regarded, Kingsbury suggested, as an area determined to produce the number of arches typical of the species, but that the arches so produced do not possess an intrinsic morphogenetic individuality. In this view it is meaningless to describe arches as being lost or added to some hypothetical ancestral number. Kingsbury applied a similar argument to the metotic somites. So far as topographical and chronological correspondence between arches and somites is concerned, this is closest, amongst gnathostomes, for the mandibular arch and somite, but even here the arch frequently becomes

associated with the hyoid somite as well. A separately identifiable arch corresponding to the premandibular somite is totally lacking while the hyoid arch, as well as being related to the hyoid somite, is also joined to the paraxial mesoderm over the second gill pouch. The branchial arches develop so late relative to the metotic somites that any topographical correspondence between them is unidentifiable. Similarly in cyclostomes, correspondence between myomery and branchiomery appears to be lacking. More recently, Jollie (1971), on the basis of study of the development of the cranium of *Squalus*, has questioned the assumption that there was ever an arch (i.e. premandibular) associated with the first prootic somite.

(3) The allocation of the cranial nerves into pairs each consisting of a dorsal and ventral element, which has become an essential part of the classical interpretation of head segmentation, has no morphological reality. Kingsbury's grounds for this assertion are several. First, the nerves innervating the muscles derived from the prootic somites do not have the central relations typical of somitic myotomal nerves and fail to pair with the profundus, maxillomandibular trigeminal and facial nerves as the classical hypothesis requires. The trochlear, in particular, is aberrant in that, as described above, it always leaves the brain dorsally and, further, that its fibres decussate. Secondly, the peripheral distribution of the cranial nerves is similarly inconsistent with the segmental plan of the head. Thirdly, the development of the vagus nerve gives no evidence of the amalgamation of segmental roots but the nerve is expanded or contracted in correspondence with the branchial region which it supplies. Moreover, the intestinal ramus of the vagus does not fit into any segmental scheme.

Kingsbury's rejection of the classical scheme of segmentation has received support from modern studies of primitive fossil chordates. Jefferies (1968), for example, in his work on the Cornuta (fossil forms with affinities to echinoderms but more probably belonging to the chordates to which they may be ancestral) describes his findings as flatly contradicting the idea of a completely segmented vertebrate ancestor with premandibular and mandibular gill pouches. Perhaps at this point it is worth emphasising, with Jefferies, exactly what rejection of the classical scheme of head segmentation implies. There is, of course, no doubt that metotic and prootic segments and visceral arches are real morphological entities. The parts of the classical scheme that are cast in doubt are: (1) that the prootic segments of the paraxial mesoderm are serially homologous with the metotic segments; (2) that there was an ancestral or typical number of metotic somites from which the numbers found in later forms can be derived by assuming the loss of certain members of the series; (3) similarly, that there was an ancestral number of visceral arches from which later arrangements can be derived by assuming loss of the caudal units (and the disappearance of a premandibular arch as a separate morphological entity); (4) that the mandibular arch, while serially homologous with the arches behind it, was ever a typical branchial arch or was associated with a mandibular gill pouch, and similarly that the hyoid arch and spiracle were once typical branchial arch and gill pouch, respectively; (5) that the nerves innervating the muscles derived from the prootic segments are equivalent to the ventral nerves supplying the myotomes of the metotic and trunk somites; (6) that the branchial nerves are equivalent to the dorsal nerves of the trunk and in the ancestral form were paired, on a one to one basis, with the ventral nerves of the head. While further evidence, especially from early fossil chordates, is required before a final decision can be reached on the true nature of head segmentation, it is clear that rejection of the classical scheme would, as just described, avoid a number of awkward morphological inconsistencies.

ORIGIN OF HEAD MESODERM

In the latter years of the nineteenth century, evidence began to accumulate which suggested that the mesenchyme of the head region is derived, at least in part, from an ectodermal source in the region of the neural crest rather than from mesoderm of primitive streak origin (Platt, 1893). This idea, which seemed at variance with the germ layer theory, was at first strongly contested, but subsequent work, in a wide variety of vertebrate types, has indicated that much of the head mesenchyme does indeed arise from the neural crest. Tissue so derived is frequently termed ectomesenchyme to distinguish it from mesenchyme arising from primitive streak mesoderm. Neural crest appears to be a unique vertebrate character and the ectomesenchymal derivatives to which it gives rise are mostly, if not entirely, peculiar to vertebrates.

The fate of neural crest cells can be traced by employing a variety of techniques, including extirpation, transplantation, explantation, vital staining or marking, and grafting of crest cells between donors and hosts of different species where the donor cells are either marked or naturally clearly distinguishable from the host cells. Experiments using such methods (see Hörstadius (1950), Weston (1970), Le Lièvre & Le Douarin (1975) for comprehensive reviews and bibliographies of this work) have shown that the neural crest gives origin to pigment cells, many components of the peripheral nervous system, supporting cells of the nervous system, and the adrenal medulla. However, only part of the neural crest is used up in the formation of these tissues, the remainder giving rise to ectomesenchyme. In the trunk this ectomesenchymal contribution is small (especially in tetrapods) but in the head it is much more extensive.

The ectomesenchymal derivatives of the head region have been analysed in greatest detail in amphibian and avian embryos because of their accessibility for experimental procedures. Amongst the studies of amphibians, those by Hörstadius (1950) and Chibon (1967) are particularly noteworthy. Hörstadius worked principally with the axolotl (*Ambystoma mexicanum*), using the techniques of extirpation, rotation and transplantation of cranial plate and neural crest. Chibon chose the newt (*Pleurodeles waltlii*) as his experimental

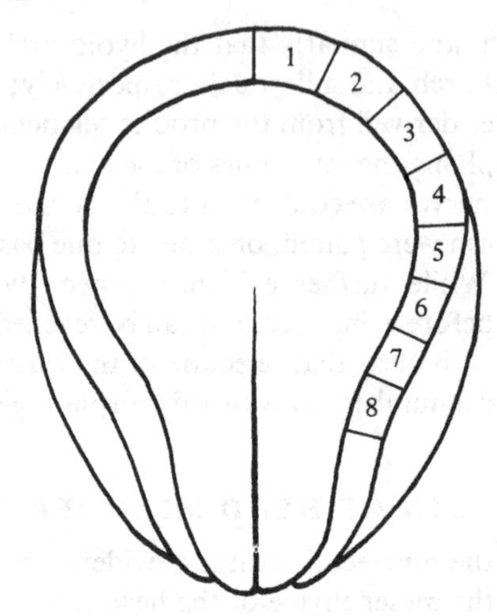

Fig. 3. Open neural plate stage of *Ambystoma mexicanum* to show zones of neural crest. (After Hörstadius, 1950.)

animal and transplantation of neural crest from a host labelled with tritiated thymidine to an unmarked host as his experimental technique. Their findings are essentially similar, and can be summarised together.

In both species, neural crest cells migrate and disperse about the time of closure of the neural tube without at first exhibiting any marked cytodifferentiation. The migration takes place laterally and ventrally between the ectoderm and the neural tube and then continues ventrally to reach, in the head region, the visceral arches. Here, the ectomesenchyme derived from the neural crest cells surrounds the mesoderm derived from the lateral plate, leaving the latter as a core (the muscle plate), running through the arch, which will eventually give rise to the branchial muscles of the arch.

The ectomesenchyme that migrates to the visceral arches gives rise to the whole of the visceral arch skeleton, apart from the second basibranchial element. Ectomesenchyme gives rise also to part of the trabecular and parachordal regions of the cartilaginous cranium. The presumptive fates of the cranial neural crest cells seem to be precisely regionalised at the stage of the open neural plate (i.e. before migration takes place, although contact with the pharyngeal mesoderm appears to be essential for further differentiation of the ectomesenchyme; see Hall (1978) for a discussion of the mechanism involved in this interaction). The crest along the most anterior part of the neural plate (zones 1 and 2, Fig. 3) has no ectomesenchymal forming capacity, differentiating into brain tissue. Zone 3 gives rise to ectomesenchyme that contributes to the anterior trabecular region of the cartilaginous skull. Zone 4 contains cells which form mandibular arch ectomesenchyme (the palatoquadrate cartilage being derived from tissue lying somewhat anteriorly to that which produces Meckel's cartilage). The ectomesenchyme for the hyoid and branchial arches is derived from cells lying in zones 5–8.

Most recent work has concentrated on the avian embryo. Both Johnston (1966) and Noden (1973) have used tritiated thymidine labelling to follow the fate of cranial neural crest cells in the chick. In this species the neural crest cells form a distinct population during or immediately following neural tube closure. The cells, which are undergoing repeated cell division, rapidly migrate, continuing to divide as they do so. The neural crest can be divided into areas which correspond broadly to the major morphological divisions of the embryonic brain and also the zones described above for the amphibian embryo. In area 1 (corresponding to the anterior prosencephalon) no migratory neural crest cells are formed. Areas 2 and 3 (posterior prosencephalon, mesencephalon) give rise to the ectomesenchyme of the maxillary region, area 4 (metencephalon) produces the ectomesenchyme of the mandibular region and area 5 (anterior myelencephalon) provides the ectomesenchyme of the hyoid arch. The migrating cells of areas 4 and 5 pass between the developing brain and the epidermis of the lateral body wall. A proportion of these cells aggregates alongside the brain to form the trigeminal (area 4) and facial (area 5) ganglia. Most of the remaining area 4 crest cells continue their migration around the pharynx to enter the mandibular arch where they form the ectomesenchyme of the arch. Similarly, the remaining crest cells of area 5 enter the hyoid arch. Most of the neural crest cells of area 3 pass into the maxillary part of the mandibular arch to form its ectomesenchyme. A proportion of area 3 cells joins the trigeminal ganglion and a further proportion aggregates posterior to the optic cup to form the ciliary ganglion. The crest cells of area 2 move at first rostrally along the dorsal aspect of the mesencephalon before turning laterally over the region of the future optic stalks and coming to lie caudal and rostral, as well as superior, to the eye.

One of the shortcomings of tritiated thymidine labelling techniques is that the progressive dilution of the marker, which occurs with repeated cell division, renders it undetectable in the later stages of histodifferentiation – which means, in practice, beyond the stage at which cartilage begins to form. An experimental procedure which avoids this difficulty has been developed by Le Douarin and colleagues (Le Douarin & Barq, 1969; Le Douarin, 1973; Le Lièvre, 1974; Le Lièvre & Le Douarin, 1975). This involves interspecific grafting from embryos of the Japanese quail (*Coturnix coturnix japonica*) to those of the domestic fowl (*Gallus gallus*). The presence of masses of unwound chromosome material in the interphase nuclei of the quail cells renders them clearly distinguishable despite repeated cell division. As a result it has been possible to follow the fate of neural crest material as far as bone formation. Such studies have shown that the spatial relationships between mesenchymal tissues of crest origin and those derived from mesoderm change little subsequent to the completion of the neural crest migration, and that the ectomesenchyme is quantitatively a rather more extensive contributor to the structures of the head and neck than had previously been thought. It appears that the dermis and much of the connective tissue of the face and ventrolateral part of the neck are derived from neural crest cells which also make important contributions to the vascular system and glands of the mouth and pharynx. In the visceral arches the ectomesenchyme comes to surround cores of mesoderm in much the same way as in amphibians. The ectomesenchyme then forms the connective tissue elements and satellite cells of the branchial muscles (which are themselves derived from the mesodermal core) as well as contributing nuclei to the muscle cells. Of particular significance in the present context is the finding of quail cells not only in the cartilages and replacing bones of the visceral arches but also in the dermal bones that develop in association with the arches. Similarly, quail cells were found in the bones that develop in the anterior part of the cranial base and cranial vault and around the eye. Only the occipital and the bones of the otic capsule entirely lack an ectomesenchymal contribution. As in amphibians, an ectomesenchyme–epithelial interaction at the site of migration appears to be a prerequisite for the development of at least some of the skeletal derivatives of the avian neural crest (Hall, 1978).

The role of cranial neural crest mesenchyme in the later stages of the morphogenesis of the mammalian head is less certainly known, largely because of the difficulties of studying living embryos *in utero*. Advances in techniques, especially that of growing whole mammalian embryos *in vitro* (New, 1966), offer possible ways of overcoming or avoiding these difficulties in the future. Until these techniques are brought into use, however, our knowledge of neural crest development in the mammal must continue to depend upon inferences drawn from the examination of dead and fixed embryos. Despite the obvious shortcomings of such methods, their use has provided a considerable amount of information of the early stages of neural crest development and migration in a wide variety of therian mammals.

Descriptions for the early human embryo have been given by Bartelmez (1922, 1923, 1960) and Bartelmez & Blount (1954). Migration of neural crest cells is first evident at the eight-somite stage (approximately 22 days old) when neural tube closure has progressed cranially as far as the caudal region of the developing hind brain. Most of the region of the future brain is thus still in the form of a neural plate. Lateral thickenings can be observed in the neural folds, beginning rostrally in the forebrain at the level of Rathke's pouch and extending caudally into the hindbrain (Fig. 4). Bartelmez (e.g. 1922) distinguished, in the thickenings of each side, a larger rostral (forebrain) part separated from a narrower caudal (midbrain and hindbrain) part by a constriction at the level of the rostral margin of the midbrain. By following the fate of these thickenings through progressively older stages (up to 28 somites) it is clear that they represent, in part, neural crest tissue.

The narrower, caudal part of the thickening gives rise to cells which enter into the formation of the trigeminal ganglion and to ectomesenchyme which

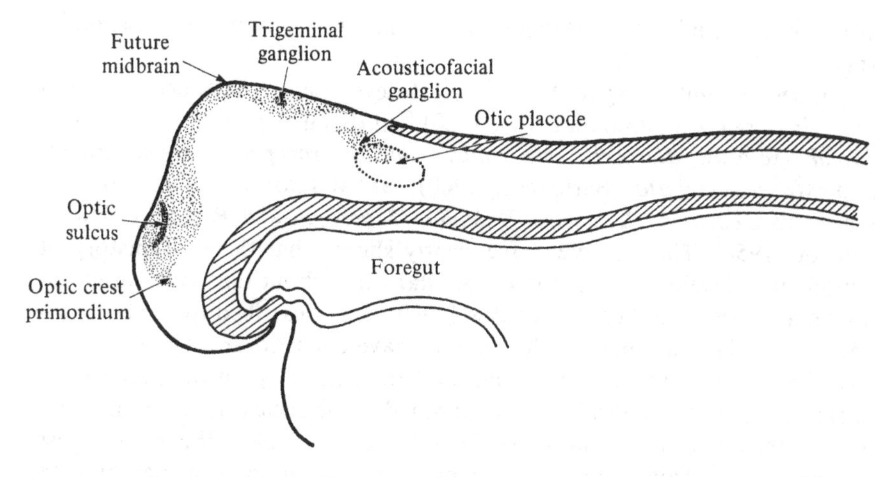

Fig. 4. Sagittal section of an eight-somite (approximately 22-day) human embryo to show neural crest (stippled) of head region. Hatching represents cut surface of neural tube. In the head region the neural plate is not yet fused to form neural tube. (Based on Bartelmez, 1922.)

migrates ventrally around the pharynx, and so behaves entirely like neural crest tissue in amphibians and birds. The mandibular and hyoid arches first become discernible at about the time of this migration. Further caudally, neural crest tissue contributes to the sensory ganglia of the facial, glosso-pharyngeal and vagus nerves. The larger rostral part of the thickening appears to give rise to the optic vesicle as well as to further ectomesenchyme which forms the sheath of the optic vesicle (Bartelmez & Blount, 1954). As in amphibian and avian embryos, no neural crest is formed from the most rostral part of the neural plate.

A very detailed investigation of neural crest development in the rat (*Rattus norvegicus*) was undertaken by Adelmann (1925) whose findings are in a number of respects at variance with those just described for man. The two principal discrepancies are the reported absence of neural crest cells from the whole of the forebrain region and failure to observe any evidence of major ectomesenchyme formation in the rat. Adelmann described the neural crest as appearing in three proliferations which give rise to the trigeminal, acousticofacial and glossopharyngeal + vagal ganglia, respectively, with little or no placodal contribution. In more recent study, however, Bartelmez (1962) has provided convincing evidence, based on observation of fixed and sectioned rat embryos, that migratory neural crest cells arise from the ventral lamina of the lateral lip of the neural plate in the region of the optic primordium of the forebrain and from the optic primordium itself. A proportion of these crest cells probably produces neurilemma cells. A further proportion migrates into the future maxillomandibular area, suggesting that it gives rise to visceral arch structures. The migration of these cells reaches its height in the 6–14-somite stages.

Studies of similar stages of neural crest development have been made in the guinea-pig (Celestino da Costa, 1920); cat (Schulte & Tilney, 1915); pig, *Ericulus setosus*, *Microcebus myoxinus*, *Cebus macrocephalus*, *Cebus gracilis* and *Macaca mulatta* (Bartelmez, 1960); as well as in the metatherian mammals *Dasyurus*, *Perameles*, *Didelphis*, *Macropus* and *Petrogale* (Hill & Watson, 1958). These studies have clearly shown that the early history of neural crest development is broadly similar throughout this wide variety of mammalian types and compares closely with that described for amphibians and birds. The only major differences to have emerged are: (1) as already described in the comparison of man and the rat, in our own species crest proliferation in the forebrain region is limited to the optic primordium, where it forms the mesenchymal sheath of the optic vesicle, while in the rodent, crest also proliferates from the ventral lamina of the lateral lip of the neural plate and migrates to form ectomesenchyme over a wide region, and (2) the exceptional finding by Schulte & Tilney (1915) of no evidence of ectomesenchyme formation in the cat.

Unfortunately, the limitations of the methods of investigation so far used in mammals have prevented elucidation of the subsequent fates of the ectomesenchyme derived from the neural crest migrations from cranial levels. Nonetheless, these methods have allowed the neural crest migrations to be traced to the visceral arches where, in the majority of mammals studied, they appear to make significant contributions to the mesenchymal component. The general similarity between these early stages of crest development and those in amphibians and birds lends confidence to the suggestion that the later stages too may be similar and that neural crest cells eventually contribute to the facial skeleton and anterior part of the cranial base in mammals as they do in the lower vertebrate forms.

2

THE ENDOSKELETAL SKULL

It was traditionally held by comparative anatomists and palaeontologists that cartilage was the first skeletal tissue to evolve in chordates and that bone appeared later but then, because of its superior mechanical properties, replaced cartilage as the principal supporting tissue of the adult stage of all gnathostomes except the cartilaginous fishes. The initial embryonic development of the endoskeleton as cartilage, and its subsequent replacement by bone, were taken to be examples of ontogeny recapitulating phylogeny. The fact that cartilage forms the skeleton of cyclostomes and that a cartilage-like material is found in the gills and mouth parts of amphioxus lent credence to the notion that cartilage was the original, primitive skeletal tissue. However, as the palaeontological evidence increased, it became clear that the role of bone as an adult supporting tissue is extremely ancient. Amongst the earliest chordates of the Ordovician several types of mineralised skeletal tissue are already present, including aspidin, dermal bone, dentine, enamel and calcified cartilage. Perichondral bone is also sometimes present and is presumed to have been developed in association with uncalcified cartilage. Endochondral ossification, which gives rise to cartilage replacing bone, appears to have evolved somewhat later in the Ordovician.

The interrelationship between these ancient skeletal tissues is still obscure (although the subject of several excellent reviews, e.g. Moss, 1968; Halstead, 1969; Hall, 1975*b*, 1978) but the findings of recent research (see Chapter 8) leave little doubt that the cells concerned in producing their modern equivalents are very closely related and that quite minor changes in the cellular environment can produce major changes in the type of tissue produced. If the ancient skeletogenic cells possessed the potential for a similar functional plasticity, then the emergence of new skeletal tissues would have represented a less radical departure than is sometimes assumed.

Whatever its evolutionary history should eventually prove to be, the widespread occurrence amongst later vertebrates of an endoskeleton composed of cartilage replacing bone clearly points to the possession of this tissue as having great adaptive value, possibly because of the advantage conferred during the growth period by the ability of the preceding cartilage to grow interstitially as well as by accretion. The presence of a skeleton composed exclusively, or almost exclusively, of unreplaced cartilage in the adult stage of some modern vertebrates is probably due to the retention of the embryonic condition throughout life, the adult bony stage having been secondarily lost, rather than representing the persistence of an ancestral condition as previously believed.

Beginning a description of the development of the skull with the cartilage and cartilage replacing components may no longer possess, as was once assumed, a phylogenetic rationale. Nevertheless, this order is retained as a descriptive convenience since in gnathostomes the cranial part of the endoskeleton forms much of the deeply lying part of the skull on which the dermal bones are overlaid.

THE CARTILAGINOUS STAGE

The development of the cartilaginous stage of the skull was studied in great detail by comparative anatomists of the nineteenth and early twentieth century (notably Bolk, de Beer, Fawcett, Gegenbaur, Goodrich and Voit amongst many others), using a variety of methods, including dissection, sectioning and wax model (or similar) reconstruction and selective staining for cartilage or bone. Their findings have been critically reviewed and synthesised by Goodrich (1930) and de Beer (1937), both of whom include a very full bibliography of earlier work. More recently, Starck (e.g. 1974) has provided a series of excellent studies of the mammalian chondrocranium. From this extensive body of work a fundamental plan of the cartilaginous stage of the endoskeletal skull common to most, if not all, gnathostomes can be derived. As seen in its most highly developed form in the cartilaginous fishes, the endoskeletal skull comprises a complete chondrocranium, and a jaw skeleton made up entirely of visceral arch cartilages. The chondrocranium consists of the braincase and sense capsules, the latter enclosing, with varying degrees of completeness, the olfactory, optic (where the capsule is represented by the sclerotic cartilage of the eyeball) and auditory sense organs. The cartilages of the jaws are derived from the mandibular and hyoid arches. Although not so readily apparent in the mature stages of other vertebrate groups, where the cartilage is largely replaced by bone and overlaid and partly fused with the dermal elements described in Chapter 3, the same fundamental plan of the cartilaginous skull can be readily discerned in the early stages of embryonic development.

BRAINCASE

The cartilaginous braincase can be visualised as being shaped somewhat like a trough lying beneath the brain. It possesses a floor and side walls but the roof is largely deficient (Fig. 5), being completed by dermal bones, except in the cartilaginous fishes where a more extensive roof of cartilage is present. The rostral part of the notochord becomes embedded in the floor of the posterior part of the braincase, and, in most forms, subsequently atrophies. Immediately in front of the notochordal tip lie the developing pituitary gland

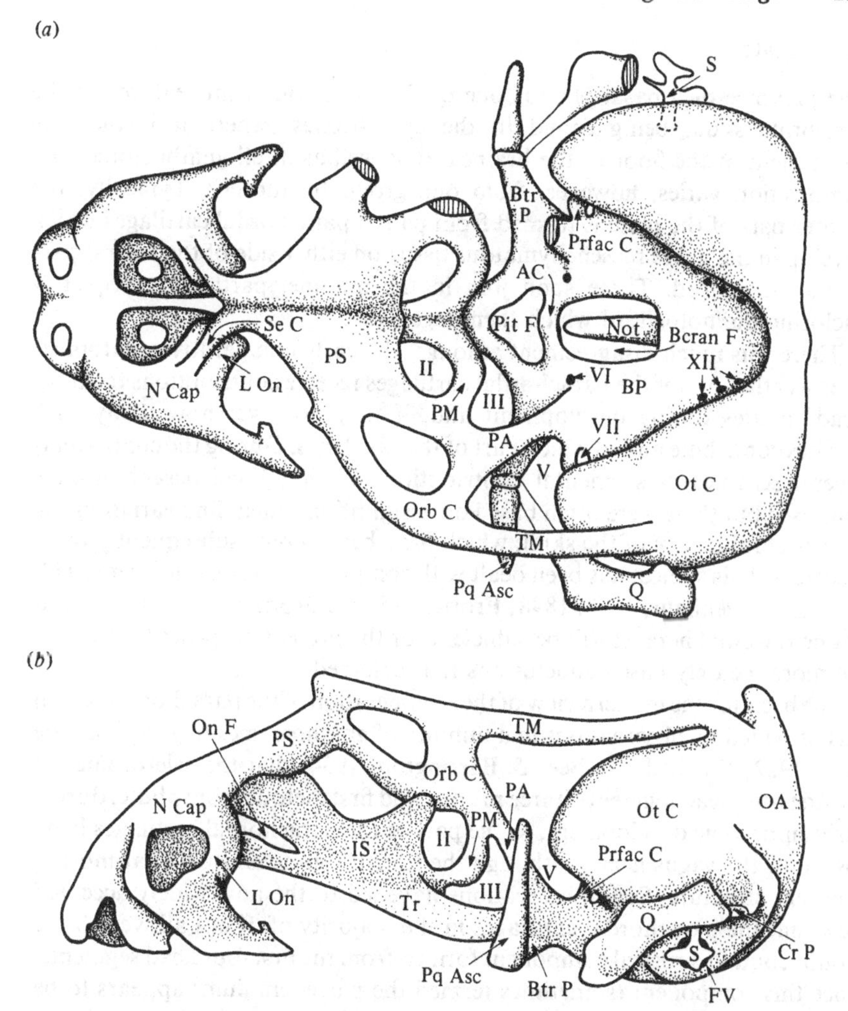

Fig. 5. Cartilaginous cranium of *Lacerta* (*a*) Dorsal view (on the right side the taenia marginalis and pila antotica have been removed; (*b*) left lateral view. For labelling see list of abbreviations. (Based on Goodrich, 1930, and Romer, 1956.)

and internal carotid arteries. This point marks the division of the braincase into prechordal and parachordal regions. Huxley (1867) suggested that this division is of more than descriptive convenience, marking a point that is truly homologous in all gnathostomes. The floor of the parachordal region is formed by the basal plate, while that of the prechordal region is composed of the paired trabeculae cranii.

Basal plate

The presence of a basal plate is one of the most consistent features of the vertebrate skull, being found in the cyclostomes (where it is the sole constituent of the floor of the braincase) as well as in all gnathostomes. Its composition varies, however, from one group to another. Typically, the greater part of the plate is formed from paired parachordal cartilages which develop in the dense mesenchymatous tissue on either side of the rostral part of the notochord. These soon join in the midline, partly or completely enclosing the notochord which then degenerates.

There was much disagreement amongst the early comparative anatomists over whether or not the parachordal cartilages receive contributions from the head somites during development and, if they do, over how many such somites contribute and to which part of the cartilages. During the controversy several schemes of segmental contributions to the chondrocranium were proposed but these were often based on assumptions, including variations on the vertebral theory of the skull, which have been shown subsequently to be incorrect. This subject has been dealt with comprehensively by de Beer (1937) (see also Gegenbaur, 1878, 1898; Froriep, 1887; Fürbringer, 1897) and need not be reviewed here. It will be sufficient for the present purposes to describe the more securely based conclusions that emerged.

The basis for the modern view of the segmentation of the parachordal region was provided by observations in a number of gnathostomes (e.g. *Squalus*, de Beer, 1922; the duck, de Beer & Barrington, 1934) that the sclerotomes of the first four head segments (three prootic and first metotic) contribute, during early embryonic development, to the part of the parachordal cartilages lying rostral to the vagus nerve, although the fourth somite is usually, at most, a transient structure. Such observations have led to the now widely accepted view that the parachordal cartilages in the majority of, if not all, vertebrate groups contain a rostral component formed from the first four head segments. Since this component (sometimes termed the palaeocranium) appears to be a very ancient acquisition, possibly antedating the vertebral column, it is not surprising that evidence of its segmentation has been lost in many higher vertebrates.

Behind the vagus is a newer region (the neocranium) to which material from a number of segments has been added. The analysis of the number of segments contributing to this region has been confused by false assumptions about the morphology of the twelfth cranial nerve. The addition to the basal plate of segments caudal to the fourth would be expected, according to the classical scheme of head segmentation, to result in the vagus and the corresponding components of the hypoglossal nerve becoming included within the skull, but to extend this argument by attempting to use the number of roots of the twelfth nerve, and the presence or absence of corresponding dorsal roots, as a precise guide to the number of such added segments, or to distinguish between protometameric and auximetameric cranial regions, is almost certainly unjustifiable even if the classical scheme is correct in all its assumptions.

Even without this guide, however, it has proved possible to determine the segmental composition of the neocranium in a variety of vertebrate types by study of their embryonic development. Two methods of analysis are available. First, it is possible in a few forms to observe directly which segments donate sclerotomes to the basal plate. Secondly, a knowledge of the segmental number of the myomere immediately preceding or succeeding the caudal limit of the basal plate will allow the number of segments included in the plate to be inferred.

In *Squalus*, for example, the first postoccipital nerve supplies the fifth permanent myomere (from the tenth segment of the whole series, assuming the fourth myotome to be lost and the fifth to be transitory as depicted in the scheme of segmentation shown in Table 1), indicating that sclerotomes five to nine have contributed to the basal plate (de Beer, 1922). In *Scyllium* the first postoccipital nerve innervates the fourth permanent myomere, indicating that sclerotomes five to eight (again according to the classical segmentation scheme) are included in the basal plate (Shute, 1972).

Although the mammalian chondrocranium has been studied in many species, the number of included cranial segments has been analysed in only a few, largely because the details of basal plate development have become obscured, thus preventing in most instances the application of the methods described above. In the mouse Dawes (1930) found that the basal plate receives contributions from five metotic somites. Since Dawes based his identification of somites on their myomeres, it seems likely that he failed to observe the first metotic somite which, according to the traditional segmentation scheme, does not contribute a permanent myomere. The myomere derived from his first metotic somite soon broke down, a behaviour usually taken as being typical of the true second metotic segment. It is possible, therefore, that Dawes's five metotic somites are in reality the second to sixth metotic segments or fifth to ninth in the complete series. The caudal limit of the basal plate in *Felis* also appears to coincide with the ninth segment (Kernan, 1915; Terry, 1917). The occipital region of the skull in birds and reptiles appears similarly to include the fifth to ninth segments.

As Shute (1972) has observed, the variable element seems to be the ninth segment as classically defined. In amniotes, as in *Squalus*, the basidorsal (the element formed from the posterior sclerotomite) of this segment is included in the occipital region, where it grows up behind the last hypoglossal root to form the occipital arch and extends posteriorly to form the exoccipital portion of the occipital condyle. In *Scyllium*, holostean and teleost fish the corresponding segment is excluded from the skull. In some fish (e.g. *Acipenser*) several spinal (occipitospinal) vertebrae are added to the occipital region extending its caudal limit beyond the ninth segment.

The anterior ends of the parachordal cartilages are separated from each other, in the majority of gnathostomes, by an unchondrified region termed the basicranial fenestra (Fig. 5). This fenestra becomes divided into anterior (pituitary) and posterior parts by the development of the transverse acrochordal bar of cartilage which stretches from the base of one pila antotica to the other. Later in development the parachordal cartilages fuse with the acrochordal cartilage, which thus becomes incorporated into the basal plate to form its rostral edge. In the process the posterior part of the basicranial fenestra may be obliterated. The acrochordal cartilage lies in a plane dorsal to the main parts of the trabeculae and parachordals. With the flooring in of the pituitary foramen the acrochordal cartilage comes to overhang the posterior part of the pituitary fossa and now constitutes the dorsum sellae.

Trabeculae cranii

The trabeculae usually begin to chondrify in the mesenchyme beneath the forebrain, one either side of the developing pituitary gland, and soon extend anteriorly. In the flat (platybasic) type of skull, found in early actinopterygians and in crossopterygians and their derivatives, the primitive amphibians, the caudal parts of the trabeculae remain wide apart, becoming united by an intertrabecular plate which bridges the gap between them. In the high vaulted (tropybasic) skulls of most modern actinopterygians and amniotes the trabeculae fuse with each other in the median plane. The anterior extensions of the trabeculae lie between the nasal capsules and become involved in forming the median nasal septum. The trabeculae fuse posteriorly with the basal plate so enclosing the basicranial fenestra. The fusion may involve initially separate polar cartilages which develop, one either side, lateral to the internal carotid arteries and close to the rostral edge of the parachordal cartilages. The development of the acrochordal cartilage divides this fenestra into an anterior part, the pituitary foramen, transmitting the adenohypophysis and internal carotid arteries, and a posterior part whose later fate has been considered above. In many vertebrate groups the pituitary foramen is floored in by cartilage which grows across from the trabeculae or, as in mammals, from separate centres, leaving patent just the two openings for the internal carotid arteries.

The more posterior regions of the trabeculae become attached to the orbital cartilages, which form the side walls of this region of the braincase, by a series of pillars (see below). Projecting laterally from the most caudal part of each trabecula (the polar cartilage when this chondrifies separately) is the basitrabecular process which typically makes contact with the basal process of the palatoquadrate cartilage to form the palatobasal articulation.

The developmental history of the trabeculae in therian mammals has proved difficult to interpret. Typical paired trabeculae have been described in the pig (Parker, 1874), but the more usual arrangement according to de Beer (1937) is for paired cartilages to appear, one on each side of the pituitary gland, and also a median rod which extends forwards from the pituitary gland between the nasal capsules. The paired cartilages spread towards the midline as the hypophysial plate, so flooring the pituitary fossa, and unite anteriorly with the median rod and posteriorly with the basal plate. These paired cartilages may represent the trabeculae or their detached posterior extremities, the polar cartilages. However, the internal carotid arteries pass lateral to these cartilages whereas in other vertebrates, as just described, they pass medial to the trabecular elements. This apparent discrepancy can be explained by assuming that in mammals either (1) the carotid arteries have cut through the posterior extremities of the trabeculae (polar cartilages) as the latter migrated medially following narrowing of the basal plate consequent upon enlargement of the otic capsules or (2), as suggested by de Beer (1937), the paired cartilages are really new elements, best termed hypophysial cartilages, whose appearance has separated and pushed laterally the internal carotid arteries. In the latter case the alicochlear commissures, which pass on either side from the base of the ala temporalis (see below) to the otic capsules lateral to the internal carotid arteries, may be equated with the posterior parts of the trabeculae (or polar cartilages). If this interpretation is correct the relationships of the internal carotid arteries are essentially the same in mammals as in other gnathostomes (an alternative interpretation of the alicochlear commissures is considered in the next section). In monotremes the otic capsules are still widely separated and the anterior region of the basal plate is unreduced as in more primitive vertebrates. The trabeculae lie lateral to the internal carotids and attach to the basal plate in the typical vertebrate manner.

The trabeculae may be derived from the sclerotomes of the most rostral (premandibular) somites or alternatively, as first suggested by Huxley (1874), may represent visceral arch skeletal elements taken up into the neurocranium. Allis (1931) extended Huxley's suggestion by equating the trabeculae and polar cartilages with the pharyngeal elements of the premandibular and mandibular arches, respectively. The demonstration by experimental methods in several vertebrate species that the trabeculae are formed from neural crest (see Chapter 1) appears to be consistent with a visceral arch origin. Recently, Jollie (1971) has produced evidence, from a study of development in *Squalus*, that, while the trabeculae do contain a visceral arch component, they do not represent part of a premandibular arch (the existence of which this author denies) or a formed element of the mandibular arch.

Side walls and roof

The side walls of the caudal part of the cartilaginous braincase are formed by what are, in effect, the neural arches of the segments included in the occipital (or neocranial) region of the skull. The most posterior of these arches is termed occipital, those preceding it preoccipital. The number of such arches varies, as would be expected, with the number of included segments. Thus, in mammals there are typically four (three preoccipital and an occipital) derived from the sixth to ninth sclerotomes, the fifth sclerotome being crushed by the otic capsules.

The preoccipital and occipital arches of each side fuse to form the definitive occipital arch or exoccipital cartilage (Fig. 5). The arches extend upwards on either side of the brainstem but do not meet dorsally. Anteriorly, the occipital arch joins with the otic capsule which forms the side wall of the braincase further forward. A gap, termed the metotic fissure, is left between them through which emerges the ninth, tenth and eleventh cranial nerves. From their phylogenetic history it might be expected that the roots of the hypoglossal nerve would emerge separately through a series of foramina located between the preoccipital and occipital neural arch elements. In fact, this regular arrangement is not usually seen. The number of hypoglossal foramina varies widely between species and even at different stages of development in a single species and provides, therefore, no guide to the number of neural arches included in the skull. These departures from the expected pattern are due to variations in the degree of chondrification in the arches.

A small part of the side wall of the braincase, immediately anterior to the otic capsule, is formed by the prefacial commissure. This is an upgrowth from the rostral part of the basal plate which fuses with the otic capsule, thus separating off the facial from the prootic foramen (see below). In mammals the enlargement of the cochlear part of the otic capsule has resulted in the prefacial commissure becoming relocated superiorly where it is more usually termed suprafacial (Fig. 6*d*).

Further forward, in the trabecular region, the side walls of the braincase are made up of paired orbital cartilages (also called orbitosphenoids). Each becomes connected to the floor of the braincase by a number of cartilaginous pillars (Figs. 5, 6*a*). In non-mammalian tetrapods there are typically three. In anteroposterior sequence these are: the pila prooptica (or preoptic root) attached inferiorly to the trabecula; the pila metoptica (metoptic root) also attached to the trabecula; and the pila antotica attached to the anterolateral corners of that part of the basal plate formed from the acrochordal cartilage. These pillars help marginate four foramina leading from the cavity of the braincase. Anteromedial to the preoptic root (i.e. between it and the nasal septum) is the olfactory foramen for transmission of the olfactory nerve. Between the preoptic and metoptic roots is the optic foramen for the optic nerve. Between the metoptic root and the pila antotica is the metoptic foramen for the oculomotor nerve. Behind the pila antotica is the prootic foramen through which passes the trigeminal (including profundus) and abducens nerves (the trochlear nerve usually emerges through a separate small foramen in the orbital cartilage). The superior edge of the orbital cartilage grows posteriorly as the taenia marginalis to join the otic capsule and complete the

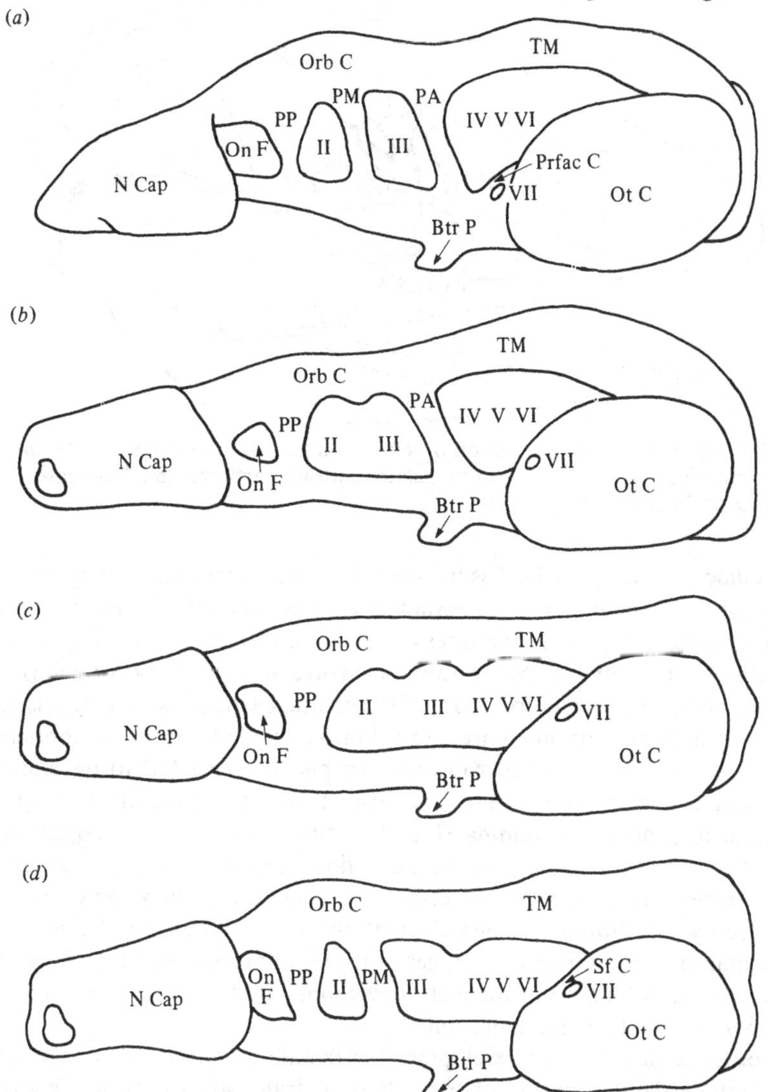

Fig. 6. Diagrammatic representation of constitution of side wall of cartilaginous braincase in various groups of amniotes. (*a*) Non-mammalian tetrapods; (*b*) monotremes; (*c*) metatherians; (*d*) eutherians. For labelling see list of abbreviations. (After Goodrich, 1930.)

margin of the orbit and prootic foramen. Anteriorly the orbital cartilage becomes connected with the lamina orbitonasalis of the nasal capsule by means of the sphenethmoidal commissure. The orbitonasal fissure is located between the orbital cartilage and its preoptic root posteriorly, the lamina orbitonasalis anteriorly, the sphenethmoidal commissure superiorly and the

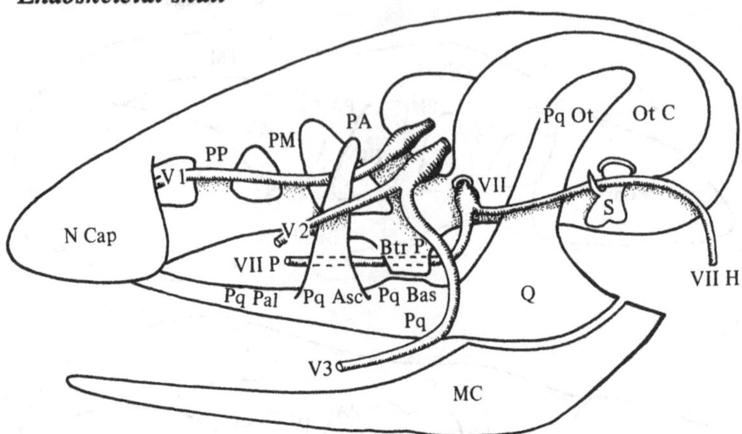

Fig. 7. Diagrammatic representation of left lateral view of cartilaginous cranium to show primitive connections between palatoquadrate cartilage and braincase. For labelling see list of abbreviations.

trabeculae inferiorly. This fissure does not lead from the cavity of the braincase, as does the olfactory foramen, but transmits the profundus nerve from the orbital region to the interior of the nasal capsule (see Fig. 13 for the relationship between the orbitonasal fissure and olfactory foramen).

In monotremes the metoptic root of the orbital cartilage is missing and the optic and metoptic foramina are merged into a single foramen postopticum (Fig. 6*b*). Both the metoptic root and the pila antotica fail to develop in metatherian mammals so that one large foramen represents the optic, metoptic and prootic foramina (Fig. 6*c*). In eutherians the preoptic and metoptic roots are present, closing the optic foramen, but the pila antotica fails to develop (Fig. 6*d*). The arrangement of these foramina becomes further modified by the addition of a new element, the ala temporalis, to the side wall of the mammalian braincase. Vestiges of the pila antotica are found in many therians as cartilaginous fragments developing in the dura mater in the neighbourhood of the ala temporalis.

In order to describe the development of the ala temporalis it is necessary to anticipate the description of the palatoquadrate cartilage which is given in the section dealing with the visceral arch skeleton. This cartilage, part of the mandibular arch, is connected primitively with the braincase by three processes: (1) anteriorly, by its palatine process with the nasal capsule; (2) posteriorly, by its otic process with the otic capsule; (3) in its middle part, by its basal process with the basitrabecular process of the trabecula (Fig. 7). Close to the basal process a fourth or ascending process may be present which connects with the orbital region of the braincase. The ascending process is well developed in the fossil crossopterygian fishes *Osteolepis* and *Eusthenopteron* (Westoll, 1943*a*), both members of the suborder Rhipidistia from which the tetrapods were probably derived. It is also present in many tetrapods,

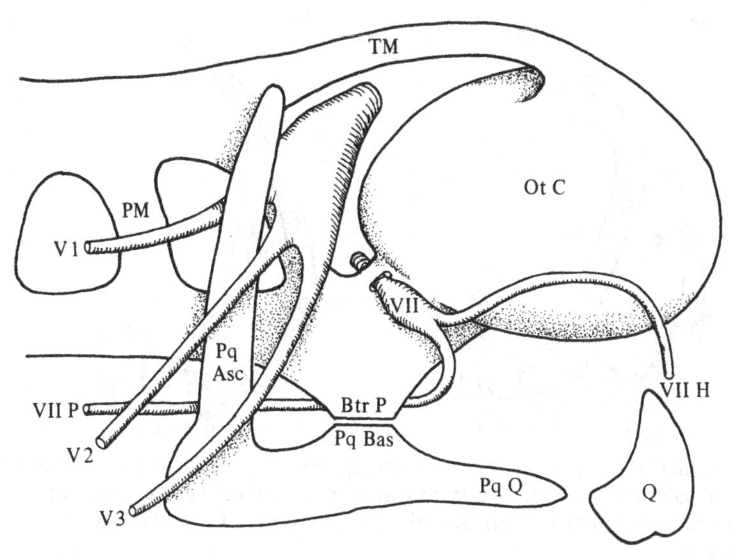

Fig. 8. Diagrammatic representation of left lateral view of the cartilaginous cranium of a primitive reptile to show structure of cavum epiptericum. For labelling see list of abbreviations.

including amphibians (although it is reduced or lost in some modern forms), many extinct reptiles (including the mammal-like reptiles) and *Sphenodon* and lizards among extant reptiles.

In the primitive reptilian condition, the palatoquadrate cartilage has been reduced to just its posterior (quadrate) and intermediate parts. The intermediate portion, which bears the ascending and basal processes as well as a quadrate ramus passing posteriorly towards the quadrate, provides the lateral boundary of an extracranial space, termed the cavum epiptericum, the medial boundary of which is the side wall of the braincase in the region of the pilae metoptica and antotica (Fig. 8). The floor of this space is formed by the basal process and its articulation with the basitrabecular process. Ossification in the intermediate part of the palatoquadrate produces the epipterygoid bone which, like its cartilaginous precursor, possesses ascending and basal processes and a quadrate ramus. The basitrabecular process is ossified from the basisphenoid as its basipterygoid process. The cavum epiptericum contains the trigeminal ganglion from which the profundus division leaves the cavum anterior to the ascending process while the maxillary and mandibular divisions leave posterior to this process. The palatine branch of the facial nerve runs forwards immediately ventral to the basitrabecular process. The dorsal extremity of the ascending process may end freely or may fuse with the orbital cartilage, pila metoptica, pila antotica, or otic capsule.

In mammals, in association with the enlargement of the brain, the cavum epiptericum has been taken into the endocranial cavity, the original side wall

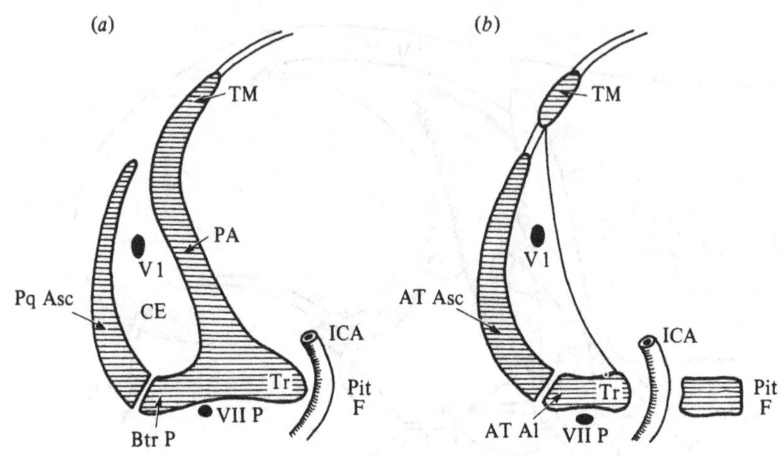

Fig. 9. Coronal section of one half of cartilaginous cranium in region of ascending process of palatoquadrate to illustrate traditional view of incorporation of cavum epiptericum into braincase. (*a*) Reptile; (*b*) mammal. For labelling see list of abbreviations.

of the braincase in this region, the pila antotica, failing to chondrify. It has been generally accepted (see, for example, Broom, 1911; Gregory & Noble, 1924) that the intermediate part of the palatoquadrate cartilage has persisted to form the new side wall of the cartilaginous braincase in this region, where it is represented by the ala temporalis. In this view the lamina ascendens of the ala is homologised with the ascending process and the alar (horizontal) process of the ala with the basitrabecular process (Fig. 9). Ossification in the ala temporalis and neighbouring membrane produces the alisphenoid bone.

Presley & Steel (1976) have recently put forward convincing evidence, derived from examination of developmental stages of principally the fruit bat but also of the rat, rabbit, pig and man, that this conventional view is incorrect in several important respects. They found, as had been noted by some earlier workers (see de Beer, 1937), that the lamina ascendens of the ala temporalis develops between the maxillary and mandibular divisions of the trigeminal nerve, a finding which throws doubt upon the supposed homology of the lamina ascendens with the ascending process, since the latter lies anterior to both these divisions. They suggest instead that the ascending process is represented by that part of the fascial layer which forms the lateral boundary of the braincase immediately anterior to the point where this boundary is pierced by the maxillary division. The body of the ala temporalis, which is almost certainly homologous with the body of the intermediate part of the palatoquadrate, is connected at the blastemal stage with the incus (the mammalian equivalent of the quadrate) by a condensation of mesenchyme which fails to chondrify but which represents, presumably, the quadrate ramus of the epipterygoid. The lamina ascendens is more probably, Presley & Steel believe, a neomorphic upgrowth from the intermediate part of the palatoquadrate cartilage in the region of the junction of the ascending process and quadrate ramus. A further small process which grows forwards from the body of the ala, anteromedial to the maxillary division of the trigeminal nerve, could represent the root of the ascending process.

The body of the ala temporalis is connected to the otic capsule by the alicochlear commissure which lies in the floor of the cavum epiptericum. Presley & Steel (1976) found that the commissure appears at the blastemal stage as an upgrowth of the otic capsule which runs close to the ala but extends on to connect with the tip of the alar process (i.e. the basitrabecular process of the trabecula), an appearance which leads these authors to suggest that the commissure is a vestige of the lateral flange of the otic capsule seen in advanced mammal-like reptiles (see the section on trabeculae cranii for an alternative interpretation of the alicochlear commissure). Later in development the commissure connects with the ala temporalis.

Presley & Steel (1976) describe ossification in this region as being initiated lateral to the tip of the ala in the layer of membrane immediately outside the fascia helping to form the lateral boundary of the incorporated cavum epiptericum, in the same plane in which the dermal pterygoid bone begins to ossify. On this basis they consider, but reject, the possibility that the intramembranous part of the alisphenoid may represent the intertemporal or quadratojugal elements of the dermal shield. They prefer instead to regard this part of the alisphenoid as being derived from the anterior lamina of the periotic which, they suggest, was probably a neomorphic, intramembranously ossifying structure in the advanced mammal-like reptiles from which the mammals evolved. As will be described in Chapter 4, the anterior lamina, together with the ascending process of the epipterygoid, marginates the trigeminal foramina in this group of reptiles. A somewhat similar situation is found in recent mammals where the intramembranous part of the alisphenoid meets the cartilage replacing bone of the ala temporalis on each side of the emerging maxillary and mandibular nerves to enclose foramina rotundum and ovale (Fig. 10). Presley & Steel (1976) thus regard the mammalian alisphenoid as a composite bone formed around the maxillary and mandibular nerves by a cartilage replacing component ossifying in the ala temporalis (only a small part of which is homologous with the ascending process of the epipterygoid), fused with an intramembranously ossifying component representing the anterior lamina of advanced mammal-like reptiles which has become detached from the periotic. In modern mammals the intramembranous component may overgrow the ala temporalis to meet the dermal bone of the lateral pterygoid plate.

In the conventional view that the alisphenoid (both cartilage replacing and intramembranous parts) is entirely homologous with the epipterygoid, it had to be assumed that the ascending process, which originally lay anterior to the maxillary and mandibular nerves, had spread posteriorly to enclose these

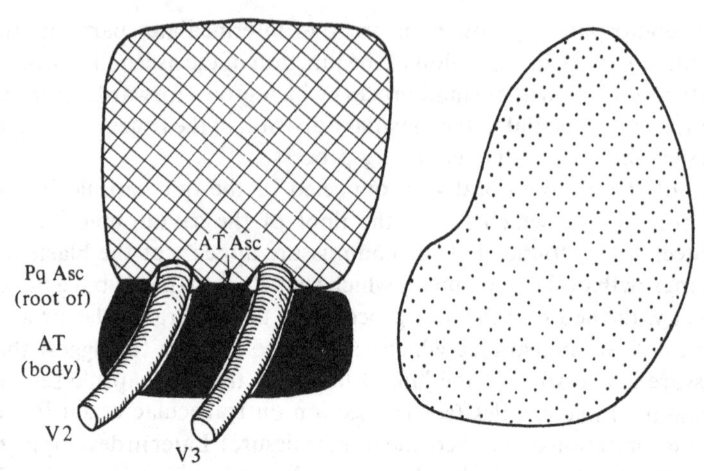

Fig. 10. Simplified representation of the structure of the side wall of the mammalian braincase according to Presley & Steel. Cross-hatching = intramembranously ossifying part of alisphenoid (i.e. anterior lamina); solid black = cartilage replacing bone of alisphenoid; broad stipple = periotic. For labelling see list of abbreviations. (After Presley & Steel, 1976.)

structures in foramina. As Presley & Steel (1976) point out, the evidence indicates that the phylogenetic trend has been rather for the bone of the alisphenoid to move in an anterior not posterior direction, the posterior margin of foramen ovale, for example, tending to be thinned and lost in recent mammals (see also below).

Presley & Steel (1976) found support for their suggestions in the situations encountered in developing stages of the marsupial *Didelphis* and the monotreme *Tachyglossus*. In *Didelphis* the ala temporalis does not grow out very far between the maxillary and mandibular nerves but a thick bar runs forwards from the ala, medial to the maxillary nerve, to join the fascia caudal to the optic nerve. This fascia probably represents the unchondrified metoptic root, the latter being absent from the cartilaginous braincase of *Didelphis*. The anteriorly projecting bar from the ala might, therefore, be regarded as a short ascending process terminating close to the metoptic root. The condition in eutherian mammals could then be regarded as a more complete regression of the ascending process than occurs in the marsupial. In *Tachyglossus* the ala temporalis is only weakly developed and the pila antotica chondrifies, unlike in the therian skull. Despite this persistence of the original side wall of the cartilaginous braincase, the cavum epiptericum is taken into the endocranial cavity of the bony skull by an extension of ossification, the anterior lamina (lamina obturans), from the periotic mass (the pila antotica undergoing ossification to form a slender spicule in the region of the sella turcica, Wilson, 1906). The anterior lamina remains attached to the otic capsule in monotremes so that its detachment in eutherians could be regarded as further evidence for the phylogenetic trend, mentioned above, for the alisphenoid to undergo a forward movement.

Presley & Steel (1976) also cite evidence from the morphology of the skull of the mammal-like reptiles in support of their views. This is considered in Chapter 4.

Immediately posterior to the cavum epiptericum, in the otic or posterior orbitonasal region, a further extracranial space is added to the mammalian cranial cavity by the formation of the tegmen tympani (see below). This new addition is termed the cavum supracochleare and lodges the facial ganglion and origin of the palatine branch, both of which were originally situated extracranially. The palatine branch runs forwards out of the cavum supracochleare into that part of the cranial cavity formed from the cavum epiptericum, thus preserving its fundamental primitive relationships.

Following the formation of the new side wall of the braincase in the region of the cavum epiptericum, the original foramina for the third to sixth cranial nerves are replaced by a new set of foramina related to the cartilage of the ala temporalis and subsequently to the alisphenoid bone. Anterior to the ala is the anterior lacerate foramen (orbital fissure), transmitting the profundus nerve and nerves to the extraocular muscles (although the latter nerves still pierce the dura mater in their original relationships and not directly opposite their exits from the cranium); posterior to the ala is the middle lacerate foramen which lies close to the entrance of the internal carotid artery into the cranial cavity. The foramina for the maxillary and mandibular nerves have been described above.

In those vertebrates with large eyes there is a tendency for the orbital cartilages and preoptic roots of the two sides to become approximated in the median plane to form an interorbital septum (Fig. 5). This may occur ventral or anterior to the brain. An interorbital septum is present in many teleosts, reptiles and birds where it frequently fails to chondrify or ossify so that, in the adult, it consists completely or partly of membrane. Above the septum a pair of cartilaginous plates may develop to form the anterior part of the floor of the cranial cavity as the planum supraseptale. Amongst mammals the tendency for extension of the nasal capsule, as part of the dominance of the olfactory system, has led to the interorbital septum becoming included in the ethmoidal region as part of the nasal septum. The interorbital septum has, however, reappeared in those mammals, most notably the primates, where there has been a secondary reduction of the nasal apparatus and an increase in the size of the eyes.

The roof of the cartilaginous braincase varies greatly in its degree of development between the different vertebrate groups. It is generally complete in cartilaginous fishes and in chondrosteans and holosteans. Amongst teleosts the roof may be complete or partial. In anuran amphibians the roof consists typically of a tectum synoticum, stretching between the otic capsules, and a tectum transversum, between the posterior parts of the orbital cartilages, but in urodeles it comprises a tectum synoticum only. Birds and most reptiles also have a roof composed of the tectum synoticum. In mammals there is generally a tectum posterius often arising from paired supraoccipital cartilages situated dorsal to the occipital arches. In some forms there is also a tectum anterius located forward of the otic capsules. Both of these structures are connected to the otic capsules only indirectly.

OTIC CAPSULES

The otic capsules arise from several (variable between species) independent centres of chondrification which appear in the mesenchyme enveloping the otic vesicles (for details of chondrification of the auditory capsule in various vertebrate groups, see for example Parker, 1874 – pig; Goodrich, 1918 – *Scyllium*; Pehrson, 1922 – *Amia*; de Beer, 1928 – *Salmo*; and Terry, 1917 – cat). Cartilage from these centres spreads around the developing inner ear, leaving foramina on the medial aspect of the capsule for the acoustic nerve and endolymphatic duct, and becomes attached to the basal plate.

In land vertebrates the tympanic membrane, tympanic cavity and ear ossicles (stapes in non-mammals; malleus, incus and stapes in mammals) have evolved to pick up the faint vibrations transmitted through the air and convey them to the inner ear. Associated with the development of these structures, two further apertures have appeared in the wall of the otic capsule – the fenestra vestibuli (ovalis) and the perilymphatic foramen. The fenestra vestibuli is located on the lateral side of the capsule and is closed by a membrane against which fits the footplate of the stapes. Deep to the fenestra a canal filled with perilymph is present by which the deflections of the membrane are conveyed to the organ of hearing. The vibrations in the perilymph are released to the exterior of the otic capsule through the perilymphatic foramen which faces into the recessus scalae tympani (see below).

The otic capsule is separated from the basal plate by a gap termed the basicapsular fenestra (Fig. 11*a*). This is continued posteriorly between the otic capsule and occipital arch, where it is known as the metotic fissure and transmits the glossopharyngeal, vagus and accessory nerves. The fenestra is bounded anteriorly by the anterior basicapsular commissure which connects the otic capsule with the basal plate. A second, posterior, basicapsular commissure may develop. Its position is variable, lying between the glossopharyngeal and vagus nerves in some forms but in front of both nerves in others. The latter condition appears to be the more usual one in amniotes.

The position of the anterior basicapsular commissure is more constant lying always behind the facial nerve, although in mammals its relationships are obscured somewhat by the development of the cochlear part of the otic capsule. This new addition to the capsule chondrifies in continuity with the older canalicular part and is, at first, separated from the basal plate by the

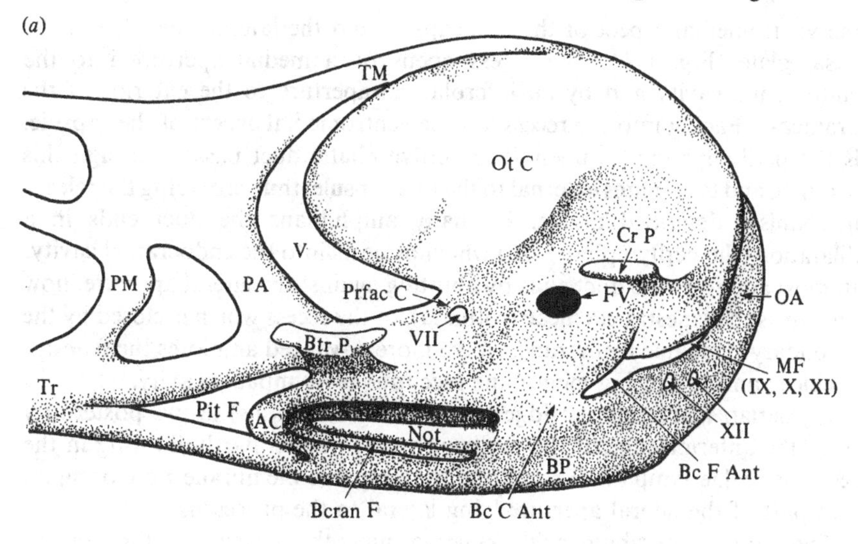

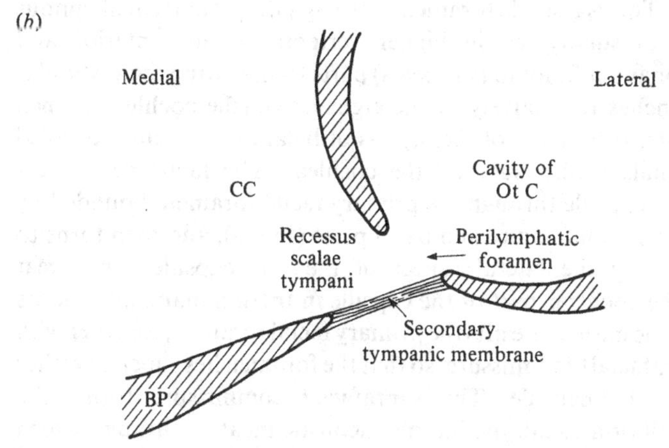

Fig. 11. To show recessus scalae tympani. (*a*) Left lateral view of cartilaginous braincase of lizard tilted so that the view is also slightly from below; (*b*) coronal section through anterior part of basicapsular fenestra to show formation of recessus scalae tympani and fenestra cochleae. For labelling see list of abbreviations. (*a* after Romer, 1956.)

anterior portion of the basicapsular fenestra. Its subsequent rapid growth in therian mammals causes it to encroach upon the anterior region of the basal plate which is consequently narrowed while the prefacial commissure is displaced superiorly to its suprafacial position.

Most of the basicapsular fenestra disappears by a process of fusion of adjacent parts leaving only the foramen for the ninth to eleventh cranial nerves and, in tetrapods, the recessus scalae tympani. The latter is a space formed from the front end of the basicapsular fenestra and lying, therefore, between the ventromedial aspect of the otic capsule and the lateral free edge of the basal plate (Fig. 11*b*). The recess opens by a medial aperture into the endocranial cavity and by an inferolateral aperture to the exterior of the braincase. Facing into the recess, on the ventromedial aspect of the capsule, is the perilymphatic foramen. The perilymphatic duct passes through this foramen and terminates external to the otic capsule, thus providing the release mechanism described above. In many amphibians the duct ends in a dilatation (the perilymphatic sac) which projects into the endocranial cavity. In most amniotes the duct ends by abutting against the lateral aperture, now known as the fenestra cochleae (rotunda), of the recess which is closed by the secondary tympanic membrane. In the more advanced amniotes the fenestra cochleae and its closing membrane face into the tympanic cavity.

In therian mammals a bridge (the processus recessus) grows posteriorly from the anterior edge of the perilymphatic foramen partly flooring in the recessus scalae tympani. The secondary tympanic membrane now occupies that part of the lateral aperture lying lateral to the processus.

The acoustic (vestibulocochlear) nerve enters the otic capsule through the acoustic foramen. This is a single foramen in the agnathous and cartilaginous fishes but becomes subdivided in higher vertebrates into anterior and posterior (superior and inferior in mammals) parts transmitting the vestibular and cochlear branches respectively of the eight nerve (the cochlear branch supplying the posterior portion of the organs of balance and being so called because in mammals it also supplies the cochlea). The facial nerve runs anterior to the otic capsule through the primary facial foramen, bounded by the capsule, prefacial commissure and basal plate (Fig. 5), and then turns to run posteriorly along the lateral aspect of the otic capsule. The great development of the cochlear part of the capsule in therian mammals and its encroachment on the basal plate lifts the primary facial foramen, together with the prefacial (suprafacial) commissure, so that the foramen lies superior rather than anterior to the capsule. The suprafacial commissure forms the anterosuperior wall of a canal, the internal acoustic meatus, through which the vestibulocochlear and facial nerves pass in company. At the distal end of the meatus the vestibulocochlear nerve gains access to the interior of the otic capsule through the acoustic foramina, while the facial nerve enters a canal in the lateral wall of the capsule (see below).

A remarkably constant feature of the lateral surface of the otic capsule is a ledge, termed the crista parotica. This lies in the horizontal plane in fishes, amphibians and reptiles (Figs. 5 and 11), but slopes posteroinferiorly in mammals (Fig. 12). Abutting against, or in close proximity to, the crista is the dorsal process of the hyomandibula (or columella auris) and the otic process of the palatoquadrate or, to use mammalian terminology, the tympanohyal and short process of the incus. In mammals the crista parotica (more usually termed the crista facialis in this group) is produced anteriorly to form the neomorphic tegmen tympani, while its intermediate part helps

(*a*)

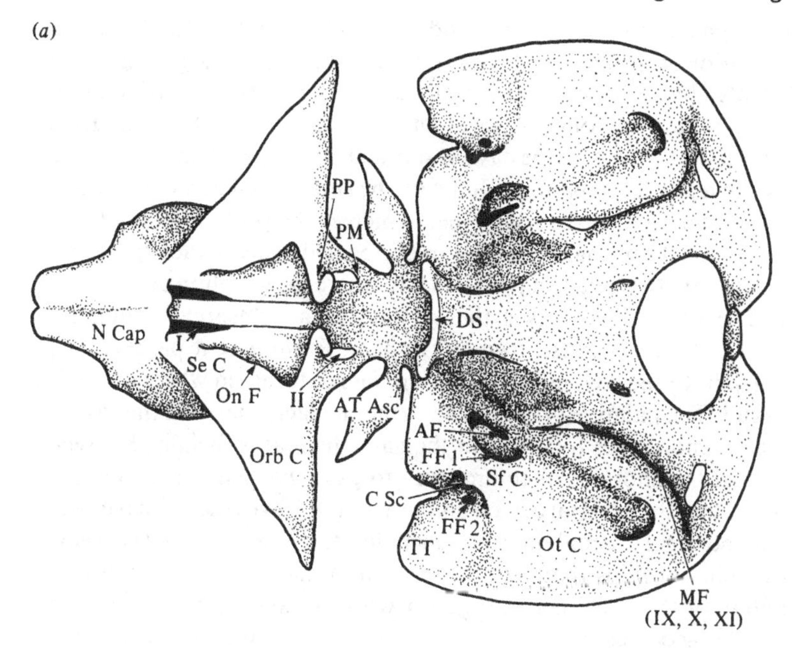

(*b*)

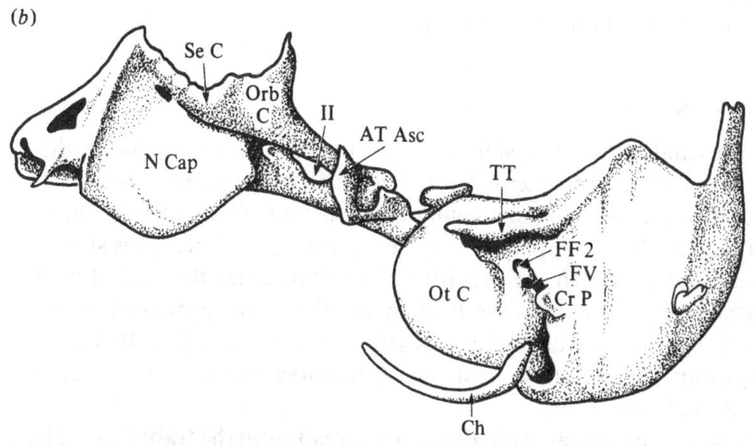

Fig. 12. Cartilaginous cranium of the New World monkey *Alouatta*. (*a*) Dorsal view; (*b*) left lateral view. For labelling see list of abbreviations. (After Starck, 1967.)

marginate a groove, the sulcus facialis, housing the facial nerve in its course along the lateral wall of the otic capsule (i.e. medial wall of the tympanic cavity).

The tegmen tympani helps to roof over the epitympanic recess which is another new feature in mammals acquired with the inclusion of the quadrate and articular (as incus and malleus) in the middle ear. As can be seen from

Fig. 12, the tegmen also forms a new side wall of the cranial cavity, in the otic and posterior orbitotemporal region, located lateral to the primary side wall (represented by the suprafacial commissure and a number of cartilaginous rudiments which may be found in this region). As a result of the formation of this new side wall, a space, the cavum supracochleare, is formed bounded laterally by the tegmen tympani, medially by the suprafacial commissure and floored by the otic capsule. The facial nerve pierces the primary cranial wall by passing through the primary facial foramen, located beneath the suprafacial commissure, to gain the cavum supracochleare where the ganglion of the nerve is lodged. The nerve then leaves the cavum supracochleare through the secondary facial foramen in the tegmen tympani. From here it passes along the sulcus facialis to the stylomastoid (tertiary facial) foramen where it finally leaves the skull. The palatine branch arises from the facial nerve in the cavum supracochleare and passes forwards in the gap (the hiatus facialis) between the tegmen and the suprafacial commissure to gain that part of the cranial cavity representing the original cavum epiptericum. It leaves the cranial cavity through the pterygoid canal in the sphenoid, thus preserving the same fundamental relationships that it possesses in non-mammals. The inclusion of the ganglion of the facial nerve, together with its palatine branch, within the cavum supracochleare, is reminiscent of the manner in which the trigeminal ganglion has become included in the cranial cavity by the incorporation of the cavum epiptericum.

NASAL CAPSULES

As the ectodermal nasal sacs sink into the underlying mesenchyme they become partly enveloped by membranous capsules which then chondrify. In the majority of vertebrates the cartilaginous capsules are fused with the brain-case but originally they were probably quite separate structures, as is still the case in agnathous fishes. In the majority of gnathostomes the medial walls of the capsules are formed by the median nasal septum produced by the anterior prolongation of the trabeculae (the original medial walls having presumably disappeared apart from the vestiges represented by the paraseptal cartilages; see below).

The walls of the capsules are chondrified by spread from the trabeculae. The degree of development of the anterior wall, or cupola anterior, depends upon the position of the external nostril, being best developed in those species where the nostril opens ventrally, and least developed, or totally absent, in those species where the nostril opens anteriorly. The posterolateral wall of the nasal capsule, separating the nasal cavity from the orbital region, is termed the lamina orbitonasalis (also termed cupola posterior or planum antorbitale). It is attached posteriorly to the preoptic root of the orbital cartilage, leaving a fissure for the passage of the nasal branch of the profundus nerve. In amniotes the lamina orbitonasalis chondrifies separately. The roof, or tectum, of the nasal capsule forms a bridge from the median septum to the lateral wall and lamina orbitonasalis.

In lower vertebrates the floor of the nasal capsule is at most rudimentary, being formed by a lateral projection from the ventral edge of the median septum. In tetrapods this projection, termed the anterior lamina transversalis, meets the ventral edge of the lateral wall of the nasal capsule and separates the fenestra narina anteriorly from the fenestra basalis (for the internal nostril) posteriorly. A posterior lamina transversalis is present in some mammals, being produced by bending forward of the ventral edge of the lamina orbitonasalis.

The extended development of the palate in mammals (described in Chapter 4) has carried the internal nostrils posteriorly, their original position being marked by the opening of the duct of the vomeronasal organ (of Jacobson). This structure is part of the olfactory region of the nasal sac which is in communication with the oral cavity. It is well developed in amniotes, lying near the nasal septum and opening in front of the nasopalatine canal. Associated with the vomeronasal organ is the paraseptal cartilage which develops in the floor of the nasal capsule alongside the ventral edge of the septum. In monotremes the paraseptal cartilage forms a capsule around the vomeronasal organ but in the majority of therian mammals it merely underlies the organ (for further details of the vomeronasal organ in mammals see Chapter 7).

The middle region of the lateral wall of the nasal capsule in amniotes is formed by the paranasal cartilage which chondrifies from a separate centre. The lateral wall has thus three components: anteriorly, that part of the wall which chondrifies in continuity with the cupola anterior; posteriorly, the front part of the lamina orbitonasalis; and in the middle region, the paranasal cartilage. The ventral edge of the lateral wall is inrolled in some reptiles and in birds and mammals to ossify as the maxilloturbinal. The rear edge of the anterior component of the lateral wall overlaps the front edge of the paranasal cartilage medially, projecting into the cavity of the nasal capsule as the concha nasalis or, in mammals, the crista semicircularis. The front edge of the lamina orbitonasalis in mammals similarly overlaps the back edge of the paranasal cartilage to produce the first turbinal element attached to the lateral ethmoidal plate. Further ethmoturbinals develop posterior to the first, as infoldings of the lateral wall of the nasal capsule, while the nasoturbinal develops anterior to the crista semicircularis.

The olfactory nerve leaves the cranial cavity by the olfactory foramen (evehens) which is bounded posterolaterally by the preoptic root of the orbital cartilage, anteromedially by the nasal septum, ventrally by the trabecula, and dorsally by the sphenoseptal commissure which bridges the preoptic root and the nasal septum. Primitively, the olfactory foramen opens directly into the nasal cavity (Fig. 13*a*). Between the preoptic root and the lamina orbitonasalis is the orbitonasal fissure, bounded above by the sphenethmoidal commissure

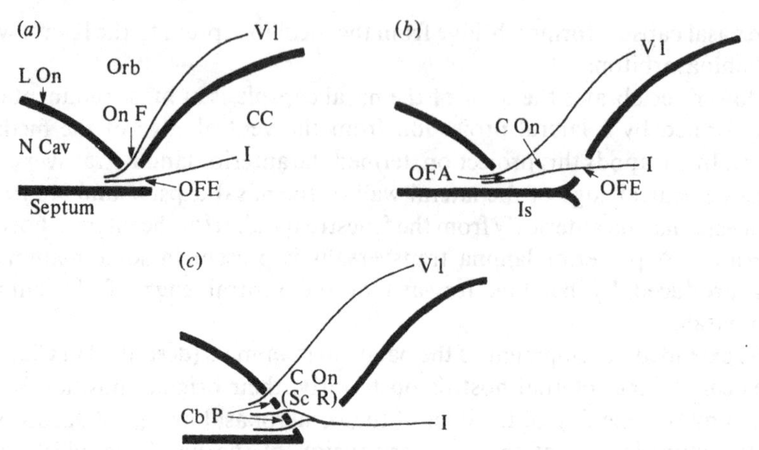

Fig. 13. Diagrammatic representation of olfactory foramen in cartilaginous skull: (*a*) with no interorbital septum; (*b*) with interorbital septum; (*c*) in mammals. For labelling see list of abbreviations.

and below by the trabecula, through which the nasal branch of the profundus nerve passes from the orbit into the nasal cavity to lie lateral to the olfactory nerve. These relationships are changed in many vertebrates by the forward displacement of the lamina orbitonasalis relative to the preoptic root. This is especially pronounced in those forms with a well-developed interorbital septum. The olfactory nerve still leaves the cranial cavity through the foramen olfactorium evehens but now enters the anterior part of the orbit, which, since it may be looked upon as an addition to the orbit gained at the expense of the nasal cavity by the forward displacement of the lamina orbitonasalis, is termed the cavum orbitonasale (Fig. 13*b*). The olfactory nerve gains access to the nasal cavity from the cavum orbitonasale by passing through the foramen olfactorium advehens which is bounded by the inner edge of the lamina orbitonasalis and the nasal septum. The nasal branch of the profundus nerve, after being transmitted through the enlarged orbitonasal fissure, has similarly to pass through the cavum orbitonasale before gaining access to the nasal cavity.

The foramen olfactorium advehens in therian mammals becomes subdivided into numerous fine foramina by cartilaginous struts which extend from the dorsal edge of the nasal septum to the corresponding border of the side wall of the capsule to form the cribriform plate. The olfactory nerves pass through the cribriform plate as a series of fine filaments. Immediately above the cribriform plate is the supracribrous recess. This is roofed over by dura mater which occupies the foramen olfactorium evehens. The supracribrous recess thus represents the cavum orbitonasale (Fig. 13*c*). The orbitonasal fissure (ethmoidal foramen) still transmits the nasal branch of the profundus nerve from the orbit to the cavum orbitonasale or supracribrous recess. Therein the

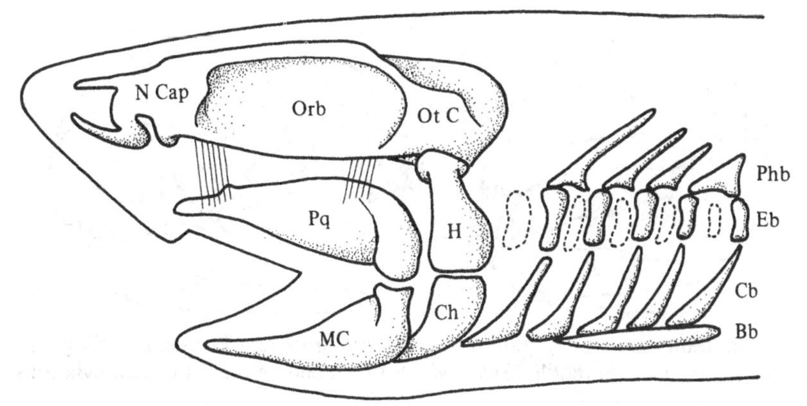

Fig. 14. Simplified left lateral view of braincase and visceral arch skeleton of *Scyllium*. For labelling see list of abbreviations.

nasal branch runs across the cribriform plate before entering the nasal cavity through the cribroethmoidal foramen. The olfactory nerve also retains its original relationships. The olfactory tract and bulb lie internal to the dura mater. The olfactory nerve filaments pass from the bulb through the dura mater in the roof of the supracribrous recess (i.e. through foramen olfactorium evehens), across the supracribrous recess to enter the nasal cavity through the foramina in the cribriform plate (i.e. foramen olfactorium advehens).

VISCERAL ARCH SKELETON

The typical branchial arch skeleton in fishes is composed of two major elements, a dorsal epibranchial and a ventral ceratobranchial (Fig. 14). In addition, a pharyngobranchial element is usually present at the dorsal end of each arch while small hypobranchials may be present ventral to the ceratobranchials. The segments of the two sides are joined together in the midventral line by an unpaired basibranchial element (possibly derived from the ventral parts of the preceding pair of branchial arches). These skeletal elements undergo great modifications in their number, arrangement and function from one vertebrate group to another and their exact homologies are often difficult to trace. According to the traditional view of the evolution of the jaws, the mandibular arch was originally a typical branchial arch, but at the beginning of gnathostome evolution it underwent modifications to give rise to the skeletal elements of the jaws. Also at an early stage the hyoid arch underwent modification to form a hyomandibular support for the jaws. As described in Chapter 1, this view no longer gains universal acceptance and it is possible that the mandibular and hyoid arches were new gnathostome acquisitions that were never serially homologous with the branchial arches. In land vertebrates the respiratory function of the posthyoid branchial arches

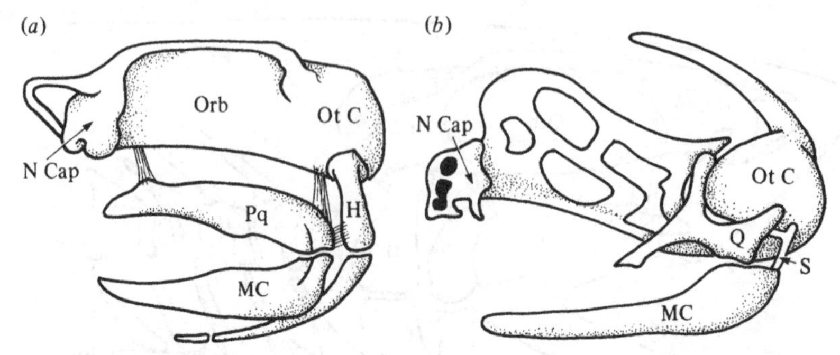

Fig. 15. Diagrammatic representation of palatoquadrate and Meckel's cartilages in: (*a*) cartilaginous fish; (*b*) reptile (*Sphenodon*). For labelling see list of abbreviations.

is lost and these too have become much modified, in this case to form pharyngeal and laryngeal structures. The possibility that the trabecular cartilage of the cranial base is also of visceral arch origin has been discussed earlier in the chapter.

The cartilages of the mandibular arch

The mandibular arch bends around the dorsal corner of the mouth to form its upper as well as its lower border. Two cartilages develop in the arch in all gnathostomes, the palatoquadrate above the mouth and Meckel's cartilage below. On the assumption that the mandibular arch was originally a typical branchial arch, these cartilages have been homologised with original epimandibular and ceratomandibular elements, respectively.

In cartilaginous fishes the palatoquadrate and Meckel's cartilages meet their fellows of the opposite side in the ventral midline to form the whole of the skeletal supports for the upper and lower jaws (Fig. 15*a*). The jaw joint is located between the caudal extremities of the two cartilages. Although the cartilages become ossified in bony fishes and tetrapods, the bulk of the skeleton of the jaws is provided by dermal bones which develop in the mesenchyme adjacent to the cartilages. It seems likely that the presence of dermal bones in the jaws represents the ancestral condition and that their absence in the cartilaginous fishes is a secondary loss.

In vertebrates with a dermal jaw skeleton the palatoquadrate cartilage does not extend to the midline, its anterior (palatine) part becoming involved in roofing over the palate or, as in amniotes, undergoing reduction (Fig. 15*b*). Its posterior (quadrate) part forms the articulation with the posterior (articular) part of Meckel's cartilage. In mammals both the quadrate and articular elements have become included in the middle ear cavity as the incus and malleus, respectively, while the intermediate part of the palatoquadrate is, as already described, involved in the formation of the lateral wall of the braincase.

In those forms in which it is well developed, the palatoquadrate cartilage is rod-like in shape and disposed anteroposteriorly, ventral to the braincase. Although the palatoquadrate may initially have had no direct skeletal connection with the braincase, articulations between these two elements probably developed at an early stage of vertebrate evolution. Primitively, three articulations are present (Fig. 7): (1) anteriorly, the palatine part of the palatoquadrate articulates with the lamina orbitonasalis of the nasal capsule; (2) the posterior part of the palatoquadrate possesses an otic process which articulates with the otic capsule; (3) the intermediate part of the palatoquadrate possesses a basal process which connects with the basitrabecular projection from the cranial base at the palatobasal articulation. An additional connection is formed in crossopterygian fishes and tetrapods between an ascending process, developed from the intermediate part of the palatoquadrate, and the side wall of the orbital region of the braincase. Between the quadrate region of the palatoquadrate complex and the side wall of the braincase is the cranioquadrate passage for the transmission of the carotid artery and head veins. The palatine branch of the facial nerve runs anteriorly, immediately ventral to the palatobasal articulation.

This fundamental plan has tended towards modification in all the major vertebrate groups. In actinopterygian fishes there has been a trend for the posterior and intermediate direct connections between palatoquadrate and braincase to be lost. Thus in modern teleosts the only direct articulation between the palatoquadrate and braincase is far anteriorly, the main support of the quadrate region being provided by the hyomandibula. This hyostylic mode of suspension allows the palatoquadrate complex to be swung laterally, so increasing the width of the mouth. The connections between the palatoquadrate and braincase in the crossopterygian fishes follow the fundamental plan more closely, but with the addition of a hyomandibular support for the quadrate region. In primitive land vertebrates the arrangements of these connections is clearly derived from that of the crossopterygian, although the hyomandibular support has been lost. There has been a tendency amongst amniotes for the anterior part of the palatoquadrate cartilage to be greatly reduced or, as in mammals, to be lacking altogether (although a vestige of it may be represented by a small piece of cartilage found transiently in the maxillary process of the embryos of some mammalian species), and for the continuity between the intermediate and quadrate parts to be lost. The further evolutionary history of the intermediate part has been dealt with in the section on the side wall of the braincase, and that of the quadrate region is considered below in the section on the ear ossicles.

Meckel's cartilage develops in the lower jaw of all gnathostomes. It first appears as a single centre but rapidly extends during embryonic life to form a rod-like structure. It comprises the whole of the lower jaw skeleton in cartilaginous fishes (Fig. 15*a*) but in other gnathostomes its anterior part fails to ossify completely (or nearly completely) and the lower jaw is made up

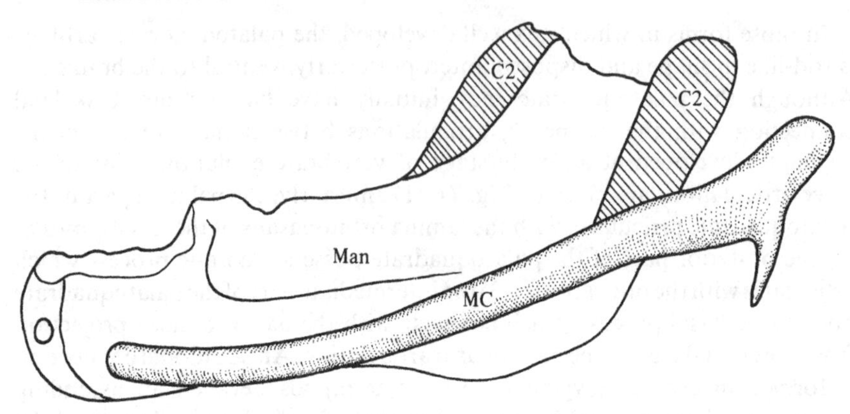

Fig. 16. The developing human mandible (approximately 12 weeks *in utero*) to show Meckel's cartilage. Secondary cartilages indicated by hatching. For labelling see list of abbreviations.

largely of dermal bones. The posterior part of the cartilage, in contrast, undergoes ossification and persists as the functionally important articular bone of the bony fishes and non-mammalian tetrapods and the malleus of mammals.

The shape of Meckel's cartilage varies greatly from one vertebrate species to another and may possess processes, such as the coronial and retroarticular, which become the site of development of independent centres of ossification. In mammals, Meckel's cartilage is usually well developed during early embryonic life. For example, in man the cartilage is present as a complete bar by about the sixth week of intrauterine life but subsequently regresses except for its posterior part which gives rise to the malleus and its extreme ventral tip which may become ossified to form part of the chin region of the mandible (Fig. 16). The fibrous covering of the cartilage persists, however, as the sphenomandibular ligament and anterior ligament of the malleus.

The cartilages of the hyoid arch

The hyoid arch in cartilaginous fishes has two skeletal elements, the hyomandibula (possibly representing the epihyal) and the ceratohyal. The hyomandibula is of considerable size and runs from the otic capsule to the posterior extremity of the palatoquadrate cartilage to which it is bound by ligaments (Fig. 15*a*). It thus braces the region of the jaw joint against the braincase. As already noted, this hyostylic condition is also the usual means of jaw suspension in the bony fishes. Amongst the actinopterygians there has been a tendency for the ventral end of the hyomandibula to become transposed anteriorly and for the jaws to shorten. Coupled with an increased mobility of the maxilla and premaxilla, this has allowed adaptation to take place to an immense range of diets.

In the rhipidistian fishes the palatoquadrate is joined to the braincase anteriorly and at the palatobasal joint, as well as being supported posteriorly by the hyomandibula (amphistylic suspension). In land vertebrates the hyomandibula plays no part in suspending the palatoquadrate (autostylic suspension), having become modified to form the columella auris or stapes of the middle ear (see section on ear ossicles and also Chapter 4).

The cartilage of the hyoid arch, apart from the hyomandibular or columellar component, is closely associated with the branchial arches to form the hypobranchial skeleton which in fishes and larval amphibians is part of the respiratory apparatus but which in adult amphibians and amniotes is involved principally in supporting the tongue and upper parts of the alimentary and respiratory tracts. In developing mammals four segments of the extrastapedial part of the hyoid cartilage (Reichert's cartilage) can usually be recognised which may chondrify and usually ossify separately. In dorso-ventral sequence these are termed the tympanohyal, stylohyal, ceratohyal and hypohyal (Fig. 78). The tympanohyal, which may be homologous with the dorsal process of the hyomandibula or columella auris, fuses at its dorsal extremity with the crista facialis (parotica) of the otic capsule and ossifies to enter into the formation of the auditory bulla (see Chapter 6). The stylohyal segment usually atrophies but its dorsal part may ossify to form a styloid process. This may fuse with the ossified tympanohyal or the two segments may remain connected by cartilage. The more ventral segments and the basihyal give rise to the body and anterior cornu of the hyoid bone. The anterior cornu may be attached to the styloid region by a stylohyoid ligament which represents a remnant of the hyoid cartilage which originally occupied this position.

CARTILAGE REPLACING BONES

In all gnathostomes except the cartilaginous fishes and a few aberrant forms such as the recent chondrostean fishes most of the cartilage of the braincase is replaced during the developmental period by bone. The sites of ossification by which this replacement takes place are extremely variable, even between closely related forms, in both their location and number. In some cases ossification follows the outline of the preceding cartilaginous stage, in others it spreads out into the surrounding membrane; in yet other instances the cartilaginous stage appears to have been repressed so that ossification begins directly in membrane. Cartilage bones and dermal bones may fuse during ontogeny or phylogeny to produce bones of composite origin. Whole bones may be lost or new bones added. Finally, there is the possibility that certain 'cartilage' bones have arisen from the conversion of dermal bones which have sunk into and invaded the underlying cartilage (this possibility is still disputed; see, for example, Jollie (1975) and also Patterson (1977), who come to the conclusion that the majority of, if not all, supposed instances of this

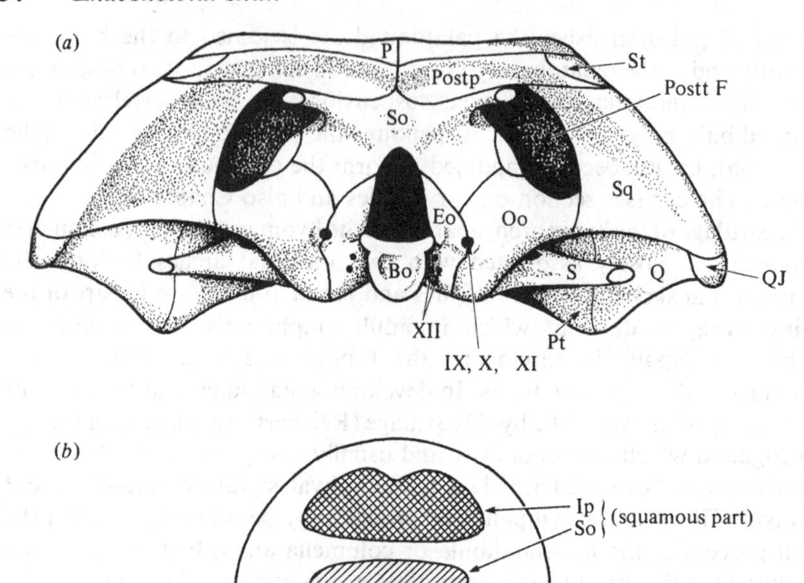

Fig. 17. To show the bones of the occipital region of the reptilian and mammalian skull. (*a*) Posterior view of skull of the stem reptile *Captorhinus*; (*b*) diagrammatic representation of posteroinferior view of the constant ossification centres of the human occipital bone (hatching = cartilage replacing bone; cross-hatching = dermal bone). For labelling see list of abbreviations. (*a* after Romer, 1956.)

happening are really due to fusion between originally separate dermal and cartilage replacing elements). Despite these difficulties – specific examples of which will be met in the following account – a fundamental plan of the cartilage replacing bones of the skull can be traced throughout the gnathostomes.

BRAINCASE

The occipital region of the braincase of the bony fishes consists typically of several cartilage replacing bones which surround the foramen magnum. These are the basioccipital, paired exoccipitals and, in teleosts, the supraoccipital. The basioccipital ossifies in the basal plate and usually extends well forward between the otic capsules. It forms the ventral margin of the foramen magnum and has its posterior surface hollowed out for articulation with the first

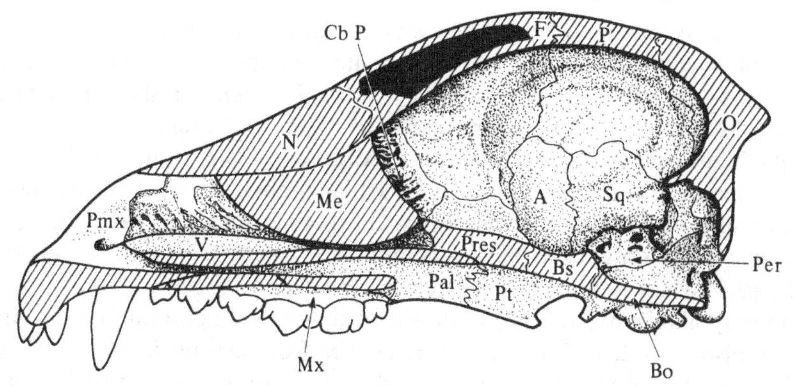

Fig. 18. Sagittal section of the cranium of the dog. Hatching = cut surfaces. For labelling see list of abbreviations.

vertebra. The exoccipital bones ossify in the occipital arches to provide the lateral margins of the foramen magnum and are pierced by the hypoglossal foramina. They enter into the superolateral corners of the articular surface for the first vertebra. The supraoccipital bone of teleosts may be a new development in these fishes, possibly having been formed from a sunken dermal bone which has invaded the cartilage (Jollie, 1975). If this interpretation is correct the supraoccipital bone of teleosts is probably not homologous with the similarly named element of the tetrapod skull, which appears to be a regular cartilage replacing bone ossifying in the tectum forming the dorsal margin of the foramen magnum in the chondocranium.

These occipital elements can be traced in most land vertebrates (Fig. 17), although the amphibians show a tendency, complete in living types, to lose the supraoccipital and basioccipital bones, the articulation with the vertebral column being provided by the exoccipitals. In mammals the occipital elements, although still ossifying from separate centres, fuse to form a single occipital bone (the superior part of the squamous occipital may ossify from a fifth element of dermal origin) and the basioccipital becomes largely excluded from the vertebral articulation so that, as in amphibians, paired occipital condyles of exoccipital origin are present. Projecting laterally and superiorly from the occipital plate of tetrapods is the paraoccipital process. Much of this process may be ossified by extension from the opisthotic centre of the otic capsule into the crista parotica. In lower forms it articulates with the tabular bone of the dermal shield, so helping to marginate the posttemporal fenestra (which communicates with the extracranial space posterolateral to the otic capsule and transmits a head vein). In mammals a similarly named process, a projection of the exoccipital, usually articulates with the mastoid region of the ossified otic capsule. A precursor of the paraoccipital process has been described as being present in the rhipidistian occiput (Romer, 1937).

The floor and side walls of the cartilaginous braincase in the orbitotemporal region, anterior to the otic capsules, ossify to form the basisphenoid bone.

The form and location of this bone vary in fishes, depending upon whether or not an interorbital septum is formed, being U-shaped in transverse section in those fishes where no septum is developed, but being T-shaped when a septum is present, as would be expected from the morphology of the cartilaginous stage in the latter case. Although there is no doubt that phylogenetically the basisphenoid is a cartilage replacing bone (being such, for example, in the palaeoniscoid fishes, Gardiner & Bartram, 1977), in most teleosts it arises ontogenetically in membrane (Daget, 1965). The trabecular cartilage, in which it might be expected to form, fails to ossify and disappears.

In tetrapods the basisphenoid extends posterior to the pituitary fossa (Fig. 18; see also Figs. 27*b*, 28*c* and 36). It ossifies from centres in the cartilage of the anterior region of the basal plate (derived from the hypophysial plate), crista sellaris and posterior part of the trabeculae (the actual number and location of centres varying from one group to another) and in non-mammals subsequently fuses with the dermal parasphenoid bone. The basitrabecular process, in those forms where it is developed, ossifies as the basipterygoid process of the basisphenoid. The palatobasal articulation, originally between the basipterygoid process and the basal process of the epipterygoid, extends in many forms to include an articulation between the parasphenoid and pterygoid, both of these dermal elements being closely applied to the cartilage replacing bones of the region. The extension of the basisphenoid posterior to the pituitary results in this bone totally enclosing the openings for the internal carotid arteries (as occurs in typical amphibians and reptiles as well as in monotremes, marsupials and some insectivores and bats) or helping partly to marginate these openings at the middle lacerate foramina (in more advanced eutherians). The variations in the internal carotid circulation and in its relationships with the base of the skull are considered in Appendix 1, Chapter 6.

In therian mammals the lateral wall of the braincase in the orbitotemporal region receives an important contribution from the alisphenoid bone which ossifies in the ala temporalis and neighbouring membrane. The details of this process and the possible homologies of the alisphenoid have already been discussed.

Immediately anterior to the basisphenoid (and alisphenoid in therian mammals), the side walls of the braincase are provided by the orbitosphenoid bones. Each of these ossifies from a centre which first appears in the orbital cartilage but which may spread into the neighbouring cartilages. The orbitosphenoid thus closes the cranial cavity anteriorly and surrounds the olfactory nerves as these pass forwards to the orbit (cavum orbitonasale) and thence to the nasal cavity. In the early labyrinthodont amphibians the two orbitosphenoids are fused across the midline to produce the single sphenethmoid bone which makes up the anterior extremity of the braincase. Its ventral portion forms a low interorbital septum. Amongst the modern amphibians the anurans possess a single sphenethmoid while the urodeles have paired orbitosphenoids but in neither case is an interorbital septum a prominent feature because of the platybasic nature of the skull. A sphenethmoid bone is present in the cotylosaurs and pelycosaurs (Fig. 27) but is much reduced in theriodonts. In the remaining reptiles the interorbital septum is tall and thin and usually incompletely ossified and the sphenethmoid reduced or lost. When it is present this bone contains a passageway (equivalent to the foramen olfactorium evehens) through which the olfactory nerves pass forwards from the cranial cavity towards the nasal cavity.

Ossification of the anterior part of the central stem of the cranial base in mammals has been the subject of controversy. Traditionally, the central stem has been regarded as ossifying as four separate elements – the basioccipital and basisphenoid in the basal plate and the presphenoid and mesethmoid (together representing the sphenethmoid) in the caudal and intranarial parts, respectively, of the trabecular region. However, Broom (1926) pointed out that in marsupials and artiodactyls only three elements are present which he equated with the basioccipital, basisphenoid and presphenoid. In a subsequent paper (Broom, 1927) he described a similar condition as existing in monotremes, perissodactyls, edentates and some insectivores (Chrysochloridae) and probably in cetaceans and proboscideans. The intranarial (mesethmoid) region of the central stem in these groups is ossified by extension from the most anterior (presphenoid) centre. It had previously been assumed that the large bone occupying both the presphenoid and mesethmoid regions was formed by fusion of initially separate presphenoid and mesethmoid ossification centres. In Broom's view, the mesethmoid is a neomorph in those mammalian groups (rodents, carnivores, primates, bats and many insectivores) where it is found. It ossifies to form the perpendicular plate of the ethmoid and spreads into the cribriform plate to fuse with the ossifications in the ethmoidal labyrinths.

On the basis of study of *Eremitalpa*, a chrysochlorid insectivore in which only three elements are present in the central stem, Roux (1947) has challenged the importance that Broom attached to the number of these ossifications. He found that the posterior part of the central stem in *Eremitalpa* ossifies from basioccipital and basisphenoid centres. The latter extends forwards into the posterior part of the trabecular region where it separates the orbitosphenoid ossifications. The remainder of the trabecular region is ossified from the single centre which Broom termed the presphenoid but which, in position, corresponds more closely to the mesethmoid of higher mammals and the sphenethmoid of mammal-like reptiles. Broom's description of ossification in the central stem of marsupials, artiodactyls and perissodactyls indicates that this is similar to that of *Eremitalpa*, except that the anterior centre (Broom's presphenoid), instead of arising far forward in the trabecular region, appears between the orbitosphenoids and is reinforced laterally by them.

The new element in those mammals with four ossifications in the central stem is not, according to Roux, the most anterior (i.e. the element traditionally termed the mesethmoid), this being the same as the anterior centre (presphenoidal of Broom) of those mammals with three ossifications although shifted forward somewhat, but is that ossification appearing in the caudal part of the trabecular plate intercalated between the anterior centre and the basisphenoid. The anterior ossification remains restricted to the intranarial part of the trabecular region while the basisphenoid does not spread forwards of the hypophysial plate. The new element arises by spread from the two orbitosphenoid ossifications into the caudal extremity of the trabeculae or, alternatively, from two separate presphenoidal centres which coalesce with the orbitosphenoids before they establish continuity with each other in the median plane and possibly represent, therefore, secondary orbitosphenoid centres.

If these suggestions are correct the mesethmoid is homologous with the presphenoid of mammals with only three ossifications in the central stem and the classical presphenoid arises by secondary invasion of the stem. In spite of these considerations, the terms mesethmoid and presphenoid are too firmly entrenched to be replaced by less confusing names.

OTIC CAPSULES

The otic capsule in bony fishes ossifies from prootic and opisthotic centres. The former arises anteroventrally and extends to surround the anterior semicircular canal and facial foramen. The opisthotic centre arises posteroventrally and gives rise to the bone of the posterior part of the capsule. In many fishes additional centres develop anterodorsally (sphenotic), posterodorsally (pterotic) and posteromedially (epiotic). It is possible that the sphenotic and pterotic centres are further examples of sunken parts of dermal bones which have invaded the cartilage of the otic capsule but there is not general agreement on this point (see de Beer, 1937; Patterson, 1977).

The otic capsule of non-mammalian tetrapods also ossifies from prootic and opisthotic centres, although the latter has tended to be lost in modern amphibians, as well as by extension from the supraoccipital bone. The mammalian otic capsule ossifies from centres whose number varies between and within species and whose homologies are accordingly uncertain (the number of centres in the human capsule, for example, may be as many as 14, but several of these centres are small, inconstant and soon fuse with adjacent centres). The bone from these centres fuses firmly to produce the periotic mass which tends to be situated quite deeply on the undersurface of the braincase rather than in its side wall as in lower tetrapods. It is usually possible to recognise distinct petrous and mastoid components of the mammalian periotic. The petrous part is composed of very hard bone and encloses the inner ear structures. It forms part of the floor of the cranial cavity but is usually hidden from view ventrally by the development of the auditory bulla (Chapter 6). The mastoid part lies posteriorly, is composed of less dense bone and is visible on the ventral surface of the skull, lying just in front of the paroccipital process.

The glossopharyngeal, vagus and accessory nerves emerge from the cartilaginous braincase through the metotic fissure between the otic capsule and occipital arch. With ossification, this fissure (now termed the posterior lacerate, vagus or, in mammals, the jugular foramen) lies between the opisthotic part of the otic capsule and the exoccipital bone (Fig. 17*a*). The relationships of the facial nerve to the bony otic capsule are considered in the section on lower jaw and ear ossicles, and also in Chapter 6.

NASAL CAPSULES

The cartilaginous nasal capsules remain partly or completely unossified in the majority of vertebrates. In the bony fishes a large number of small bones may be formed in this region (Starks, 1926; Harrington, 1955) but probably none is homologous with any of the bones found in the ethmoidal region of land vertebrates. The nasal capsules of amphibians and reptiles remain completely unossified. In mammals, cartilage replacing bones are formed in the nasal septum (mesethmoid, or equivalent ossification, and vomer) and in the lateral walls of the capsule (ethmoidal labyrinths and turbinals). Further details of the ossification of the mammalian nasal capsule will be found in Chapter 7.

VISCERAL ARCH SKELETON

Palatoquadrate complex

In the rhipidistian fishes the palatoquadrate complex appears to have been completely ossified (Fig. 19). The anterior (palatine) part of the ossified complex is attached (often fused) to the posterior face of the ossified lamina orbitonasalis. The intermediate (epipterygoid) part bears a basal process which articulated movably with the braincase at the palatobasal joint, and an ascending process. Behind the epipterygoid the dorsal margin of the palatoquadrate is closely applied to the side wall of the braincase in the form of an incipient otic process. The posterior (quadrate) part has connections with the squamous and quadratojugal bones of the dermal shield and is also supported by the partly ossified hyomandibula. The role of the palatoquadrate in the skull kinetism of rhipidistians is discussed, together with the general phenomenon of intracranial mobility, in Chapter 3.

In modern actinopterygians there are typically three cartilage replacing bones in the palatoquadrate complex, the palatine (autopalatine), metapterygoid (epipterygoid) and quadrate. The palatobasal articulation between the epipterygoid and basisphenoid has been lost. The direct connection between quadrate and otic capsule is also missing and the quadrate is supported by the hyomandibular and symplectic bones, both of which ossify in the hyomandibular cartilage of the second arch.

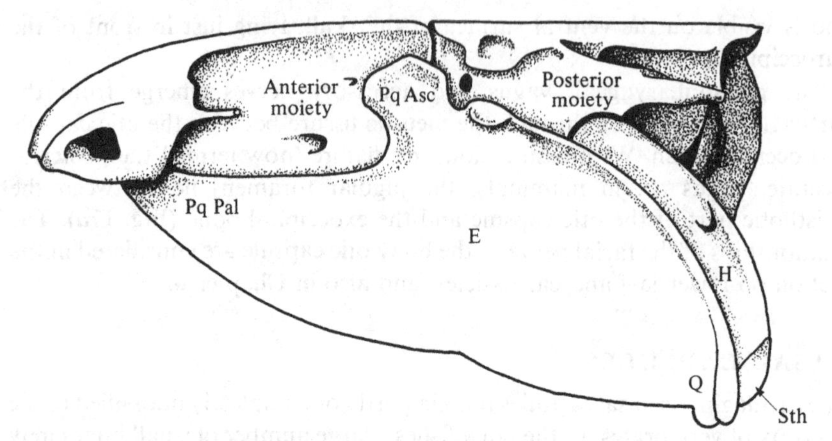

Fig. 19. The palatoquadrate complex and hyomandibula of the rhipidistian skull. For labelling see list of abbreviations. (Modified from description of *Eusthenopteron* in Moy-Thomas & Miles, 1971.)

The palatoquadrate bony complex of tetrapods is readily derived from that of rhipidistians, although the palatine part of the cartilage and its ossification have tended towards reduction and loss, especially in amniotes. The intermediate region of the palatoquadrate cartilage has persisted and its replacing bone, the epipterygoid, possesses typically, like the preceding cartilage, a basal process which articulates with the basipterygoid process of the basisphenoid, an ascending process (columella cranii), and anterior and posterior (quadrate) processes. The latter may reach and connect with the quadrate part of the palatoquadrate complex. The epipterygoid has undergone great modifications and in some forms (for example, turtles) is lost altogether. In early amphibians and reptiles, including the early mammal-like reptiles, it retained a movable palatobasal articulation but in later amphibians and many reptiles the articulation has become fused. The posterior part of the palatoquadrate cartilage and its ossification, the quadrate bone, also persists to provide, in the non-mammalian groups, the articulation for the lower jaw.

In mammals the epipterygoid, as already described, has become involved in the formation of the lateral wall of the braincase while the quadrate has been modified to form the incus (see next section).

Lower jaw and ear ossicles

The ossifications in Meckel's cartilage are somewhat variable. There is almost always a posterior ossification to form the articular bone (Fig. 26), although this may fuse with the neighbouring dermal angular bone or, less frequently, fail to ossify. The articular is continuous anteriorly with the unossified anterior part of Meckel's cartilage which continues forwards in the floor of the meckelian canal. In reptiles the articular is produced posteriorly as the retroarticular process into which the depressor mandibulae muscle is inserted. The retroarticular process in mammal-like reptiles tends to project downwards, although some authors (e.g. Romer & Price, 1940; Romer, 1956; Watson, 1951) have preferred to regard this structure as a neomorph into which part of the pterygoid musculature was inserted, the original retroarticular process, in their view, having disappeared (see Chapter 4). Other cartilage replacing ossification centres which may appear include an anterior mentomeckelian (often fused either in ontogeny or phylogeny with the dermal dentary) near the anterior tip of Meckel's cartilage, and coronomeckelian and retroarticular in correspondingly named processes of the cartilage. In mammals there is always an articular (malleolar) ossification in Meckel's cartilage. The fate of the remainder of the cartilage varies from one mammalian group to another, largely disappearing in some but undergoing calcification or ossification in others (Friant, 1960, 1966). The history of the bones of the lower jaw is dealt with further in Chapter 3.

The quadrate and articular bones in mammals have become included in the chain of ear ossicles by means of which vibrations are transmitted across the tympanic cavity from the tympanic membrane to the inner ear. The original quadratoarticular jaw joint has been replaced by a new jaw articulation between two dermal elements, the dentary and the squamous. In order to describe how the change in function of quadrate and articular came about it is necessary to digress briefly to the evolution of hearing. The inner ear of jawed fishes consists of a membranous labyrinth, comprising three semicircular canals and the sack-like saccule and utricle, filled with endolymph. The membranous labyrinth is separated from the wall of the otic capsule by a thin layer of perilymph. The sensory cells in the maculae of saccule and utricle and in the ampullae of the semicircular canals, believed to be modified neuromasts of the lateral line system, function to detect changes in orientation, acceleration and turning movements. The sensory cells in the saccule are probably also involved in hearing. In amphibians and reptiles the organ of hearing is located in the basal papilla region of the saccule, while in birds and mammals it is situated in a new organ, the cochlea, developed from the saccule (this condition also obtained in the mammal-like reptiles).

Hearing in fishes must be accomplished by the sound vibrations being passed from the water medium to the inner ear through the structures of the head (although some fishes have swimbladders modified to help in the transmission of vibrations). In land vertebrates, where hearing depends upon the detection of air-transmitted vibrations, which are much weaker than those passing through water, the tympanic membrane and structures of the tympanic cavity have been developed to amplify and transmit (impedance match, see Chapter 6) these vibrations directly to the inner ear. In most types that have been studied it is apparent that the tympanic cavity is derived ontogenetically from the first (spiracular) endodermal pouch (e.g. Kanagasuntheram, 1967). Such observations have led to the generally accepted view

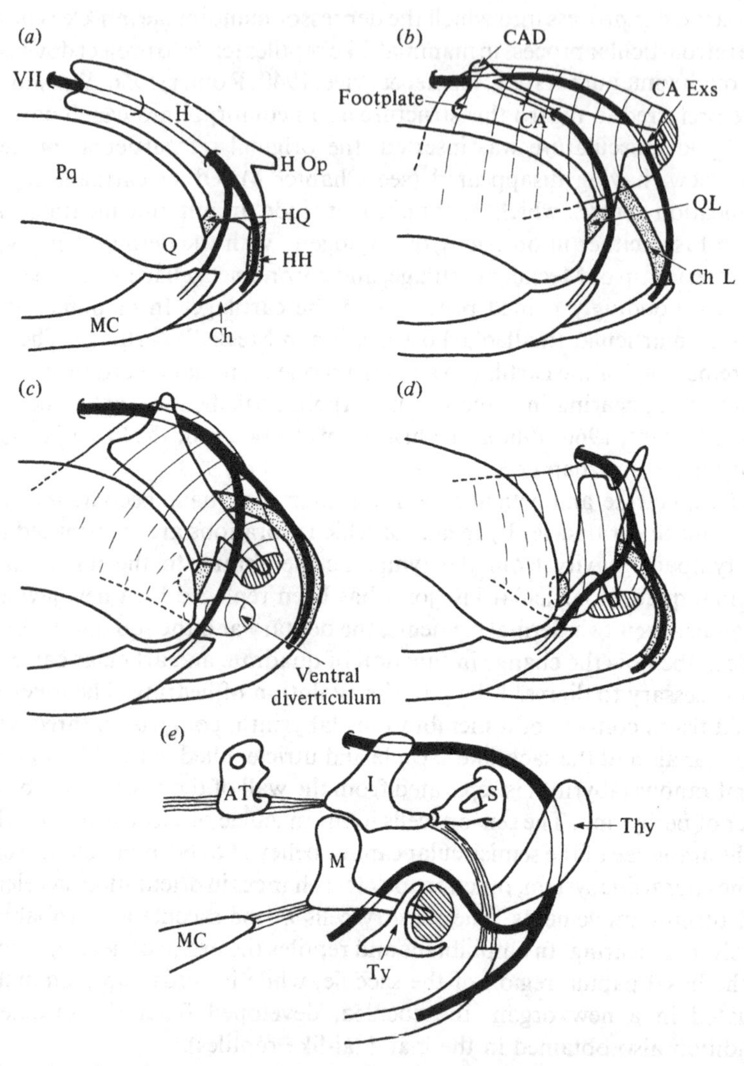

Fig. 20. The evolution of the stapes from the hyomandibula. (*a*) Rhipidistian; (*b*) primitive amphibian; (*c*) primitive reptile; (*d*) modern reptile; (*e*) mammal. In (*b*) to (*d*) the possible evolutionary changes in the tympanic cavity are shown. In (*e*) the development of the mammalian ear ossicles and their relationships to other derivatives of the mandibular and hyoid arches are depicted. Broad hatching = tympanic cavity; close hatching = tympanic membrane. The hyomandibular branch of the facial nerve is shown dividing into hyoid and internal mandibular (chorda tympani) branches. For labelling see list of abbreviations. (Based on Westoll, 1943*a*.)

that phylogenetically, too, the tympanic cavity is a derivative of the first pouch, although this assumes that the cavity is a homologous structure in all tetrapods. In ontogeny the inner part of the pouch becomes the pharyngotympanic tube which opens internally into the pharynx. Ventrally, a tympanic diverticulum from the pouch develops and enlarges to become closely applied to the surface ectoderm to form the tympanic membrane (which does not, therefore, represent an unpierced spiracular gill slit). This delicate membrane is capable of rapid vibration and in many amniotes, especially mammals, is protected by being placed at the inner end of an external acoustic meatus.

According to the traditional view the tympanic cavity appeared during the early stages of tetrapod evolution. As it expanded phylogenetically it came to surround the hyomandibula which became modified into the columella auris, stretching from the tympanic membrane to the fenestra vestibuli in the lateral wall of the otic capsule. The columella functions as a piston by which the weak pressure changes in the air produced by the sound vibrations are amplified to produce much greater pressure changes in the fluid of the inner ear (the amplification being equal to the ratio of the area of the tympanic membrane to that of the columella footplate occupying the fenestra vestibuli).

The hyomandibula of the rhipidistian fishes, from which the tetrapod columella was probably directly derived, has been described in detail by Romer (1941) and Westoll (1943*a*). It is a fairly substantial rod of bone which is attached by dorsal and ventral processes to the otic region of the braincase and by an opercular process to the bony operculum (Fig. 20*a*). Its distal unossifed part is, of course, unknown but was probably attached by a quadrate process to the quadrate bone and by a hyoid process to the ceratohyal.

Westoll suggested that the evolution of the tympanic cavity was facilitated by the increasing separation of hyoid and mandibular arches and relative forward movement of the quadrate which occurred during the transition from crossopterygian fish to primitive tetrapod. With the loss of the opercular bones the opercular process is believed to have been transformed into the lateral extrastapedial or extracolumella part of the columella auris which, in non-mammalian tetrapods, is attached to the tympanic membrane by its tympanic process. The ventral connection between the hyomandibula and otic capsule probably became the footplate of the columella auris, fitting into the newly developed fenestra vestibuli. These suggestions are supported by the structure of the columella auris in labyrinthodonts where, in addition to the footplate and extrastapes, there is a dorsal process and indications of a quadrate ligament (equivalent to the quadrate process) and ceratohyal ligament (hyoid process) – thus all five processes of the rhipidistian hyomandibula are accounted for (Fig. 20*b*). Similarly in reptiles, including modern forms such as lizards, the five processes, or vestiges of them, can often be traced (Fig. 20*c*, *d*). As would be expected from this phylogenetic history the columella auris in modern tetrapods develops ontogenetically in the dorsal part of the mesenchymal blastema of the hyoid arch, being formed first in cartilage. The medial (closest to the otic capsule) part of the cartilage is ossified to form the stapes while its lateral extrastapedial part may remain unossified.

In mammals the stapes is joined in the tympanic cavity by the articular and quadrate, now termed malleus and incus, respectively (Fig. 20*e*). These two bones are formed first in cartilage which chondrifies in the mesenchymal blastema of the first arch. The incus chondrifies separately but the developing malleus is continuous with Meckel's cartilage and may retain this continuity until quite late in development. Later, ossification centres appear in the two ossicles (the anterior process (also termed Folian, tympanic or gonial process) of the malleus, representing its point of connection with Meckel's cartilage, may have fused with it a remnant of the dermal prearticular bone) while the remainder of Meckel's cartilage undergoes complete or near complete atrophy apart from its perichondrium which may persist as the anterior malleolar and sphenomandibular ligaments. The mammalian stapes is developed from the stapedial portion of the columella auris, the extrastapedial part being largely underdeveloped so that the stapes is disconnected from the tympanic membrane and articulates instead with the incus. (Fragments of the extrastapes may be represented by two minute cartilages found in many mammals, one (Paaw's, frequently written Paauw's, cartilage) in the tendon of the stapedial muscle and the other (Spence's cartilage) close to the pars flaccida of the drum, Fig. 45*c*.) The tympanic membrane is now connected to the manubrium of the malleus and its vibrations are conducted to the stapes through the incus via the malleolar-incudal and incudo-stapedial joints. The stapes is pierced by the stapedial artery, a branch of the second aortic arch. A similar artery is found in other tetrapods but has a variable relationship to the columella auris. The mammalian tympanic membrane is supported in an incomplete ring of bone, the tympanic, which is believed to have been derived from the dermal angular (see Chapter 3). The evolution of the mammalian middle ear structures is described in detail in Chapter 4.

Recently, Tumarkin (1968) has pointed out certain difficulties that are presented by the acceptance of this classical theory of the evolution of the middle ear apparatus. He notes that, while the middle ear apparatus of most modern tetrapods is clearly adapted for the reception of air conducted sound in possessing a tympanic membrane connected with the inner ear by the columella (non-mammals) or the ossicular chain (mammals), in a number of amphibians and reptiles it is of the bone conduction type in which the inner ear is connected by the stapedial part of the columella to the quadrate (snakes and lizards), hyoid (amphisbaenid reptiles) or squamous (urodele amphibians) and in which the tympanic membrane and middle ear cavity may be completely lacking. Intermediate forms also exist. Tumarkin specifies three principal difficulties with the classical theory. First, it leads to bone conduction (and transitional) forms being regarded as the result of secondary or degenerate changes which, Tumarkin believes, are unlikely to occur in so important a sense organ. Secondly, Tumarkin argues that it is unlikely that two obsolescent jaw bones could have become interpolated into a refined air conduction system based on a single ossicle such as is held to have happened in the transition from mammal-like reptiles to mammals. Thirdly, Tumarkin believes the theory does not fit with the palaeontological facts. Some of the very primitive labyrinthodonts lacked a fenestra vestibuli without which an air conduction mechanism could not have been very effective, while in later tetrapods such as the stem reptiles and mammal-like reptiles, although a fenestra vestibuli is present, a tympanic membrane may have been lacking (certainly they lacked an otic notch which is believed to have housed this membrane – see Chapters 3 and 4). Moreover, in many early reptiles, and especially in the mammal-like group, the stapes appears to be too massive a structure to be well suited for an air conduction mechanism.

To meet these difficulties Tumarkin has suggested that the following evolutionary changes occurred. The earliest amphibians probably spent much of their time in shallow water and when on land adopted the prone position. In both cases, sound generated in the substrate (by, for example, the footsteps of predator or prey) would be more intense than that in air and would enter the body more easily because of closer impedance matching. Given the structure of the rhipidistian hyomandibula (see above), there are three ways in which a stapes efficient for bone conduction could have evolved in the early amphibians: (1) its quadrate process could remain attached to the quadrate; (2) it hyoid process could remain attached to the hyoid; or (3) its dorsal process could become attached to the squamous. Tumarkin believes that all three arrangements occurred in the early tetrapods. Thus, the three direct types of bone conduction mechanism found in modern tetrapods may represent the retention of a primitive condition rather than secondary or degenerative states.

In such bone conduction mechanisms the vibrations in the substrate could be canalised to the inner ear through the bones of the cranium and the stapes (via the fenestra vestibuli once this was present). There is, however, an alternative way in which bone conduction can work. If the whole skull (including the otic capsule) is set vibrating by the bone conducted sound and the stapes (or ossicular chain) fails to vibrate or vibrates out of phase with the skull, the sensory cells of the inner ear will be activated just as they will when it is the stapes that vibrates relative to the otic capsule. In order for this reverse route to be effective, the stapes needs to be as massive as possible (in the direct route of bone conduction the mass of the stapes is not critical while in air conduction it needs to be as small as possible).

With the evolution of tetrapods adopting a less prostrate posture, the effectiveness of all of these bone conduction mechanisms would have been reduced because the head was no longer maintained in contact with the ground. Tumarkin suggests that this reduction in efficiency may have been offset in three separate ways. In tetrapods in which the bone conduction mechanism was of the direct route variety, a new pathway from the ground was established by the inner ear becoming connected to the scapula (but by an element probably not homologous with the stapes of other mechanisms) and thus to the bones of the forelimb, the nearest parts of the body in contact with the ground. In those tetrapods with reverse-route bone conduction, the efficiency of hearing was maintained by increasing the mass of stapes. This, in Tumarkin's opinion, is the reason why the stapes of the mammal-like reptiles is so massive. Finally, Tumarkin suggests, in many groups of later tetrapods (possibly during the Permian) air conduction mechanisms evolved (including a middle ear cavity and tympanic membrane). These would have been of considerable survival value in that they would have allowed the ear to function even when very distant from the ground (i.e. in very large animals). The early mammal-like reptiles would have been at a disdvantage in that they possessed a heavy stapes adapted for reverse-route bone conduction and unsuitable for an air conduction mechanism (Tumarkin postulates that the enormous sail borne by the vertebral spines in the pelycosaur *Dimetrodon* was a substitute tympanic membrane which allowed airborne vibrations to drive the reverse-route mechanism; the more usual view is that the sail acted as a solar panel permitting the reptile to warm up quickly in the morning sun), but the more advanced mammal-like reptiles, the cynodonts and their derivatives, underwent a phylogenetic decrease in size, as a result of which the mass of the stapes must have become proportionately diminished. The stapes, which remained connected to the quadrate, may not have reached the surface at a tympanic membrane. Nonetheless, the stage would have been set for the interpolation of the quadrate and articular into the ossicular chain which (together with the development of a tympanic membrane if not already present) finally gave the successors of the mammal-like reptiles an efficient middle ear mechanism, sensitive to air conducted sound.

In fact, the phylogenetic decrease in size in the cynodonts is of modest degree only and the stapes is large but lightly built. As we shall see (Chapter 4), it is generally held that the cynodonts already possessed both a tympanic membrane and a tympanic cavity as part of an ear apparatus fully adapted for detecting air conducted sound (although in a recent analysis Allin (1975) has suggested certain modifications of this traditional view which have several points in common with Tumarkin's proposals – see Chapter 4). The critical stage in the progression postulated by Tumarkin would have come after the emergence of the first full mammals, since the earliest mammals still possessed a full postdentary complex of bones as part of the lower jaw. The early mammals were certainly of diminutive body size and the associated reduction in the mass of the stapes may possibly have preadapted it for inclusion in a triple ossicle mechanism. Unfortunately, the relevant fossil evidence is lacking.

The development of the tympanic cavity has modified the relationships of the facial nerve and its chorda tympani branch in this region. In rhipidistians the hyomandibular trunk of the facial nerve is believed to have passed posteriorly between the dorsal and ventral connections of the hyomandibular with the otic capsule, immediately alongside the lateral surface of the capsule.

It then turned ventrally and passed from the medial to the lateral apect of the hyomandibula through a short canal in the bone (Fig. 20*a*). The hyoid branch of the trunk presumably continued along the ceratohyal while the internal mandibular (chorda tympani) branch passed forwards over the hyomandibula to the medial side of the lower jaw. In living reptiles the hyomandibular trunk runs dorsal and then posterior to the columella auris. The chorda tympani passes forwards immediately superior to the lateral part of the tympanic cavity, and hence to the contained extrastapes, before turning ventrally to reach the medial surface of the articular bone (Fig. 20*d*). Westoll (1943*a*) has described a series of structural modifications by which the transition from the rhipidistian to the reptilian condition could have been achieved: (1) a change in the orientation of the axis of the columella associated with the downturning of the tabular and paroccipital process which occurred in primitive reptiles, with the result that the extrastapes became relocated in a more ventral position, close to the quadrate process (Fig. 20*b*, *c*); (2) a ventral diverticulum of the tympanic cavity pushed outwards between the hyoid and quadrate processes to reach up to and surround the extrastapes (Fig. 20*c*, *d*); (3) the hyomandibular trunk and its branches cut free from the dorsal aspect of the hyomandibula (Fig. 20*c*). The chorda tympani was thus left passing superiorly and anteriorly to the extrastapes and the lateral part of the tympanic cavity. A similar sequence of events is seen during ontogeny in lizards. The chorda tympani is both phylogenetically and ontogenetically a posttrematic (i.e. postspiracular) nerve and its apparent pretrematic relationships are misleading, resulting, according to Westoll, from the development of the ventral diverticulum.

The course of the facial nerve in mammals is, in essence, the same as in reptiles (Fig. 20*e*). After passing through the primary facial foramen (which has been displaced superiorly by the development of the cochlear part of the otic capsule) and giving off its palatine (greater superficial petrosal) branch, the nerve turns to run posteriorly and then inferiorly along the lateral surface of the otic capsule immediately beneath the mucous membrane lining the medial wall of the tympanic cavity. It passes medial to the tympanohyal, which is fused with the crista facialis of the otic capsule, to enclose the facial nerve in a deep gutter termed the sulcus facialis. The formation of the auditory bulla results in this part of the nerve becoming enclosed within a bony canal which it leaves through the stylomastoid foramen marginated primitively by the bullar floor anteriorly and by the petrous and mastoid parts of the periotic elsewhere (see Chapter 6). Similarly, the mammalian chorda tympani bears the same essential relationship to the stapes, incus and malleus as does the reptilian nerve to the columella, quadrate and articular. As the nerve passes over the roof of the tympanic cavity it lies lateral to the stapes, having no close relationship with this ossicle because of the loss of the extrastapedial portion. It then runs downwards posterior to the incus and medial to the malleus to reach the inner aspect of Meckel's cartilage and the dermal bones of the lower jaw.

3

THE DERMAL SKULL

ORIGIN OF DERMAL BONE

A body armour made up of sclerifications in the form of plates or scales in the skin is a very ancient vertebrate acquisition, being already present and well developed in the agnathans of the Ordovician. Although such armour is absent in modern jawless vertebrates and is represented by only skin denticles in the cartilaginous fishes (both of these conditions are probably degenerative rather than primitive), it is present to some extent in all other vertebrates. The principal function of the skin is protection and its protective properties are clearly enhanced by the development of skeletal structures within its layers. Several types of skeletal tissue are encountered in these sites. They may be developed predominantly from the epidermis (e.g. keratinised teeth, turtle scutes and whale baleen), from the dermis (e.g. aspidin and dermal bone) or from a combination of ectoderm and ectomesenchyme (e.g. oral teeth). According to Holmgren's delamination hypothesis (Holmgren, 1940; Jarvik, 1959), dermal sclerifications are first formed at the junction between the basal layer of the epidermis and the dermis and then tend to sink into the deeper dermal layers. Subsequent generations of sclerifications (not necessarily of the same type) may form at the epidermis–dermis junction and, in their turn, sink into the deeper tissues (for detailed discussions of the relationships between the various types of dermal sclerifications and of the delamination hypothesis see Moss, 1969; Hall, 1970, 1975*b*).

In the head and shoulder region of bony fishes the skin armour takes the form of large plates of bone rather than the small scales found in the trunk. These plates complete, together with the cartilage replacing bones already described, the cranium and jaws and also contribute to the pectoral girdle. Despite the fact that they have come to develop, in ontogeny, deep to the dermis and without a direct relationship to the epidermis, such bones are still termed dermal in reference to their phylogenetic origin.

In land vertebrates the dermal component of the skull (and, to a variable extent, of the pectoral girdle) is retained but the remainder of the skin armour is greatly reduced or lost completely. Within the skull there has been a general tendency for reduction in the number of dermal elements, the place of the missing bones being taken by enlargement of the remaining dermal bones. In most cases, the distinction between the dermal and endoskeletal parts of the skull is quite clear but, as discussed in Chapter 2, fusion between dermal and cartilage replacing bones during ontogeny or phylogeny is not uncommon. Examples of this phenomenon have been described in the previous chapter and others will be met in this. The further possibility that dermal elements may actually invade the endoskeleton and become bones which appear, in ontogeny, to replace cartilage is still debated (see Chapter 2). The dermal bones in many birds and mammals acquire areas of secondary or adventitious cartilage which may play an important part in subsequent growth, but the structure, development, growth and response to various stimuli of this tissue all indicate that it is quite separate and distinct from the primary cartilage of the endoskeleton. Its occurrence can in no way be taken as an example of a dermal bone being converted into a cartilage replacing bone. The phylogenetic and ontogenetic significance of secondary cartilage is discussed further in Chapter 8.

Because of the better preservation and more superficial (and therefore more easily observed) location of dermal bones in fossilised remains, their phylogenetic history is generally better known than that of the cartilage replacing bones. Despite the uncertainties always present in determining homologies, a common plan can be discerned for the dermal part of the skull which persists, although with great adaptive modifications, throughout the gnathostomes. The lateral line canals of the head (and trunk) have become embedded in a fairly constant manner in the dermal skeletal elements and these relationships have proved a valuable guide to the homologies of many of the cranial bones in the less advanced vertebrates.

DERMAL SHIELD, UPPER JAW AND PALATE

The dermal components of the skull in the rhipidistian fishes are well known from studies by several authors of which perhaps the most notable are those by Westoll (1936, 1937, 1943*b*) of *Osteolepis* and *Eusthenopteron*. The dorsal part of the dermal skull in these creatures is shaped like an armorial shield with its pointed end situated anteriorly and its concave surface facing inferiorly. It completes, with the cartilage replacing bones of the chondrocranium, the cranial, orbital and nasal cavities (Fig. 21). The shield is made up of numerous small bones, most of which appear to be clearly homologous with correspondingly located bones in the tetrapod skull, connected to each other by sutures except along a transverse line between the parietals and postparietals where a slightly movable fibrous joint was present. Together with the movable articulations between the anterior and posterior moieties of the braincase (Fig. 25) and between the epipterygoid and basisphenoid, this joint allowed a small amount of movement between the anterior and posterior parts of the skull. Westoll postulated that the ancestral forms from which the crossopterygians were derived possessed a dermal shield consisting of a far greater number of elements than occurs in rhipidistians, with many small

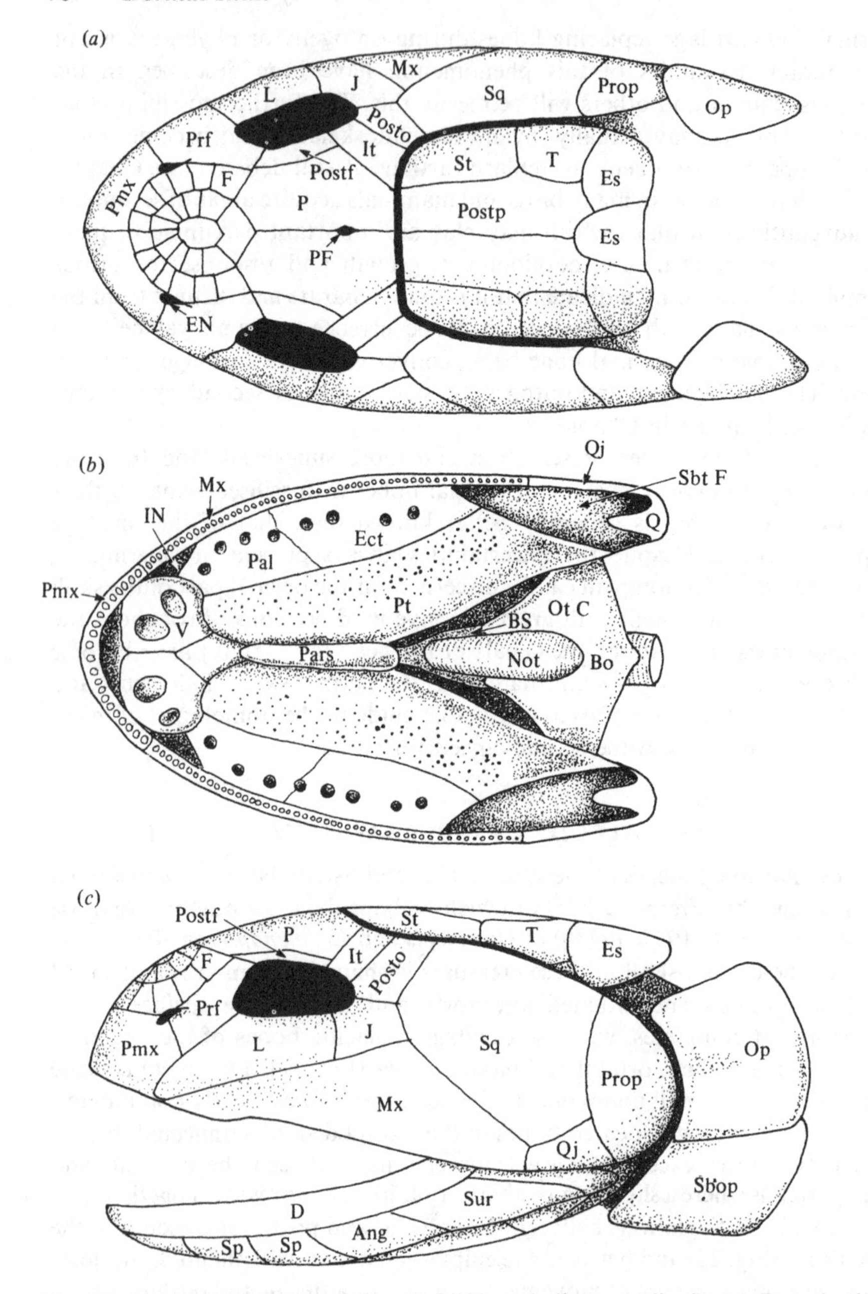

Fig. 21. The dermal shield of the rhipidistian skull (generalised). (*a*) Dorsal; (*b*) ventral; (*c*) left lateral views. For labelling see list of abbreviations. (After Westoll, 1943*b*.)

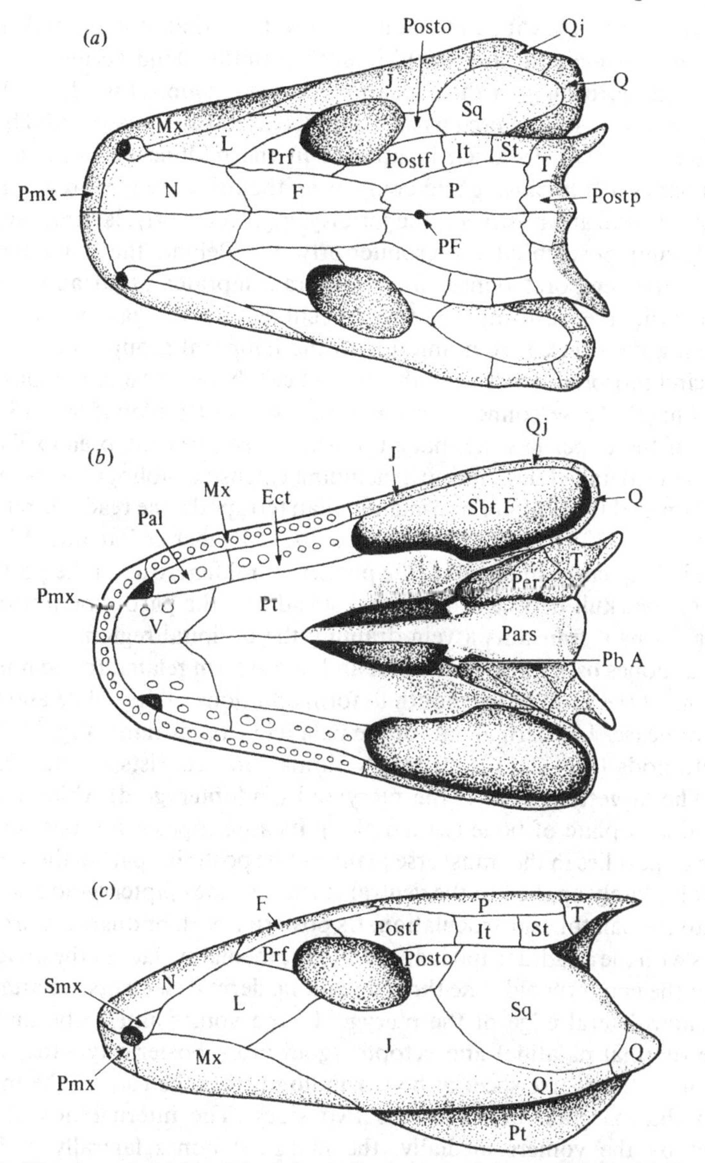

Fig. 22. The dermal shield of the labyrinthodont skull (generalised). (*a*) Dorsal; (*b*) ventral; (*c*) left lateral views. For labelling see list of abbreviations. (After Romer, 1956.)

bones related to the lateral line organs and a variable number of median bones. Primitive dipnoans display this structure.

Apart from a marked increase in the length of the snout relative to that of the postorbital region, the elimination of the transverse joint and the loss of a few elements, the structure of the dermal shield in the early amphibians is directly comparable with and readily derived from that of the rhipidistians. The bones that make up the shield in such primitive land vertebrates may be allocated, according to their location, to five groups (Fig. 22). (1) The dorsal group consists of paired nasal, frontal, parietal and postparietal bones. The parietal (pineal) foramen is situated in the midline between the two parietal bones. (2) Enclosing and completing the orbit are the circumorbital bones: prefrontal and postfrontal superiorly, jugal inferiorly, lacrimal antero-inferiorly and postorbital posteroinferiorly. (3) Behind the circumorbital group lie the temporal bones: intertemporal, supratemporal and tabular which overlie the subtemporal fossa occupied by the jaw muscles. (4) Completing the cheek region, inferior to the temporal group and posterior to the circumorbital series, are the two cheek bones: the squamous and quadrotojugal. These connect with and help support the quadrate. (5) The margins of the upper jaw are made up of the tooth-bearing premaxilla and maxilla. (For detailed descriptions, including extensive bibliographies, of the dermal bones of the skulls of non-mammalian tetrapods, the reader is referred to Goodrich, 1930; Romer, 1956; Olson, 1971; Romer & Parsons, 1977.)

There is frequently an opening, the posttemporal fenestra, on the posterior surface of the skull between the dermal shield and the paroccipital process. In living forms it transmits a vein draining the occipital region.

Dermal bones ossify also in the first arch mesoderm related to the mucous membrane of the vault of the mouth to form additions to the palate and floor of the braincase. The dermal part of the palate in rhipidistians (Fig. 21*b*) and early tetrapods (Fig. 22*b*) is strikingly similar and consists of four paired bones. The largest element is the pterygoid (endopterygoid) which can be visualised as a plate of bone twisted along its anterioposterior axis, so that its anterior part lies in the transverse plane but its posterior part in the vertical plane. It is closely applied to the ventral surface of the epipterygoid and may enter into the palatobasal articulation. Its posterior part, or quadrate process, connects with the quadrate and is applied to the medial surface of the quadrate ramus of the epipterygoid. The three remaining dermal elements are attached to the anterolateral edge of the pterygoid – the vomer anteriorly, and the palatine (dermal palatine) and ectopterygoid more posteriorly – the whole making up a transverse sheet of bone running across the roof of the mouth between the marginal bones of the two sides. The internal nostrils are bounded by the vomers medially, the marginal bones laterally and the palatines posteriorly. The posterior, more vertical, parts of the two pterygoids are separated from each other to leave a midline gap through which the ventral surface of the braincase can be seen. Between the cheek bones and the posterior part of the pterygoid is a space roofed in by the temporal bones and completed anteriorly and posteriorly by the ectopterygoid and quadrate, respectively. This is the subtemporal fossa which houses the muscles moving the lower jaw. It opens ventrally through the subtemporal fenestra which is bounded anteriorly by the pterygoid and ectopterygoid, posteriorly by the quadrate, medially by the quadrate process of the pterygoid and laterally by the cheek bones.

The dermal palatal bones are closely applied to the palatoquadrate ossifications whose connections with the braincase thus help to determine the mobility of the whole palatal complex. This topic is considered in more detail below in the discussion of skull kinetism.

Applied to the ventral surface of the braincase is the unpaired dermal parasphenoid bone. In the rhipidistian skull this bone lies entirely beneath the anterior moiety of the braincase and would not, therefore, have hindered movements at the joint between the two moieties (Fig. 21*b*). However, in tetrapods where the two moieties of the braincase are fused, the parasphenoid has extended posteriorly to underlie both parts (Fig. 22*b*). It may fuse with the adjacent basisphenoid bone and frequently possesses strong transverse processes which lie ventral to and support the basipterygoid processes of the basisphenoid. The palatine branch of the facial nerve, which runs immediately beneath the basipterygoid process, is thus enclosed in a bony (vidian) canal. The rostral part of the parasphenoid, extending underneath the sphenethmoid, is sometimes termed the cultriform process.

The modern bony fishes show numerous departures in the structure of the dermal shield and palate from the arrangement just described for rhipidistians and primitive amphibians. These are mostly specialisations for varying modes of aquatic life and have no direct bearing on the phylogeny of the tetrapod skull (for full descriptions see Harrington, 1955; Patterson, 1975). Briefly, it can be mentioned that the actinopterygian braincase contains dermal bones comparable (but not always directly homologous) with those of the tetrapod skull. The bones of the cheek region are usually deficient (the quadrate being supported by the hyomandibula), the jaws shortened in association with the forward displacement of the ventral part of the hyomandibula and the marginal bones largely freed from the remaining bones of the skull.

In the later tetrapods there has been a tendency for elements of the dermal shield to undergo degeneration and be lost. Although the primitive pattern of bones in the dermal shield was retained (except for the loss of the intertemporal) in the cotylosaurs, there have been considerable losses and modifications in the later reptiles, especially in connection with the development of temporal fenestrae. In the cotylosaurs the jaw closing muscles took origin from the undersurface of the roof of the subtemporal fossa and passed down to the lower jaw through the subtemporal fenestra. This is the anapsid condition (Fig. 23*a*). It is retained (or possibly reacquired) by the Chelonia amongst the living reptiles. Generally, however, new openings have developed in the temporal region of the reptilian skull to allow the jaw-closing muscles to bulge outwards as they thicken during contraction. In the majority of reptiles, living and extinct, two such openings are present – the diapsid condition (Fig. 23*b*). In early diapsids the superior of the two fenestrae is situated between parietal, postorbital and supratemporal bones and the

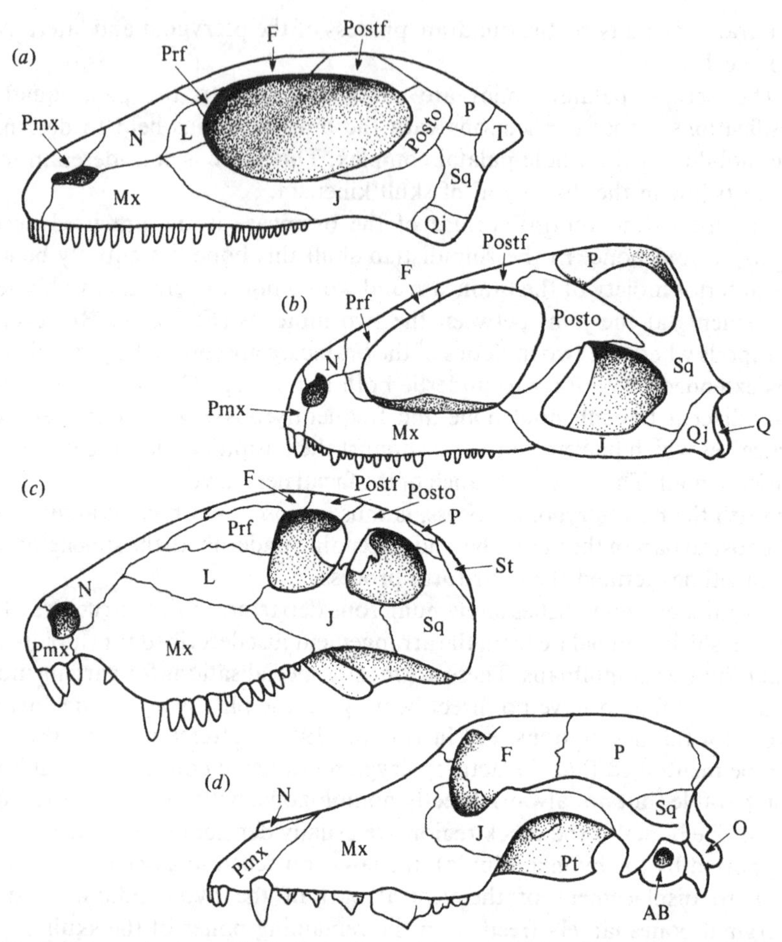

Fig. 23. Left lateral views of skull to show (*a*) anapsid (cotylosaur), (*b*) diapsid (*Sphenodon*), (*c*) synapsid (*Dimetrodon*), (*d*) mammalian (dog) structure of temporal region. Not to scale. For labelling see list of abbreviations. (*a* and *b* after Romer, 1956; *c* after Romer & Price, 1940.)

inferior between postorbital, jugal, quadratojugal and squamous bones. In later diapsids considerable divergences have occurred in the bones in the neighbourhood of the temporal fenestrae. Amongst living types *Sphenodon* (probably one of the least specialised) has lost the temporal, tabular and postparietal bones, while in lizards the inferior temporal arch and in snakes both arches have been secondarily lost. In the mammal-like reptiles only a single opening developed – the lateral temporal fenestra of the synapsid condition (Fig. 23*c*). The fenestra is located between the postorbital, squamous and jugal, with the first two elements meeting above, never below, the opening. A single temporal opening also evolved in the extinct Euryapsida but differs from the lateral temporal fenestra of the synapsids in being more dorsally placed.

The transition to the mammalian condition from that in the mammal-like reptiles has been brought about by the enlargement of the temporal fenestra and the loss of the prefrontal, postfrontal, postorbital, supratemporal and quadratojugal bones (Fig. 23*d*). The tabular bone appears also to have been lost in most mammals but may be represented by interparietal ossification centres in others (de Beer, 1937). The originally paired postparietal bones are reduced to a single median element in pelycosaurs and therapsids and may be represented in mammals by interparietal ossification centres which fuse with the supraoccipital bone. Following the progressive enlargement of the temporal fenestra through the mammal-like reptiles to mammals, the ventral part of the side wall of the dermal shield is reduced to a narrow bar, the zygomatic arch, made up of the jugal (zygomatic or malar) and the zygomatic processes of the maxilla and squamous. The zygomatic arch, though a diagnostic feature of mammals, has a very variable structure. Many mammalian orders exhibit a tendency towards diminution of the jugal. Hogben (1919) suggested that the primitive condition is that found in most marsupials and many less specialised placentals where the jugal extends from the lacrimal anterodorsally to the mandibular fossa of the jaw joint. More generally, the jugal has been encroached upon by the squamous posteriorly and by the maxilla anteriorly. The precise composition of the zygomatic arch is thus extremely variable, its constitution varying quite independently of such functional correlates as its thickness or shape (see Hogben (1919) for further details). The bar between the orbit and temporal fossa is also reduced and has been lost completely, together with the postorbital and postfrontal bones, in advanced therapsids and many mammals but has been reconstituted in primates by union of the postorbital process of the jugal with the frontal. A similar enclosure of the orbit has been effected in some specialised members of other orders, notably the ungulates, but here the postorbital bar may or may not include the jugal, depending on the constitution of the zygomatic arch (generally the jugal enters into the postorbital bar in artiodactyls but not in perissodactyls).

Although the general tendency amongst tetrapods has been for reduction and loss of elements from the dermal shield, one new addition may be the septomaxillary bone which is found in the posterior margin of the external nostril, immediately anterior to the maxilla and extending inwards to overlie the vomeronasal organ, in many reptiles (including the mammal-like reptiles), non-therian mammals (where it fuses with the premaxilla) and Edentata (and possibly some other eutherian mammals). Its phylogenetic history is obscure. Whether or not the septomaxilla is a true dermal bone or is an example of a cartilage replacing bone transformed during phylogeny into a membrane bone has yet to be decided. Goodrich (1930) described it as having been derived from one of the dermal plates bordering the external nostril in the fish ancestors of the tetrapods, in which case the septomaxilla would be a retention of an ancient bone, not a new addition.

The parietal foramen has been lost in modern amphibians and many reptiles, as well as in birds. Although the foramen is still present in pelycosaurs it is reduced or absent in the more advanced mammal-like reptiles and missing entirely in mammals.

A new feature of land vertebrates is the tympanic membrane. In the labyrinthodonts this is generally believed to have occupied the otic notch between the tabular and the bones of the cheek (Fig. 22). In other early amphibians the membrane may have been located more ventrally (the possibility that a tympanic membrane was lacking at this stage has been considered in Chapter 2; see also Panchen (1972) for a discussion of the tympanic membrane in early tetrapods). In primitive reptiles it is generally believed to have been located behind the quadrate bone and to have lain flush with the external surface of the head. This condition is found, for example, in modern turtles but in some reptilian groups, in birds and in mammals the ear drum has come to occupy a position at the base of a short external acoustic meatus developed by a thickening of the surrounding bones. An external acoustic meatus may also have been present in advanced mammal-like reptiles (see Chapter 4).

The tympanic membrane of mammals is attached around its perimeter to the tympanic bone. The fossil evidence indicates that this element is derived from the angular, a dermal bone of the lower jaw which lies close to the presumed position of the tympanic membrane in mammal-like reptiles (Fig. 29). In its most primitive form the tympanic is just a ring of bone, still having this structure in monotremes and marsupials. During its early ontogeny in eutherian mammals the tympanic has at first a ring-like structure but it subsequently expands to form part or, in some forms, all of the auditory bulla, a capsule of bone which encloses and protects the tympanic cavity. The periotic bone and auditory bulla remain separate in some mammalian types but in others they fuse with each other and with the squamous to form the temporal bone. The main (hyomandibular) trunk of the facial nerve, which originally left the braincase through the facial foramen located just anterior to the otic capsule, emerges in mammals behind the bulla through the stylomastoid foramen. A detailed account of the evolution of the tympanic bone will be found in Chapter 4 and of the auditory bulla in Chapter 6.

The palatal dermal bones of advanced tetrapods, like the dermal shield, have undergone considerable modifications from the arrangement seen in the early amphibians. This is especially true of the later amphibians, where both the cartilage replacing and dermal components of the palate are much modified and elements are lost. In primitive reptiles, including the early mammal-like reptiles, the palatine bones follow essentially the same plan as that found in primitive amphibians but again there is considerable divergence

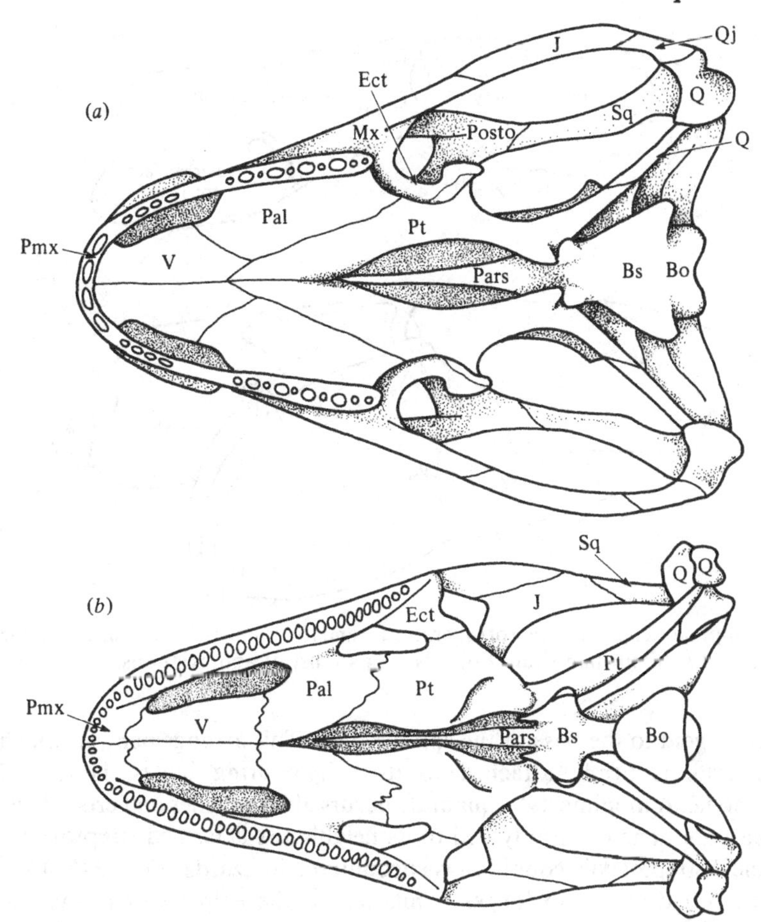

Fig. 24. Ventral views (somewhat simplified) of skull of (*a*) *Sphenodon* and (*b*) *Iguana*. Not to scale. For labelling see list of abbreviations.

in the later reptilian groups. A characteristic of the reptilian palate is the presence of a transverse ridge, the lateral or pterygoid flange, on the pterygoid bone which is often tooth bearing.

Amongst living reptiles the least modification to the palate has occurred in *Sphenodon* (Fig. 24*a*). Here, the parasphenoid is small and becomes fused to the basisphenoid during ontogeny, its transverse processes fusing with the basipterygoid processes to enclose, on each side, the palatine branch of the facial nerve in the vidian canal. The basipterygoid processes have movable articulations with the pterygoids. The pterygoids diverge posteriorly to reach and support the quadrate bones which gain further support from their sutural connections with the quadratojugal and squamosal elements of the dermal shield. The quadrate has a broad, anterior prolongation which reaches lateral

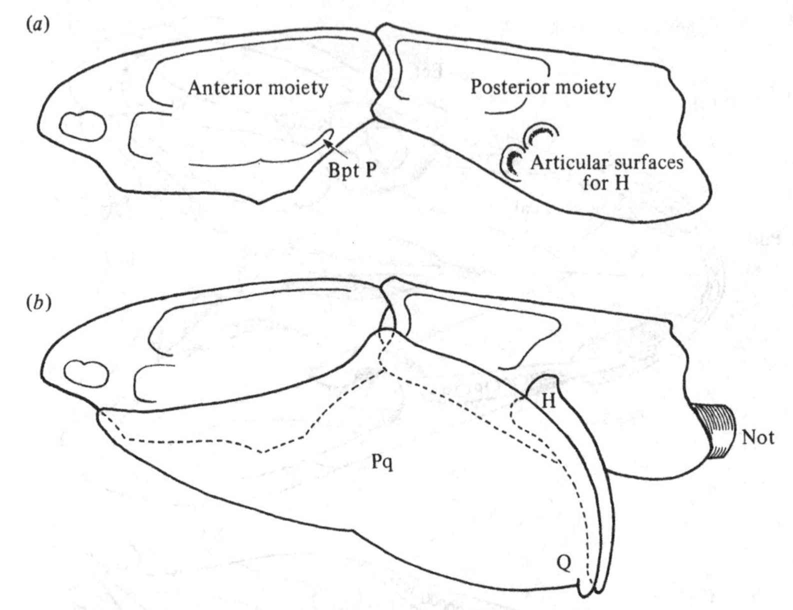

Fig. 25. Simplified left lateral view of (*a*) braincase and (*b*) braincase and palatoquadrate complex of the rhipidistian skull. For labelling see list of abbreviations.

to the pterygoid to the base of the epipterygoid. This arrangement, in which the quadrate is firmly attached to its neighbouring bones, is termed monimostylic. A monimostylic quadrate occurs also in turtles and crocodiles.

A quadrate bone not firmly tied to its neighbours is termed streptostylic. The typical streptostylic condition is encountered in lizards (Figs. 24*b*, 40*a*), where the quadrate is no longer connected to the epipterygoid, and the quadratojugal bone has been lost as part of the reduction of the inferior temporal arch. The quadrate articulates by movable joints at its upper end with the squamous and at its lower end with the pterygoid. It thus possesses a considerable degree of mobility.

The mobility of the quadrate is related to the condition of skull kinetism. In the skull of the rhipidistian crossopterygian fishes the braincase consists of two separate moieties, an anterior, or oticooccipital, and a posterior, or ethmosphenoid (Figs. 21, 25). Although descriptions of the braincase (e.g. Romer, 1937; Jarvik, 1954; Thomson, 1965, 1967) are necessarily of the fossilised (i.e. bony) adult stage, it is probable that, developmentally, the posterior and anterior moieties were related respectively to the basal plate and trabeculae of the chondrocranium. The cephalic portion of the notochord is unconstricted and passes forwards through a large canal in the floor of the posterior moiety to end in the floor of the anterior moiety. Immediately above this break in the continuity of the braincase is the transverse hinge in the dermal shield already described. The flexibility of the notochord clearly allowed the anterior moiety of the braincase, together with the associated anterior segment of the dermal shield, to move in a dorsoventral direction relative to the posterior moiety and segment.

In addition to the notochord the two sections of the skull are connected by the hyomandibula and palatoquadrate complex (Fig. 25). The palatoquadrate is attached to the anterior moiety by a firm union anteriorly and by the movable palatobasal joint somewhat further back. The hyomandibula has a double articulation with the posterior moiety dorsally and a single articulation with the palatoquadrate ventrally. As Thomson (1967) has pointed out, the orientation of the articular facets of the hyomandibula is such that when the anterior section of the skull moved dorsoventrally the quadrate part of the palatoquadrate would have moved in a more horizontally orientated plane (the joints between the cheek bones of the rhipidistian skull being arranged in a manner that would have allowed this mediolateral movement to take place). The functional significance of this type of kinetism is uncertain, but Thomson has argued convincingly that it was related to the exigencies of feeding in shallow water.

It was for long believed that, because of the jointed structure of the braincase, the skull of the rhipidistians could not have been directly antecedent to that of the land vertebrates. It is now known, however, that very early amphibians (such as the Ichthyostegalia from the Upper Devonian) retain a line of division across the ventral aspect of the braincase, suggesting that the rhipidistian structure may have become modified by a process of fusion to give the non-jointed structure typical of the tetrapod skull. Despite this loss of intracranial kinetism the mobility between the palate and the braincase was retained and is a feature of the skull of many reptiles, both extinct and living.

In *Sphenodon*, for example the quadrate, palatal complex and dermal shield (collectively termed the maxillary segment) are rigidly connected to each other but their connections with the braincase (the occipital segment) are somewhat looser – the palatobasal articulation is incompletely obliterated, the occipital region of the braincase is joined to the parietal bones of the dermal shield by small amounts of cartilage or ligament, and the anterior part of the braincase and the interorbital septum are incomplete and partly membranous. Thus, the maxillary and occipital segments can move slightly relative to each other under the influence of muscle action. The degree of kinetism is much greater in lizards where the palatobasal articulation is freer, the connections between the braincase and parietal bones are very loose (the metakinetic condition) and the quadrates, as already described, are markedly streptostylic. In some lizards there is, in addition, a transverse, movable joint across the dermal shield in the form of a loose suture between the frontal and parietal bones (the mesokinetic condition – this is a secondary development not homologous with the transverse joint in the rhipidistian dermal shield which is situated between the parietal and postparietal elements). When the lower end of the quadrate is moved forwards by muscle action the palate and anterior part of the dermal shield (including premaxillae and maxillae) tilt upwards as a unit about the axis of the joint. This action plays an important part in feeding (for a discussion of the functional aspects of kinetism, see Frazzetta (1962) and Bellairs (1969)). Still greater degrees of streptostylism and skull kinetism are seen in the extremely mobile skulls of snakes (Frazzetta, 1966). It should be noted that while, in general, streptostylism is associated with and complements high degrees of kinetism, this is not always so: in mammals, for example, the quadrate (as the incus) is freely mobile but the skull is akinetic. Conversely, as we have seen, the quadrate in *Sphenodon* is monimostylic but the skull still possesses a measure of kinetism; indeed this was probably the primitive tetrapod condition, being found, amongst other early reptiles, in the pelycosaurs. The subsequent phylogenetic changes in skull kinetism in the progression from the mammal-like reptiles are considered in Chapter 4.

Two reptilian groups, the crocodiles and the mammal-like reptiles, exhibit modifications of the maxillary and palatine bones, leading to the formation of a secondary palate, a continuous shelf of bone situated beneath the primary palate. In both cases, medially projecting flanges of bone, termed the palatine processes, are developed from the laterally disposed elements of the primary palate. These processes grow across and eventually fuse in the midline to produce the secondary palate (Fig. 36). The part of the oral cavity enclosed between the primary and secondary palates becomes an extension of the air passages which, in consequence, now have their internal openings located far posteriorly. The development of a secondary palate facilitates breathing, while the mouth is being used to seize and masticate food. In the crocodiles this is clearly of advantage in that it allows breathing to be continued whilst prey is held below the waterline, while in the mammal-like reptiles the greater metabolic requirements of the warmblooded state (generally assumed to have evolved in this group) may well have demanded continuous respiration whilst eating. The complete secondary palate of mammals is directly derived from that of the mammal-like reptiles. A detailed account of the evolution of this structure and the associated modifications of the pterygoid and parasphenoid bones is more conveniently delayed until Chapter 4.

LOWER JAW

It was seen in Chapter 2 that a variable number of ossifications may occur in Meckel's cartilage. In all gnathostomes, except the Chondrichthyes, these cartilage replacing bones are supplemented by dermal elements which ossify in the first arch mesoderm of the lower jaw. As elsewhere in the skull, there has been a trend for the number of dermal elements to be reduced in the more advanced vertebrates (the reader is again referred to Goodrich (1930), Romer (1956), Olson (1971), Romer & Parsons (1977) for detailed accounts of these elements in non-mammalian vertebrates).

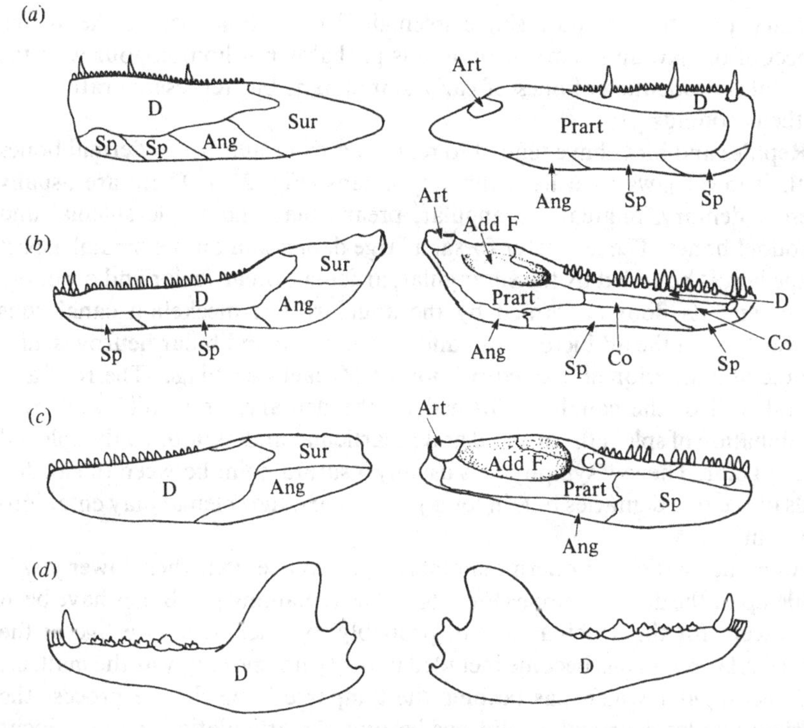

Fig. 26. Lower jaws of (*a*) rhipidistian, (*b*) primitive amphibian, (*c*) primitive reptile and (*d*) mammal (dog). Lateral views to left; medial views to right. Not to scale. For labelling see list of abbreviations. (*a*, *b* and *c* after Romer, 1962.)

In early fossil bony fishes the outer surface of the jaw is formed by the dentary, angular, surangular and a variable number of infradentary bones. The dentary is the marginal, tooth-bearing bone. It makes a sutural connection in the anterior midline with its fellow of the opposite side. Medially, the sheathing bones are the prearticular and a number of coronoid bones. In modern actinopterygians the number of dermal bones in the lower jaw has been greatly reduced and the homologies of the remaining bones are confused. There is usually present a dentary (containing a fused cartilage replacing bone, probably representing the mentomeckelian) and an 'angular' bone, although this may not represent the original angular and may itself be lost. In addition, the cartilage replacing articular bone may be replaced by, or possibly fused with, a dermal component which can possibly be equated with the true angular (Harrington, 1955).

As might be expected, the lower jaws of the rhipidistian fishes and early amphibians closely resemble each other, and are much nearer to the primitive condition than is the case in the modern actinopterygians. Both possess a fairly complete complement of dermal bones (Fig. 26*a*, *b*). In the modern amphibians, however, the number of dermal bones is reduced, as a rule, to dentary, prearticular and a single splenial. The last bone lies on the medial aspect of the jaw and bears teeth, and is probably not homologous with the splenial (infradentary) bones of early amphibians but represents rather one of the coronoids.

Reptiles and birds have tended to retain a greater number of dermal bones in their lower jaws than have the amphibians (Fig. 26*c*). There are usually present dentary, angular, surangular, prearticular and single splenial and coronoid bones. The adductor fossa, a large depression on the medial aspect of the jaw, is bounded by the surangular, articular, prearticular and coronoid bones and its floor is formed by the angular. The meckelian canal runs forwards from the adductor fossa and houses the mandibular neurovascular bundle and anterior unossified portion of Meckel's cartilage. The roof and lateral wall of the canal are formed by the dentary, its medial wall by a combination of splenial, coronoid and prearticular and its floor by the splenial and angular. The jaw symphysis is usually a sutural joint between the medial ends of the two dentaries but, in long-jawed types, the splenial may enter into the joint.

A characteristic of modern mammals is, of course, that their lower jaw is made up of the dentary alone (Fig. 26*d*). The remaining jaw bones have been lost except for the articular which (probably together with a vestige of the prearticular bone) has become included in the tympanic cavity as the malleus, and the angular which has become the tympanic bone. In the process the original quadrate–articular joint has become the articulation between incus and malleus and a new jaw joint has been established between the dentary and squamous bones. The stages involved in the evolution of the mammalian lower jaw and ear ossicles from the reptilian structures are known, or can be inferred with some confidence, from the fossil record of the mammal-like reptiles and are fully described in the next chapter.

Although the dentary is usually regarded, phylogenetically, as a dermal bone, it may receive minor cartilage replacing contributions from centres of endochondral ossification in Meckel's cartilage. In some mammalian species, notably man, Meckel's cartilage undergoes little or no ossification and may then persist as a rod of cartilage or atrophy (see Friant (1960, 1966) for accounts of the fate of Meckel's cartilage in a wide variety of mammals). Further endochondral bone is added to the dentary by the development within it of secondary cartilages (see Chapter 8 for a full description of this tissue). The close proximity of the dentary and other dermal bones of the lower jaw to a primary cartilage is an arrangement not encountered elsewhere in the body, and, as pointed out by Hall (1975*b*), raises the intriguing possibility that these dermal elements represent perichondral bone which originally developed in the perichondrium of Meckel's cartilage in a manner akin to that in which intramembranous bone forms around the diaphysis of an endochondrally ossifying long bone of the postcranial skeleton.

SECTION II

EVOLUTION OF THE MAMMALIAN SKULL

4

EVOLUTION OF THE MAMMALIAN SKULL

ORIGIN OF MAMMALS

While the purpose of this chapter is not to discuss in detail the evolution of the mammals from their reptilian ancestors, a brief consideration of this complex and still controversial subject is necessary as a background against which to trace the major steps in the development of the characteristic features of the mammalian skull (for comprehensive and critical treatments of this subject the reader is referred to Olson (1959), Simpson (1959, 1961, 1971), Kermack (1963), Hopson & Crompton (1969), Parrington (1971), Kermack, Mussett & Rigney (1973) and Jenkins & Parrington (1976); many of these studies include consideration of the dental and postcranial parts which is generally excluded from the present account).

The stem reptiles (cotylosaurs), from which it is believed that the later reptilian groups were derived, first appear in the fossil record in the Carboniferous period. Their skulls lack temporal openings. At a very early stage, still during the Carboniferous, the cotylosaurs gave rise to the pelycosaurs, the earliest of the Synapsida (mammal-like reptiles). The pelycosaurs, which became widespread during the Permian period, were generally rather primitive but in certain skull features, such as the presence of a single temporal opening, they foreshadowed the mammalian condition. Three major groups of pelycosaurs are known: the ophiacodonts, the carnivorous sphenacodonts and the herbivorous edaphosaurs.

During the Early Permian the sphenacodont pelycosaurs (or a related group) gave rise to the progressive therapsids. The therapsids diversified early into two major groups: the anomodonts, consisting of a large variety of carnivorous and herbivorous forms usually of large body size; and the early theriodonts, a less variable group of small to medium sized carnivores with a skeletal morphology generally similar to that of their pelycosaur ancestors. By the end of the Triassic, other reptilian groups, notably the dinosaurs, had become prominent and the anomodonts disappear from the fossil record. During the Middle Permian more advanced theriodonts, the gorgonopsians and therocephalians, appeared. The gorgonopsians remained morphologically conservative apart from a tendency to enlargement of the incisor and canine teeth and reduction in size and number of the postcanine teeth. The therocephalians were a more progressive group which, even in the primitive pristerognathid forms, possessed cranial and postcranial features tending towards the mammalian type – notably, an expanded epipterygoid and a mammalian phalangeal formula. The later therocephalians had also begun to develop a secondary palate but tended to become specialised in their dentition. Neither the therocephalians nor the gorgonopsians survived the Permian.

At an early stage in their evolution the therocephalians gave rise to the bauriamorphs, a still more advanced theriodont group. The more primitive members of this group were small, probably insectivorous animals. Even at this early stage, however, they had acquired several progressive features, including a secondary palate, reduction of the postorbital bar and loss of the parietal foramen.

Also derived from an early theriodont stock were the cynodonts, generally of small to medium body size (i.e. up to the size of a large dog). Their exact derivation has been much debated, having been attributed to the gorgonopsians (Watson, 1921) and to the therocephalians (Broom, 1938; Brink, 1960). More recently Kemp (1972*a*) has suggested, principally on the basis of cranial evidence, a close relationship to the whaitsiid therocephalians (the whaitsiids are a therocephalian family from the Upper Permian with a specialised dentition). The earliest known cynodonts, the procynosuchids, date from the end of the Permian. The later members of this group possessed numerous cranial features approaching the mammalian condition. These include a complete secondary palate, an expanded epipterygoid, a lateral flange on the prootic, a large dentary and features of the lateral part of the dermal shield and of the lower jaw which suggest that the external adductor muscle of the jaw was differentiating into the typical temporal and masseter muscles of mammals.

From the Early and Middle Triassic a variety of cynodont families is known, including the galesaurids, cynognathids and traversodontids amongst others. The members of these families have, in addition to the cynodont characteristics already described, a still further enlarged dentary, with coronoid and angular processes and an extensive fossa for the insertion of the masseter (indicating that this muscle was of mammalian proportions), and postdentary bones that are correspondingly reduced in size. These trends are seen to be still further advanced in the late reptilian derivatives of the cynodonts, the ictidosaurs (including the tritylodontids) of the Late Triassic.

It is generally agreed that the mammals have a theriodont ancestry but which type or types of theriodont were directly involved is still debated. Many authorities (e.g. Simpson, 1959, 1961; Olson, 1944, 1959) have stressed the varying degrees of evolutionary advance towards the mammalian condition which occurred in each of the theriodont groups, especially the cynodonts, bauriamorphs and ictidosaurs, and suggest that the mammalian threshold may have been crossed independently by several lineages, probably during the Triassic. Others believe that the evidence, especially that provided by the newer mammalian finds from the Late Triassic, more strongly supports a monophyletic origin for mammals. Hopson & Crompton (1969), for example, have taken this view and, from a consideration of the dentitions of the early mammals, came to the conclusion that this origin may have involved a single family of the Cynodontia, possibly of early galesaurid stock.

Until relatively recently the earliest known mammals came from the Jurassic by which time several distinct forms were already in existence, known mainly from their jaws and teeth. Most authorities have recognised five groups placed at the ordinal level, the Multituberculata, Triconodonta, Docodonta, Symmetrodonta and Pantotheria, but there is considerable disagreement about their relationships. In more recent years the fragmentary remains of a number of mammals from the Late Triassic of Europe, Africa and Asia have been discovered. These were all very small, probably insectivorous animals and appear to fall into five genera: *Morganucodon* (there is much dispute about the correct name of this animal; *Eozostrodon* may be a synonym used prior to *Morganucodon*; this issue has no direct bearing on the present discussion and the name *Morganucodon* is adopted here merely because it is the one used by Kermack and his colleagues in their detailed descriptions of the braincase and lower jaw of this animal which form the bases of two later sections of this chapter); *Erythrotherium*, *Megazostrodon* and *Sinacodon* (all three probably closely related to *Morganucodon*); and *Kuehneotherium*.

These newer finds have, as yet, led to no greater unanimity of opinion about the relationships between the early mammals or about those between the early mammals and later types. Simpson (1971), in a review of work in this field, has offered the following tentative scheme. Of the living mammals, the Metatheria and Eutheria probably share a common ancestry which evolved from the Pantotheria. The Pantotheria appear to have been closely allied to the Symmetrodonta and both may have evolved from an ancestor like *Kuehneotherium*. The evidence provided by the teeth and the structure of the side wall of the braincase (see below) suggests that the non-therian Jurassic mammals, the Triconodonta, Docodonta and possibly the Multituberculata, were derived from a *Morganucodon*-like form. The living non-therian mammals, the monotremes, have no known fossil record but in the structure of their braincase wall resemble the Jurassic non-therians and *Morganucodon*. Hopson & Crompton (1969) from dental evidence have suggested that *Morganucodon* and *Kuehneotherium* were quite closely related to each other, both having a possible galesaurid ancestry.

The eutherian mammals began their first major radiation towards the end of the Cretaceous. This led to the establishment of the characteristic fauna of the Palaeocene which included archaic primates, ungulates, carnivores, insectivores and the multituberculates. Most of these archaic groups became extinct during the Eocene or shortly after. The beginning of the Eocene is marked in the fossil record of Europe and North America by the appearance of mammals belonging to the modern orders: Rodentia, Primates, Chiroptera, Carnivora, Artiodactyla, Perissodactyla, Cetacea and Lagomorpha. Unfortunately, fossil evidence bearing on the origin of these orders is generally lacking.

(*a*)

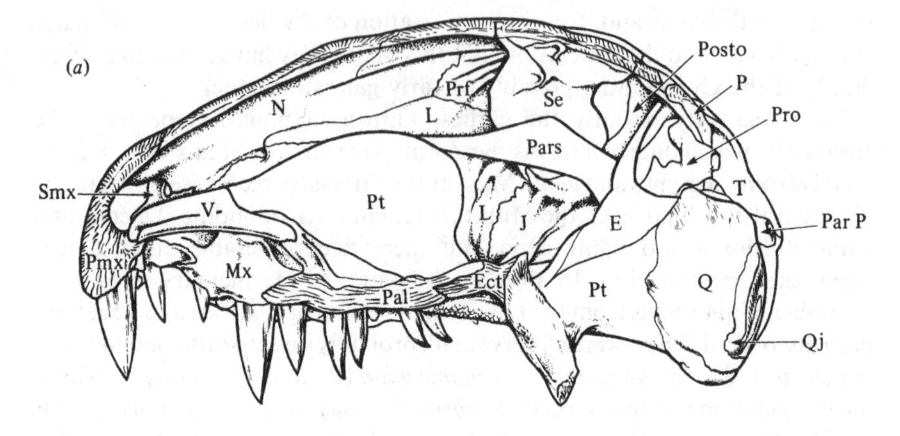

(*b*)

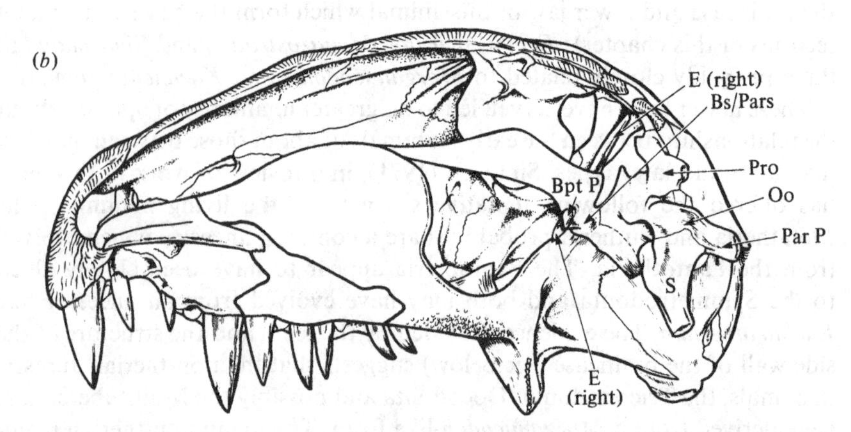

Fig. 27. To show cartilage replacing bones of skull of *Dimetrodon*. (*a*) Left lateral view with dermal shield of left side removed; (*b*) left lateral view as in (*a*) but with bones of palate and palatoquadrate complex of left side also removed. For labelling see list of abbreviations. (After Romer & Price, 1940.)

SKULL OF SPHENACODONT PELYCOSAURS

It is clear from the above account that a convenient starting point from which to trace the evolution of the mammalian skull is provided by the cranial morphology of the sphenacodont pelycosaurs. The description that follows is based largely on *Dimetrodon*, chosen because the skull of this sphenacodont is particularly well known and has been the subject of several excellent descriptions (e.g. Case, 1910; Romer & Price, 1940).

BRAINCASE AND OTIC CAPSULES

The cartilage replacing bones of the pelycosaur braincase include all the major elements typical of land vertebrates (Figs. 27, 28*b*, *c*). These elements show a strong tendency to fuse with each other. The basioccipital bone forms the ventral surface of the posterior part of the braincase but is excluded from the internal surface of the cranial cavity by the exoccipitals which cover it dorsally. The exoccipitals constitute the lateral margins of the foramen magnum and connect suturally above with the supraoccipital. Each is pierced by two foramina for the roots of the hypoglossal nerve. A basal ramus passes dorsal to the basioccipital and meets its fellow of the opposite side in the midline, the rami of the two sides together providing the posterior part of the floor of the cranial cavity. The exoccipitals form the dorsolateral parts of the single occipital condyle, the remaining, major, portion being provided by the basioccipital. In mature specimens the basioccipital and exoccipitals are fused. The supraoccipital becomes firmly fused with the bones of the otic capsules to make up a continuous occipital plate of bone in the posterior part of the roof and in the side walls of the braincase anterior to the vagus. The occipital plate provides the superior margin of the foramen magnum from which it slopes forwards as the roof of the braincase to meet and extend beneath the unpaired postparietal bone of the dermal shield. Lateral to the postparietal the supraoccipital makes contact with the tabular bones. The occipital plate is produced on each side ventrolaterally as the paroccipital process which passes beneath the tabular to contact the squamous and quadrate. Between the paroccipital process and tabular is the posttemporal fenestra. The paroccipital process probably ossified from the opisthotic as well as from the supraoccipital bone.

The otic capsules comprise opisthotic and prootic bones tightly fused with each other and with the supraoccipital. The major element appears to be the prootic, the opisthotic being confined superficially to the paroccipital process and posterior margin of fenestra vestibuli. Each capsule is shaped like a three-sided pyramid with medial, dorsolateral and ventrolateral surfaces. The two capsules pass obliquely forwards from the occipital plate, on either side of the cranial cavity, converging so that their apices meet anteriorly to form a ridge of bone (the dorsum sellae – in cotylosaurs, living reptiles and mammals this ridge is ossified from the basisphenoid). The medial surface of the otic capsule forms the lateral wall of the medullary region of the cranial cavity and is pierced by a wide internal acoustic meatus. The upper part of the ventrolateral surface provides the medial wall of the cranioquadrate passage. Posteriorly, this surface is occupied by the fenestra vestibuli for the footplate of the stapes (a fenestra cochleae appears to be lacking in pelycosaurs as in many other synapsids). The dorsolateral surface runs forwards without interruption from the occipital plate and paroccipital process to the lateral edge of the dorsum sellae. The medial and dorsolateral surfaces meet at the dorsal margin. This is notched for the escape of the root of the trigeminal nerve from the cranial cavity into the cavum epiptericum.

The triangular area of the floor of the braincase between the converging medial walls of the otic capsules is completed, as already described, by the basioccipital overlain by the exoccipitals. Medially, at the line of junction of the opisthotic with this occipital plate, is the jugular (posterior lacerate) foramen for nerves nine to eleven.

The cartilage replacing basisphenoid is intimately fused on its ventral surface with the dermal parasphenoid (Figs. 27, 28*c*). The posterior part of the dorsal surface of the basisphenoid forms the sella turcica. Behind the sella the compound bone extends towards but is usually not fused with the dorsum sellae (presumably cartilage intervened in the recent state). The parasphenoid extends further posteriorly than does the basisphenoid and overlaps the ventral surfaces of the apices of the otic capsules and anterior end of the basioccipital. Here, the parasphenoid is enlarged on each side to form the basisphenoidal tubera (presumably for attachment of prevertebral muscle).

In the region below the sella turcica the basipterygoid processes project ventrally close to the midline. On each process, there are two articular facets, one located venterolaterally, the other dorsolaterally. The facets appear to be on bone belonging to the basisphenoid but ensheathed by bone belonging to the parasphenoid. Grooves for the palatine nerves and internal carotid artery run forwards on the ventral surface of the bone between the basipterygoid processes. These grooves are not enclosed ventrally. The further course of the internal carotid artery is uncertain. It may have turned upwards anterior to the basipterygoid process and lateral to the parasphenoid–basisphenoid complex or alternatively it may have entered the cranial cavity through a bilaterally paired foramen situated on the ventral surface of the bone between the basipterygoid processes.

Anterior to the basipterygoid processes the bone becomes keel-shaped in section with the walls of the keel being formed by the parasphenoid enclosing a core of cartilage replacing bone. Its ventral margin runs anterosuperiorly to a point well above the palate in the interpterygoid space. The dorsal surface of the bone rises even more steeply before turning forwards at an abrupt angle. The basisphenoid ends at a variable point in the trough formed by the parasphenoid. The side wall of the braincase above the basisphenoid is devoid of bone and presumably consisted in the living animal of the unreplaced cartilage of the side wall (pila antotica) of the chondrocranium in this region and membrane.

The sphenethmoid bone in its posterior part is Y-shaped in section. The vertical plate lies in the median plane of the skull and forms the interorbital septum (Fig. 27). Its ventral border is received by the parasphenoid trough anterior to the termination of the basisphenoid. The dorsal expansion of the bone (i.e. the arms of the 'Y') appears to correspond to the planum supraseptale of the cartilaginous braincase. It is produced into a process on each side which probably continued in cartilage as the sphenethmoidal commissure to the lamina orbitonasalis. Anteriorly, the median vertical plate of bone continues upwards without a dorsal expansion until it meets the dermal shield. Here, it spreads out into a plate which is attached to the ventral surface of the frontal bone near the anterior margins of the orbits.

On the superior surface of the dorsal expansion are two grooves on either side. The medial groove is converted to a foramen anteriorly by a bridge of bone while the lateral groove also becomes a foramen when the articulating frontal is in place. These foramina probably transmitted lateral and medial branches of the olfactory nerve.

The nasal capsules of the pelycosaurs remained entirely cartilaginous, no cartilage replacing bone having been found anterior to the sphenethmoid.

DERMAL SHIELD

The bones of the dermal shield, unlike the cartilage replacing ossifications, tend not to fuse but remain distinct even in specimens of full maturity.

The general construction of the dermal shield is shown in Fig. 28*a*, *b*. The full complement of elements typical of primitive land vertebrates is present with the exception of the intertemporal bone. The shield is relatively narrow, especially anteriorly, its widest part being located in the postorbital region. Its dorsal surface is flat and ends posteriorly at a transverse occipital crest which runs approximately along the line of sutures between the parietal bones in front and the postparietal and tabular bones behind. The latter, therefore, form part of the occipital surface of the skull. The parietal foramen lies in the midline at the posterior end of the suture between the two parietals. The postparietal is an unpaired median bone.

There is no otic notch in the pelycosaur skull. In primitive amphibians the notch was bounded by tabular, squamous and supratemporal bones. Romer & Price (1940) suggested that an indication of the point of closure of the notch might be provided in the pelycosaur skull by a projection of the tabular, supratemporal and adjacent surface of the squamous. The supratemporal bone in pelycosaurs is small and was probably degenerating.

The side walls of the dermal shield are more or less vertical in disposition. Their ventral margins vary in shape from one pelycosaur form to another. In *Dimetrodon* there is, in the margin on each side, a small concavity below the external nostrils, a marked convexity in the maxillary region succeeded by an equally marked concavity below the orbit. The development of the maxillary convexity is associated, presumably, with the powerful dentition of this region while the suborbital concavity would have allowed free play for the coronoid process of the lower jaw and its attached musculature.

The external nostril lies well forward. It is bounded in *Dimetrodon* by the nasal, premaxillary and maxillary bones. In many other types of pelycosaur the lacrimal also enters into the boundary. The nostril is divided into a large

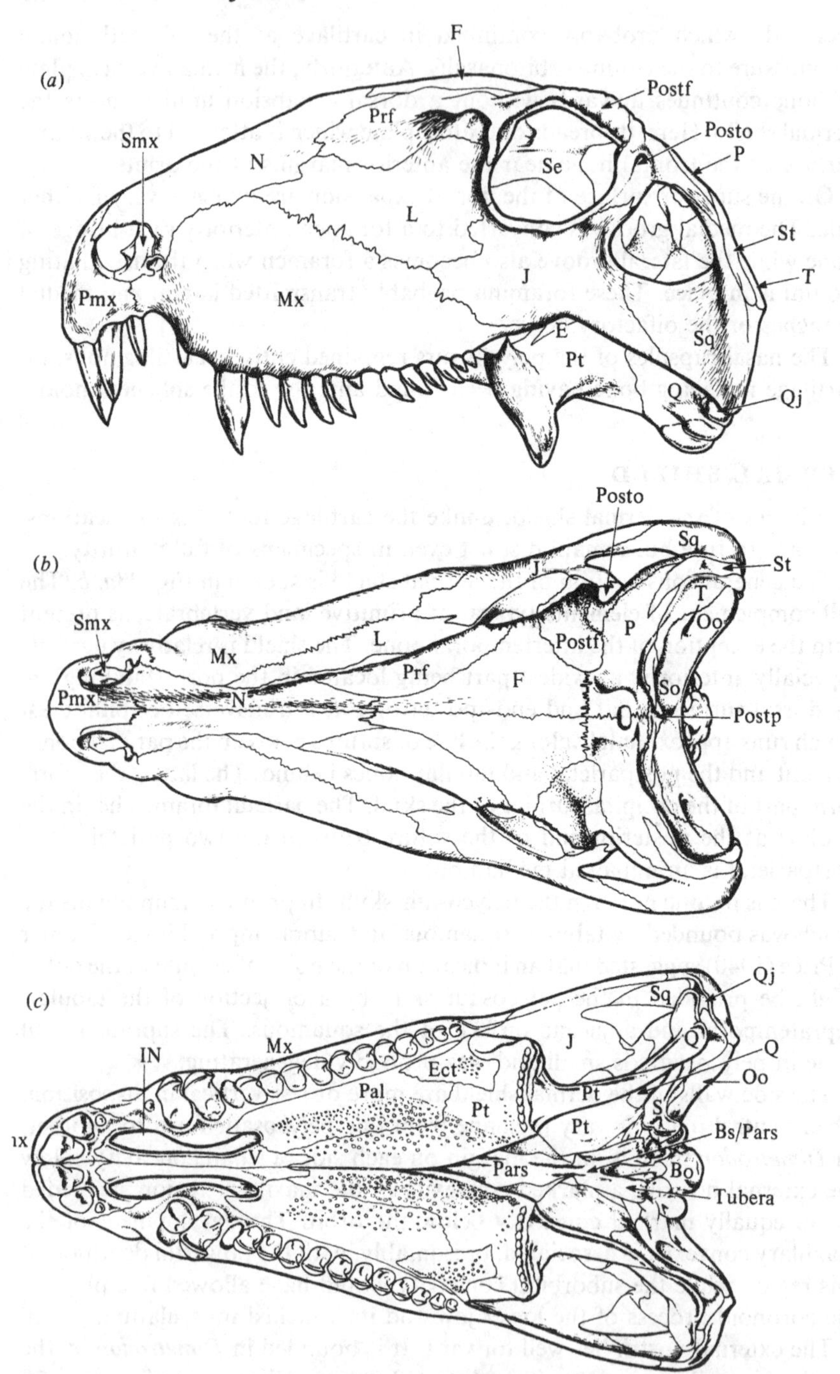

Fig. 28. To show dermal bones of skull of *Dimetrodon*. (*a*) Left lateral view of dermal shield; (*b*) dorsal view of dermal shield; (*c*) palatal view. For labelling see list of abbreviations. (After Romer & Price, 1940.)

anterior and a small posterior (septomaxillary) compartment by the septomaxillary bone which bridges the nasal opening. The functional significance of this division is unknown but it has been suggested (e.g. Watson, 1921) that the septomaxillary opening may have formed part of a separate air passage leading out from the vomeronasal organ.

Behind the orbit is the single lateral temporal opening characteristic of the pelycosaurs and their descendants. The size of the opening varies widely between the different pelycosaur groups, especially in a posterior and ventral direction. Its boundaries are formed by the postorbital, squamous and jugal elements – the first two meeting above the opening in the characteristic synapsid manner.

The squamous bone is the dominant element of the cheek region. The posterodorsal margin of the bone is thickened to form a ridge which slopes downwards and backwards to the quadrate with which it makes a long connection. Just superior to the contact with the quadrate, the squamous sends a ridge medially, ventral to the tabular and supratemporal, to make contact with the lateral border of the supraoccipital and paroccipital process. The quadrotojugal is a rather smaller element, of variable size, which occupies the notch between the most ventral parts of the squamous and quadrate in the posteroventral corner of the dermal shield.

PALATE

The dermal bones of the palate include the four paired elements – vomers, palatines, ectopterygoids and pterygoids (Fig. 28*c*) – found in primitive tetrapods. The maxillae and premaxillae also contribute to the lateral parts of the palate by means of their tooth-bearing, thickened and inwardly turned lower borders.

The anterior part of the palate is largely occupied by the apertures of the internal nostrils which are bounded anterolaterally by the maxillae and premaxillae, medially by the vomers and posteriorly by the palatines. Behind the nostrils the palate forms a continuous plate, composed of palatines, ectoterygoids and anterior parts of the pterygoids, stretching between the alveolar borders of the maxillae. The medial parts of the pterygoids are inrolled to turn upwards to meet each other and form a high ridge projecting dorsally in the median plane. Presumably this ridge was continuous with the cartilaginous nasal septum separating the nasal capsules which lay on the upper surface of the dermal palate. Posteriorly, the pterygoids diverge to leave a central interpterygoid vacuity in which can be seen the posterior part of the parasphenoid applied to the floor of the braincase. The posterior border of the palate is marked by the typical reptilian transverse flanges of the pterygoid bones. A buttress projects laterally from this region of the pterygoid to contact and brace the palate against the jugal.

Posteriorly, the pterygoid continues towards the quadrate as the quadrate process. By contrast with the more or less horizontal disposition of the palatal part of the bone, the quadrate process lies in the vertical plane, the bone being twisted, so to speak, along its anteroposterior axis so that the upper surface of the palatal part is continuous with the lateral surface of the quadrate process. The quadrate process runs posterolaterally to make a wide contact with the anteromedial surface of the quadrate.

The cartilage replacing bone of the intermediate part of the palatoquadrate, the epipterygoid, consists of a triangular shaped plate of bone closely applied to the lateral surface of the quadrate process of the pterygoid (Fig. 27*a*). At its anterior corner the epipterygoid passes forwards off the surface of the quadrate process so that its medial surface is uncovered. This is the basal process which articulates with the basipterygoid process of the basisphenoid at a movable palatobasal joint. Superiorly, a slender ascending process passes upwards and almost reaches the dermal shield. Posteriorly, the quadrate ramus of the epipterygoid in *Dimetrodon* connects with the quadrate bone but in other sphenacodonts the connection was probably cartilaginous.

The quadrate, the cartilage replacing bone of the posterior part of the palatoquadrate, is a rounded plate attached by its anteromedial surface to the posterolateral surface of the quadrate process of the pterygoid (Figs. 27, 28*c*). Posterodorsally, it also contacts the squamous and quadratojugal bones. Further contacts are made with the epipterygoid and with the paroccipital process (this contact may possibly represent the otic process of the palato-quadrate). The posteroventral part of the bone is thickened and rounded and provides the articular surface for the lower jaw. It is divided into lateral and medial condyles which face somewhat laterally as well as ventrally and which have long axes that run anteromedially to posterolaterally. On the medial surface of the quadrate, a short distance above the articular surface, is a pit which may have given attachment to a ligamentous extension of the stapes.

It will be noted that the contacts between the dermal shield and palatal complex, on the one hand, and the braincase, on the other, are few and rather loose. The basisphenoid articulates movably with the epipterygoid, the sphenethmoid gains only a small attachment to the dermal shield while the connections in the occipital region (supraoccipital with postparietal, tabular with squamous, paroccipital with squamous and quadrate) would appear to have allowed some movement. The skull, therefore, probably possessed a measure of kinetism but the quadrate is monimostylic.

LOWER JAW

The lower jaw consists of a single cartilage replacing bone, the articular, and six or seven dermal elements, the dentary, splenial, angular, surangular, prearticular and one or two coronoids (Fig. 29). The superior border is reciprocally curved to the margin of the upper jaw, having, in *Dimetrodon*, a small convexity anteriorly, then a long concavity followed by a pronounced

(*a*)

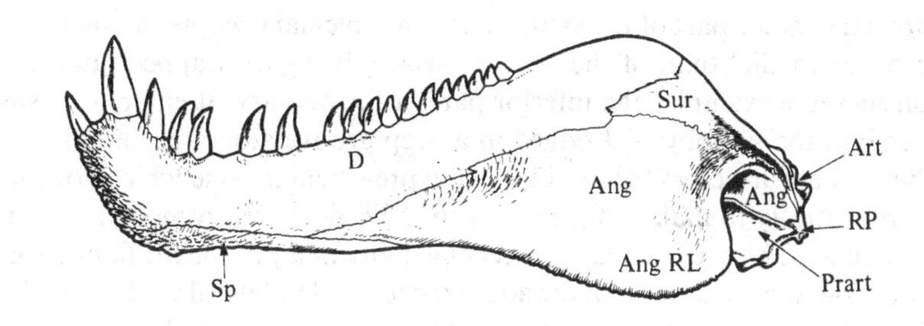

(*b*)

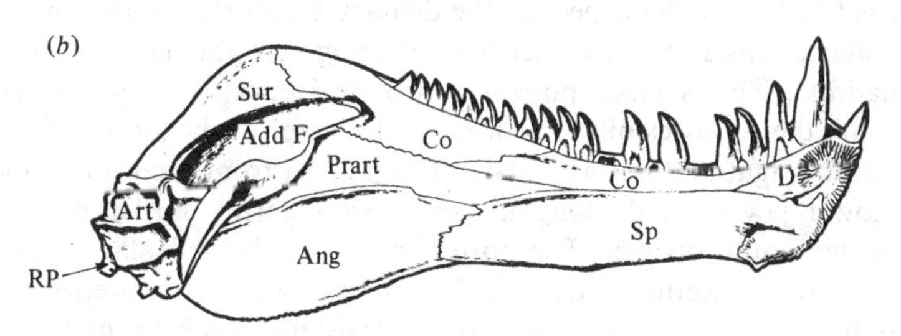

Fig. 29. To show left lower jaw of *Dimetrodon*. (*a*) Lateral view; (*b*) medial view. For labelling see list of abbreviations. (After Romer & Price, 1940.)

convexity in the coronoid region. The inferior border has a straighter outline except posteriorly where the angular region is strongly convex. The jaw is compressed from side to side so that its external and internal walls are parallel.

The external surface is made up of dentary, splenial, angular and surangular, the dentary being the largest element and forming the anterior three-quarters of the superior border, including the tooth-bearing portion, and the anterior border and peak of the coronoid projection. The splenial forms the ventral border of the jaw in front of the angular. The latter bone makes up the posteroventral part of the lower jaw. It has a thickened upper part which anteriorly overlaps the dentary and posteriorly sends a flange to overlap the upper part of the lateral surface of the articular. Projecting posteroinferiorly from the thickened upper part is a plate of bone termed the reflected lamina. This lies in a plane somewhat lateral to the remainder of the bone. Posteriorly, a notch is present between the reflected lamina and body of the angular which was probably associated with the insertion of the pterygoid musculature (see below). A reflected lamina is a characteristic feature of the angular bone of sphenacodonts and therapsids. The angular is surmounted by the surangular which occupies the posterior one-quarter of the superior border of the jaw.

The four bones on the lateral surface of the jaw also appear on its medial surface, where they are joined by the prearticular and coronoids. The dentary is exposed on this surface only along the tooth row and far anteriorly where it forms the major part of the symphysis. The splenial occupies a much larger part of the medial than of the lateral surface, being overlapped laterally to a considerable extent by the inferior part of the dentary. It enters to a small degree into the symphysis. Lodged in a deep groove along the superior part of the medial surface of the angular is the prearticular bone which continues posteroventrally to make a firm connection with the ventromedial aspect of the articular. The prearticular projects medially in a prominent flange which provides the ventral margin of the adductor fossa. The lateral wall of this fossa is formed by a concavity on the lower part of the medial aspect of the surangular. The number of coronoids in the pelycosaur lower jaw appears to vary but two are usually present in *Dimetrodon*. They are thin plates of bone attached to the medial aspect of the dentary below the tooth row.

The articular consists of a compact body bearing the articulating surface for the quadrate. This surface possesses two oval concavities which are congruent with the condyles of the quadrate. The shape of the joint surfaces suggests that the right and left articulars moved apart from each other and that each lower jaw twisted along its long axis (ventral border moving inwards) as the mouth opened. The posterior part of the articular bears a small, downward projecting retroarticular process while its anterior tip appears to have continued forwards as a persistent Meckel's cartilage. Laterally, the articular is overlapped by the angular and surangular and ventromedially it is buttressed by the prearticular.

The meckelian canal passes forwards from the adductor fossa along the main axis of the lower jaw, being bounded superiorly by the dentary and coronoids, laterally by the dentary, medially and inferiorly by the angular and splenial.

AUDITORY OSSICLE

The auditory transmitting mechanism contains a single osseous element, the stapes (Figs. 27*b*, 28*c*). It consists of a large, thick rod, passing from the footplate in a posterior, lateral and inferior direction to end medial to the lower margin of the quadrate. Its termination has an unfinished appearance, indicating that it was probably continued as a cartilaginous extrastapes to the tympanic membrane, assuming one was present. Just below the footplate the stapes is pierced by the stapedial foramen. From the lateral surface of the bone in this region the dorsal process passes superiorly to articulate with the paroccipital process. There are indications of a ligamentous attachment on the anterolateral surface of the stapes. From here, a ligament (the homologue of the internal or quadrate process) may have passed to the pit already noted on the medial aspect of the quadrate.

The position of the tympanic membrane in the pelycosaurs is not known with certainty. Romer & Price (1940) suggested that it may have been located close to the ventral part of the posterior border of the quadrate, possibly with its anterior border attached to that bone, and some distance deep to the surface of the head. This would place the membrane in line with the stapes and in its characteristic reptilian position above the depressor mandibulae muscle (assuming this was inserted into the retroarticular process of the articular bone). The precise extent of the tympanic cavity is, of course, unknown, but it is presumed to have occupied the space between the paroccipital process dorsoposteriorly, the quadrate process of the pterygoid anterolaterally and the prootic medially. As described in Chapter 2, Tumarkin (1968) has suggested that the pelycosaurs lacked a tympanic membrane and that they relied upon bone conduction for the transmission of sound vibrations, citing the massive nature of the stapes as evidence in this respect.

EVOLUTIONARY TRENDS IN THE THERIODONT SKULL

All the major morphological features which characterise the mammalian skull are foreshadowed or actually achieved in one or other of the theriodont lineages. Thus, although the precise steps by which the mammals evolved from their theriodont ancestors are not known, it is possible to trace with a high degree of probability the changes by which the mammalian cranial structure was achieved. Nonetheless, it must be borne in mind that the types from which our knowledge of these changes is derived most probably do not come from a strict evolutionary sequence but more likely represent lineages that were undergoing parallel evolution in the mammalian direction.

The ensuing account is based mainly on the work of the following authors: Broom (1911, 1936, 1938), Watson (1914, 1921, 1951, 1953), Parrington (1934, 1946, 1955, 1967), Parrington & Westoll (1940), Olson (1944, 1959), Crompton (1958, 1963*a*, *b*, 1972), Kermack & Mussett (1958), Kermack (1963), Barghusen (1968), Kermack, Kermack & Mussett (1968), Hopson & Crompton (1969), Kermack & Kielan-Jaworowska (1971), Kemp (1972*a*, *b*, 1979) and Kermack, Mussett & Rigney (1973). (In most instances, the references cited are only a small part of the authors' published work on the theriodont skull; they have been chosen because they are either key works or contain an extensive bibliography.) In order to avoid repetition, specific references from the above list will be cited only where important differences of opinion are discussed.

GENERAL SHAPE AND STRUCTURE OF THE SKULL

Most of the evolutionary changes in the theriodont skull can be related to two underlying trends – increasing power and efficiency of the masticatory apparatus, and enlargement of the braincase. The teeth of each quadrant are differentiated into incisors, a large dagger-like canine and cutting or grinding

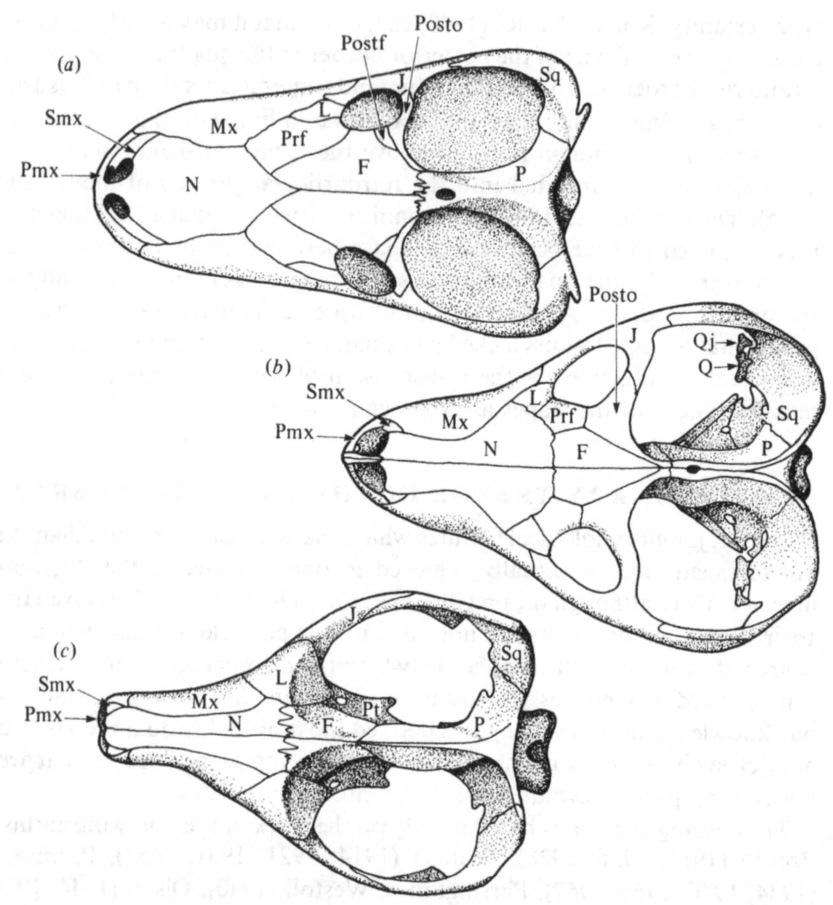

Fig. 30. Dorsal views of skulls of (*a*) therocephalian (*Lycosuchus*), (*b*) early cynodont (*Thrinaxodon*) and (*c*) ictidosaur (*Oligokyphus*). Not to scale. For labelling see list of abbreviations. (Based on several sources including Romer, 1956, and Estes, 1961.)

postcanine teeth. Although the tooth row is reduced in length in gorgonopsians and later therocephalians relative to that of the pelycosaurs, it is extended again in cynodonts. To support this powerful dentition the jaws are strongly constructed and project forwards as a well-developed muzzle. A maxillary sinus has been described in the upper jaw of some cynodonts (Kemp, 1979). In the lower jaw the dentary becomes greatly expanded and the remaining bones correspondingly diminished in size. Despite this reduction of the postdentary bones there was probably little structural weakening of the jaw since, as will later be described, there was a rearrangement of the bony elements and of the muscular forces acting upon them which preserved mechanical efficiency. In addition, some of the later cynodonts and ictidosaurs developed a subsidiary jaw articulation between the dentary and squamous bones to help support the reduced quadrate–articular joint.

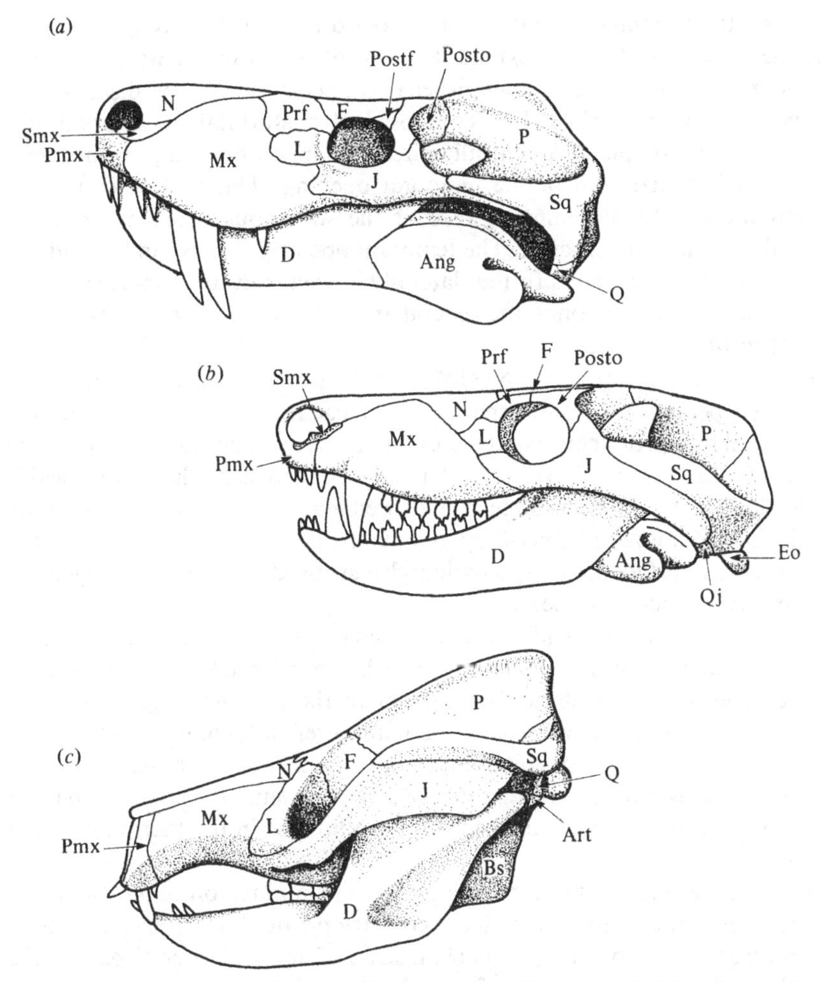

Fig. 31. Lateral views of skulls of (*a*) therocephalian (*Lycosuchus*), (*b*) early cynodont (*Thrinaxodon*) and (*c*) ictidosaur (*Oligokyphus*). Not to scale. For labelling see list of abbreviations. (Based on several sources including Romer, 1956, and Estes, 1961.)

In order to accommodate the enlarged and more efficiently structured jaw muscles (see section on lower jaw and jaw articulation), there was a progressive tendency for the temporal openings to enlarge and for the temporal (zygomatic) arches to become bowed outwards. These changes are already in evidence in the pristerognathid therocephalians (Figs. 30*a*, 31*a*). The temporal openings in these early theriodonts have increased in both anteroposterior and transverse diameters as compared to their dimensions in the pelycosaur skull. As a result of the transverse enlargement the openings approach each other over the dorsal aspect of the skull with elimination of the contact between the postorbital and squamous and often with the formation of a median sagittal crest on the parietal bones. At this stage, however, the zygomatic arches are not bowed laterally to any great degree and the squamous does not extend far forwards in its suture with the jugal. The supratemporals, already diminished in the pelycosaurs, have disappeared altogether. The postorbital bar, composed of postorbital, postfrontal and jugal bones, is complete. A ventrally directed flange from the parietal forms the vertical medial wall of the temporal opening. This wall is continued posterolaterally by the anterior face of the squamous which is extended vertically as an occipital crest. The temporal openings are expanded and the zygomatic arches extended further laterally in whaitsiids but, apart from the loss of the postfrontal bones, the general arrangement of the region is similar to that in pristerognathids.

In the cynodonts (Figs. 30*b*, 31*b*) there is a further enlargement of the temporal openings which are separated from each other on the dorsal aspect of the skull by a prominent sagittal crest on the parietal bones. The zygomatic arches are thick, strong bars bowed laterally, so increasing the relative width of the skull. The squamous bone extends well forwards in its contact with the jugal. In the more advanced cynodonts from the Middle Triassic the posterodorsal surface of the zygomatic arch is grooved, probably for lodgement of the external acoustic meatus.

In association with the still further enlarged jaw muscles, the zygomatic arch in ictidosaurs is curved strongly outwards between its anterior root in front of the orbit and its posterior attachment in the occipital region, and the postorbital bar together with postorbital and prefrontal bones has been lost (Figs. 30*c*, 31*c*). The postorbital bar is missing in most mammals. In some later forms, however, it has been reconstructed but, since the postorbital bone has been lost, the elements utilised differ from those in the theriodont skull (see Chapter 3).

The dominating evolutionary change in the palatal region of the mammal-like reptiles is the formation of the secondary palate. Its presence results in the internal nostrils opening far back, instead of into the anterior part of the mouth, and in the first parts of the airway and alimentary system being separated from each other. An incipient secondary palate is seen in the therocephalian skull where the marginal dermal bones of the palate exhibit a tendency to fold first ventrally and then medially to produce palatine processes projecting inwards beneath the primary palate (Fig. 32*a*). Complete separation of the airway may have been achieved by soft tissue structures. A similar condition exists in procynosuchids while in the later cynodonts the palatine processes have met and fused in the midline to form a complete bony secondary palate (Fig. 32*b*). Above the secondary palate a median septum, probably formed from the vomers, separates the airway into right and left passages. The secondary palate tends to become considerably elongated in the most advanced cynodonts (Fig. 36*a*) and ictidosaurs (Fig. 32*c*). The pterygoid bones, once the largest elements in the palate, become reduced to flanges forming the roof and side walls of the most posterior part of the nasal

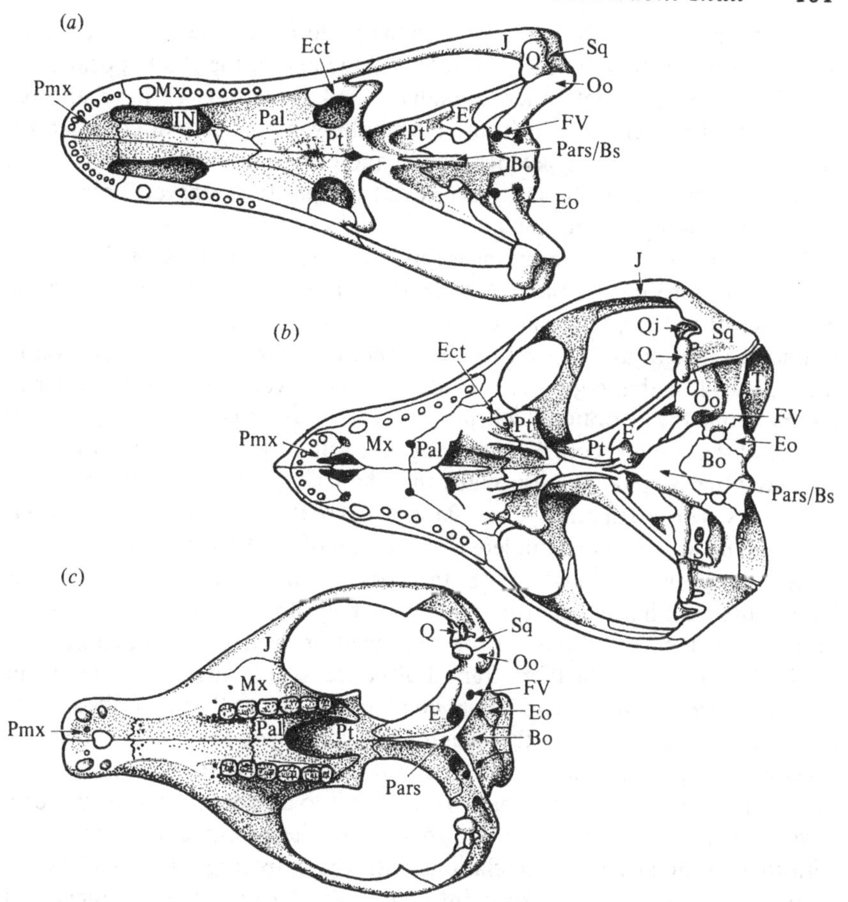

Fig. 32. Ventral views of skull of (*a*) therocephalian (*Scylacosaurus*), (*b*) early cynodont (*Thrinaxodon*) and (*c*) ictidosaur (*Oligokyphus*). Not to scale. For labelling see list of abbreviations. (Based on several sources including Romer, 1956, and Estes, 1961).)

passage. The quadrate processes of the pterygoids in both therocephalians and cynodonts have tended towards reduction in size and have moved medially to become firmly fused to the cranial base, converting the latter to a very strong, girder-like structure and, in the more advanced cynodonts, closing the interpterygoid vacuity. In the ictidosaurs the quadrate processes have completely disappeared. The quadrate rami of the epipterygoids also tend to reduction and the articulations between the basal processes and basipterygoid processes become fused.

Important changes also occur in the quadrate region. The quadrate bone of the pelycosaur skull is a large structure with broad contacts with the pterygoid and squamous bones. In theriodonts it is much reduced in size and occupies a recess in the anterior surface of the squamous lateral to the paroccipital process (Figs. 30*b*, 39). The quadrotojugal is also much diminished and remains as a small bone attached to the posterior surface of the quadrate. The connections between the quadrate and the rest of the skull appear to be sufficiently loose in both therocephalians and cynodonts to have allowed a measure of movement which, though small, was probably of considerable significance in facilitating jaw depression and elevation in those forms where the axis of rotation of the jaw joint deviated from the transverse plane (see section on lower jaw and jaw articulation).

The enlargement of the braincase involves the absorption of the cavum epiptericum into the cranial cavity. In the process, the ascending and basal processes of the epipterygoid become incorporated into the side wall of the braincase anterior and ventral to the temporal fossae. A further addition to the side wall in this region is provided by the development of an anterior lamina from the periotic which projects forwards to contact the posterior border of the epipterygoid. These changes are already in evidence in the more advanced therocephalians, especially the whaitsiids, and are completed in the cynodonts. More superiorly, the flanges from the parietal and squamous project ventrally to help complete the side wall of the braincase in the depths of the temporal openings. With the loss of the postorbital bar in the ictidosaurs a similar flange from the frontal helps to form the side wall of the braincase in the orbital part of the combined orbitotemporal opening.

The cranial base of the theriodont skull contains as its main central element a fused basisphenoid–parasphenoid complex similar to that described in the pelycosaur skull. In both therocephalians and cynodonts the cultriform process of the parasphenoid is reduced. As already noted, the pterygoid bones become fused with the cranial base to convert it to a strong, anteroposteriorly aligned girder. The basipterygoid processes of the basisphenoid persist but with their originally free articulations with the epipterygoids fused. As the cavum epiptericum is absorbed into the cranial cavity the basipterygoid processes come to form part of the alisphenoid bone in the floor of the braincase. The parasphenoid is lacking in the mammalian skull but its transverse processes may be represented by the dorsal ossification centres of the pterygoid plates (see section on palate).

The region of the sphenethmoid is not well known. The bone appears to have been absent in cynodonts but presumably in forms leading to mammals it persisted to give rise to the presphenoid and mesethmoid ossifications (see Chapter 2). The nasal capsules are believed to have remained cartilaginous throughout life.

Amongst the more significant changes in the otic capsules of the theriodonts is the development of the anterior lamina and lateral flange. The anterior lamina, as already noted, plays a part in walling in the cavum epiptericum as it becomes incorporated into the cranial cavity. The lateral flange, first seen in whaitsiids (Fig. 34), is well developed in cynodonts. It contacts the quadrate ramus of the epipterygoid and helps floor in the posterior part of the cavum epiptericum.

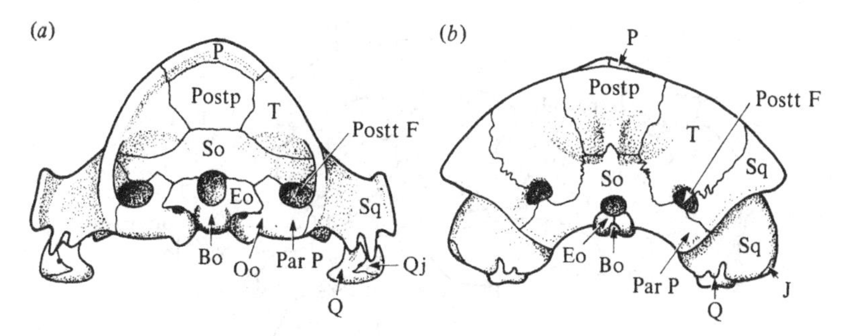

Fig. 33. Posterior views of skulls of (*a*) therocephalian (*Lycedops*) and (*b*) cynodont (*Cynognathus*). Not to scale. For labelling see list of abbreviations. (After Romer, 1956.)

The occipital region of the theriodont skull is relatively broad and low with much of its surface made up of tabular and postparietal bones. The supraoccipital is reduced in size. The posttemporal fenestrae are present but small, and are bounded inferiorly by stout, paroccipital processes. The squamous bone forms the lateral part of the occipital plate on each side, curving outwards and forwards from its contact with the tabular bone and paroccipital process. The occipital region of the therocephalian skull tends to be particularly broad and low (Fig. 33*a*). The occipital condyle is single with a median basioccipital and a bilateral exoccipital component. In the cynodont skull the occiput is somewhat higher than in therocephalians and the supraoccipital much narrowed with a compensating medial enlargement of the tabular which now more completely marginates the posttemporal fenestra (Fig. 33*b*). The occipital condyle has become a paired structure by the reduction in size of the basioccipital component, an arrangement which probably allowed a great range of flexion and extension at the craniovertebral joint.

During the course of these evolutionary changes, several elements of the dermal shield have been lost (the intertemporal is already absent at the pelycosaur stage); the supratemporal is lacking in all therapsids; the postfrontal is lost in the later therocephalians; the prefrontal and postorbital are lacking, along with the postorbital bar, in ictidosaurs.

Having considered the general phylogenetic trends in the shape and structure of the theriodont skull, a more detailed examination will now be made of certain regions of the skull where the morphological changes were critical to the achievement of the mammalian condition.

SIDE WALL OF BRAINCASE

The reptilian otic capsule consists of opisthotic and prootic ossifications tightly fused into a single periotic mass. In primitive theriodonts the periotic possesses typically a solidly built central part with a laterally projecting

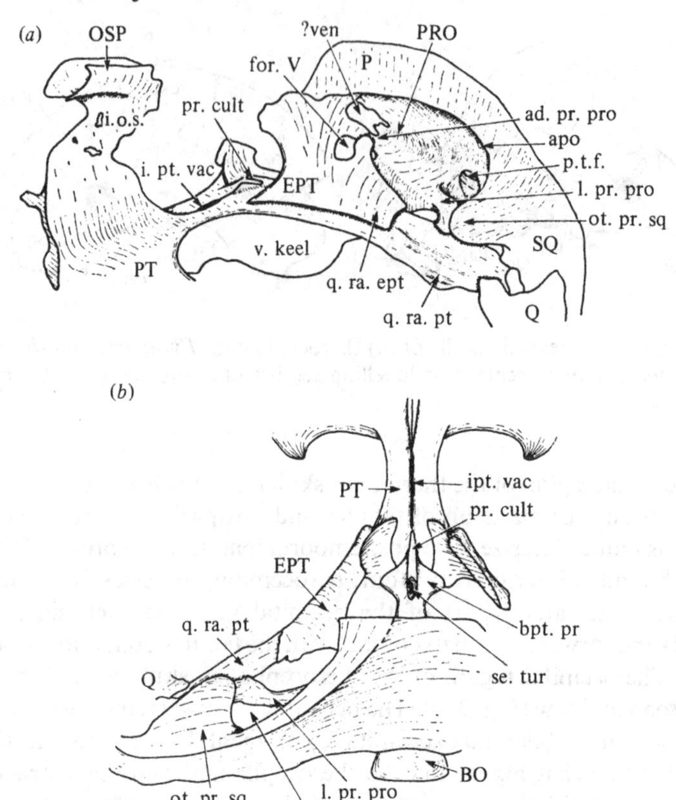

Fig. 34. Whaitsiid braincase in (*a*) lateral and slightly anterodorsal view; (*b*) dorsal view. Abbreviations: ad. pr. pro = anterodorsal process of prootic; apo = attachment of aponeurotic sheet; BO = basioccipital; bpt. pr = basipterygoid process: EPT = epipterygoid: for. V = trigeminal foramen; i.o.s. = interorbital septum; i. pt. vac = interpterygoid vacuity; l. pr. pro = lateral process of prootic; OSP = orbitosphenoid; ot. pr. sq = otic process of squamous; pr. cult = cultriform process; PRO = prootic; PT = pterygoid; p.t.f. = posttemporal fossa; pt. par. f = pterygo-paroccipital foramen; Q = quadrate; q. ra. ept = quadrate ramus of epipterygoid; q. ra. pt = quadrate ramus of pterygoid; se. tur = sella turcica; SQ = squamous; ? ven = probable foramen for vein; v. keel = ventral keel. (Reproduced with permission from Kemp, 1972*a*.)

paroccipital process and an anteroventral process which represents the ossified pila antotica. The paroccipital process makes contact with the squamous and tabular bones. The posterior part of the periotic meets and usually fuses with the exoccipital and supraoccipital. Its ventral border contacts and often fuses with the laterodorsal border of the basioccipital. The anteroventral process similarly makes contact with the posterior part of the laterodorsal border of the basisphenoid–parasphenoid complex, its lateral surface forming the medial wall of the cavum epiptericum. The anterior border

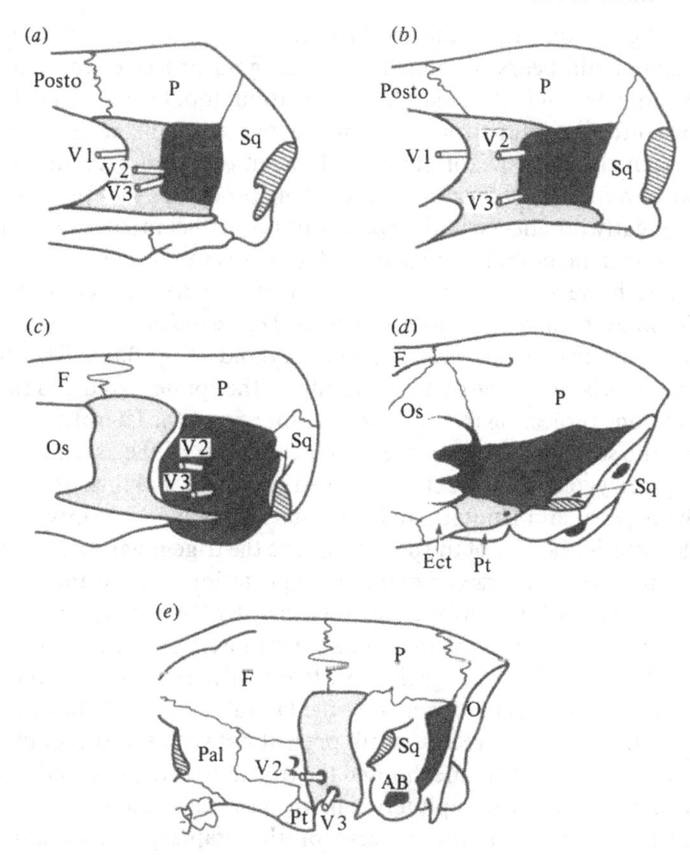

Fig. 35. Schematic representation of side wall of braincase in (*a*) early cynodont (e.g. *Thrinaxodon*), (*b*) more advanced cynodont (e.g. *Trirachodon*), (*c*) *Morganucodon*, (*d*) *Kamptobaatar* (multituberculate) and (*e*) modern therian (e.g. dog). Light shading = epipterygoid–alisphenoid; dark shading = periotic; hatching = cut surface of squamous. Not to scale. For labelling see list of abbreviations. (*c* and *d* after Kermack & Kielan-Jaworowska, 1971.)

of the periotic is marked by the trigeminal notch (prootic incisure) which lodged the root of the trigeminal nerve. The ventral border of this notch is provided by the anteroventral process of the periotic which often possesses a dorsal projection that passes upwards a variable distance anterior to the notch. The trigeminal nerve thus lay posterior to the ossified pila antotica as would be expected from their relationships in the cartilaginous braincase (Chapter 2). The dorsal border of the notch is formed by an anterodorsal process, which is probably an ossification of the taenia marginalis. The side wall of the braincase between the otic capsule and the sphenethmoid region is unossified, having been completed in life, presumably, by membrane. The ascending process of the epipterygoid lies lateral to this unossified part of the wall, the space between being the cavum epiptericum in which lay the trigeminal ganglion. The palate and braincase tended to become fused but the basisphenoid still bears prominent basipterygoid processes for articulation with the epipterygoid, although this articulation, too, tended towards fusion.

There is little departure from this typical arrangement in the therocephalians (apart from the presence of suborbital vacuities anterior to the pterygoid flanges), although in whaitsiids the ascending process of the epipterygoid has become greatly broadened and its posterior border bears a deep notch for the maxillary and mandibular divisions of the trigeminal nerve (Fig. 34). In cynodonts, however, there are several significant new developments. The periotic bone in early cynodonts such as *Thrinaxodon* makes sutural connections over two areas with the epipterygoid (Fig. 35*a*). One of these connections is between the quadrate ramus of the epipterygoid and the lateral flange of bone projecting from the ventral margin of the lateral surface of the periotic; the second is between the posterior border of the ascending process of the epipterygoid, now much broadened (and also making sutural contact with the parietal and frontal), and the anterior lamina, a process projecting from the anterior border of the periotic above the trigeminal notch. The notch is thus converted to a foramen marginated posteriorly by the anterior lamina and lateral flange of the periotic and anteriorly by the posterior border of the epipterygoid and the root of the quadrate ramus. The ventral border of the foramen, however, lies in a slightly more lateral plane than the original ventral border of the notch, being formed by the lateral flange, not the ossified pila antotica. Hence, a small recess is still present between the lateral flange and the pila antotica, which, in life, housed the trigeminal ganglion and represents the much reduced cavum epiptericum. In more advanced cynodonts (e.g. *Triarachodon*) the ascending process of the epipterygoid is still further enlarged and two trigeminal foramina may be present along the suture between anterior lamina and ascending process, the lower for the mandibular and the upper for the maxillary division (Fig. 35*b*). The ossified pila antotica is situated medial to the epipterygoid, indicating that a cavum epiptericum still persisted internal to the epipterygoid. The structure of the lateral wall of the braincase in ictidosaurs and tritylodonts is similar, in all essentials, to that of the cynodonts.

In the Late Triassic mammal *Morganucodon* (? *Eozostrodon*) the lateral wall of the braincase anterior to the otic capsule is formed predominantly by the anterior lamina of the periotic (Fig. 35*c*). The mandibular and maxillary nerves pierce this lamina through two separate foramina (pseudovale and pseudorotundum, respectively). Anterior to the anterior lamina the side wall is completed, after a small gap, by a not greatly enlarged ascending process which, since we are now dealing with an animal generally recognised as mammalian rather than reptilian, can be termed the alisphenoid. The lateral flange of the periotic is narrower than in cynodonts, indicating that the cavum epiptericum in *Morganucodon* is very small. The structure of the lateral wall of the braincase in the remaining Triassic and Jurassic mammals is incompletely known but the morphology of the periotic masses in *Triconodon* and *Trioracodon* (Kermack, 1963) suggests that, in the Triconodonta, it had a structure not greatly dissimilar to that in *Morganucodon*. The Late Cretaceous multituberculate *Kamptobaatar* (Fig. 35*d*) and the living non-therians, the monotremes, also possess lateral walls built on *Morganucodon* lines although in both cases the anterior lamina is much more extensive, and the alisphenoid correspondingly smaller, than in the Triassic mammal.

Unfortunately, knowledge of the structure of the lateral wall of the braincase is unknown for any of the therian mammals earlier than the Late Cretaceous. In all therians of this date and later the lateral wall is constructed in what appears to be a fundamentally different manner from that in non-therians. According to the conventional interpretation (Chapter 2): (1) the anterior lamina of the periotic mass is missing and the wall of the braincase is made up of a much enlarged alisphenoid (deemed to be homologous with the reptilian epipterygoid, as is the alisphenoid of non-therians) and the squamous (Fig. 35*e*); (2) the pila antotica is suppressed and the cavum epiptericum completely absorbed into the cranial cavity; and (3) the maxillary and mandibular nerves leave the braincase through foramina – rotundum and ovale, respectively – in the alisphenoid. The structure of the side wall of the cynodont skull would appear to provide a starting point from which either the therian or non-therian pattern could be derived – the therian condition being obtained by further expansion posteriorly of the epipterygoid around the trigeminal branches together with reduction of the anterior lamina while the non-therian condition resulted from forward extension of the anterior lamina and reduction of the alisphenoid.

The derivation of the therian alisphenoid has been interpreted differently by Presley & Steel (1976). They argue that the lateral situation of the cynodont anterior lamina (lying in the same plane as the ascending process) is consistent with it ossifying in membrane and being, therefore, an element which is quite separate from the chondrocranium and which has fused secondarily with the cartilage replacing bone of the periotic. They have produced evidence, from the study of ontogenetic stages in modern mammals (Chapter 2), that the therian alisphenoid is a composite bone, containing an intramembranously ossifying part (which they homologise with the anterior lamina of cynodonts) and a cartilage replacing part ossifying in the ala temporalis, the lamina ascendens of which appears to be largely a neomorphic outgrowth from the junction of the ascending process and quadrate ramus of the intermediate part of the palatoquadrate (and not, therefore, homologous with the ascending process). The line of fusion of these two parts lies approximately along the line of the foramina rotundum and ovale (which, in fact, are marginated by this fusion). If the intramembranous part of the alisphenoid is homologous with the anterior lamina, then an essential difference between the non-therian and therian mammals is that in the former the anterior lamina has remained fused with the periotic, as it is in cynodonts, while in the latter the lamina has fused with the cartilage replacing bone of the ala temporalis and has lost its fusion with the periotic.

So far as the fossil evidence is concerned, perhaps the strongest support for Presley & Steel's (1976) view comes from the position of the trigeminal foramina. The general arrangement in lower vertebrates is for the maxillary and mandibular nerves to emerge posterior to the ascending process. This arrangement appears to have been preserved in cynodonts for, as we have seen in *Trirachodon*, two separate foramina are present in the line of fusion of the anterior lamina and ascending process which are usually attributed to the two branches of the trigeminal nerve. If the conventional view that the alisphenoid is entirely homologous with the epipterygoid is correct, then the ascending process must somehow have migrated posteriorly during therian evolution to lap around the maxillary and mandibular nerves, so enclosing them in the foramina rotundum and ovale. As Presley & Steel point out, the assumed detachment of the anterior lamina from the periotic and its fusion with the cartilage replacing bone of the ala temporalis and also the tendency for the posterior margin of foramen ovale to be thinned and even lost in advanced therian mammals suggest that, if anything, the phylogenetic trend has been for the bone of the alisphenoid to move anteriorly not posteriorly. No such inconsistencies would exist if the lateral wall of the braincase had evolved as Presley & Steel postulate. The enclosure of the trigeminal foramina brought about by the fusion of the anterior lamina with the cartilage replacing bone of the ala temporalis, the detachment of the anterior lamina from the periotic and the thinning of the posterior margin of foramen ovale could all be attributed to the tendency towards anterior migration.

Presley & Steel cite the situation in *Trirachodon* as support for their suggestions. Here, the two trigeminal foramina are separated from each other by a projection from the junction of the ascending process and quadrate ramus of the epipterygoid which meets a projection from the anterior border of the anterior lamina (Fig. 35*b*). The relationships in this region are thus exactly the same as those postulated for the therian alisphenoid.

Although the evidence in favour of Presley & Steel's views is impressive, final judgement as to their correctness or otherwise cannot be made until information becomes available about the stages by which the structure of the therian braincase wall was derived from that of cynodonts. As already noted, the earliest therians for which information is presently available about this region of the braincase already have the full therian construction. If Presley & Steel's suggested homologies prove to be correct, the differences in the structure of the side wall of the braincase between therian and non-therian mammals would be less fundamental than is implicit in the conventional view.

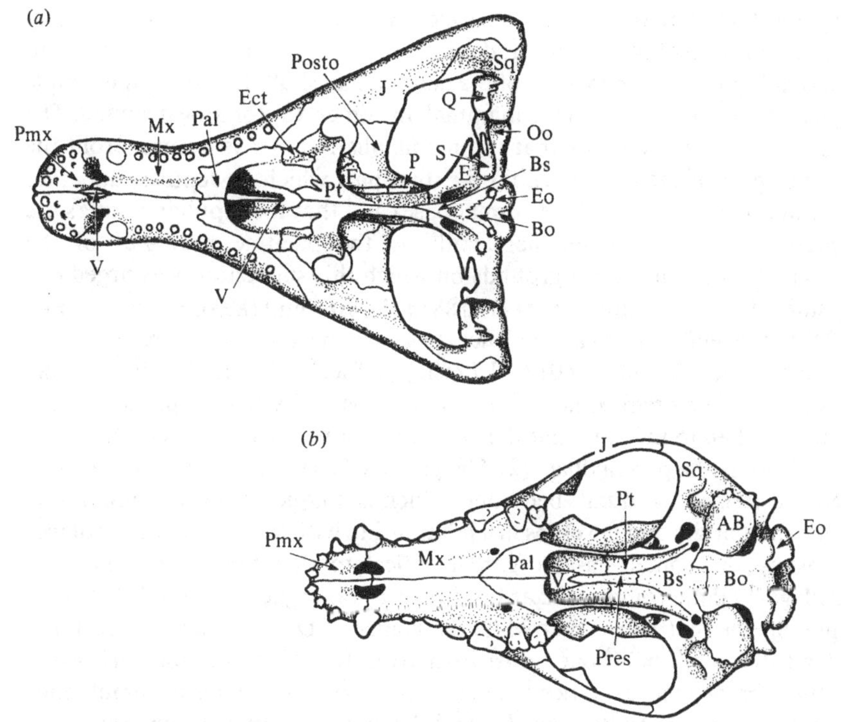

Fig. 36. Ventral views of skulls of (*a*) *Cynognathus*; (*b*) dog. Not to scale. For labelling see list of abbreviations. (*a* based on Romer, 1956.)

PALATE

The palate of the pelycosaur possesses, in all essentials, the structure believed to have been typical of the ancestral land vertebrates (Fig. 28*c*). In the Therocephalia are seen the beginnings of the development of a secondary palate (Fig. 32*a*). The vomers, which primitively are paired, flat bones, show a tendency in the more advanced members of this group to become deeper and narrower and to fuse with each other. In this manner, a low, bony, nasal septum is formed (the upper part of the septum was composed presumably of the fused, medial walls of the cartilaginous, nasal capsules). In addition, the palate becomes highly vaulted with the edges of the vault being formed by flanges (the palatine processes) projecting from the premaxillary, maxillary and palatine bones. These flanges may have supported a soft palate which completed the separation of the airway from the mouth.

In procynosuchids the palatine processes tend towards further phylogenetic enlargement and in the more advanced cynodonts and ictidosaurs meet and fuse in the midline to produce a complete secondary bony palate (Figs. 32*b*, *c*; 36*a*). Attached to the dorsal surface of the secondary palate is a vertically disposed low plate of bone which is clearly derived from the paired vomers of more primitive theriodonts and which appears, from its relationships, to be homologous with the vomer of the mammalian skull. However, this simple explanation of the origin of the mammalian vomer was for long disputed. The alternative suggestion was that the mammalian vomer is derived from the anterior part of the parasphenoid (i.e. that part which, in reptiles, underlies the trabecular segment of the cranial base) and that the reptilian vomers are represented in the mammalian skull by the palatine processes of the premaxillae. The principal grounds on which this suggestion was urged can be summarised as follows (Parker (1885*a*, *b*), Broom (1895), de Beer & Fell (1937); the embryological evidence has also been critically reviewed by Parrington & Westoll (1940) and Presley & Steel (1978)): (1) The palatine processes of the premaxillae in some mammals ossify from separate centres which is taken to indicate that they represent bones originally separate from the bodies of the premaxillae. (2) The palate of *Ornithorhynchus* contains an element, termed the dumb-bell bone, which is unique amongst mammals. It is of paired origin and develops in a relationship with the paraseptal cartilages (investing the vomeronasal organ) identical to that found between these cartilages and the vomers in lizards – hence, it was argued, the dumb-bell must represent the reptilian vomer while the vomer of *Ornithorhyncus*, which lies behind the dumb-bell bone, must be derived from the anterior part of the reptilian parasphenoid. Since the vomer of therian mammals is homologous with the vomer of *Ornithorhynchus* and the dumb-bell bone is represented by the palatine processes of the premaxillae, it follows that the vomer of the therian skull must be equated with the anterior parasphenoid of reptiles and the premaxillary palatine processes with the reptilian vomer. It was perhaps the paper by de Beer & Fell (1937) which played the crucial part in the wide acceptance that this scheme received. They described the dumb-bell bone in the platypus embryo as ossifying quite separately from the premaxillae and as being, therefore, a discrete palatal element, differing in this respect from Green (1930), who, in an earlier examination of the same material, had described thin threads of ossification connecting the dumb-bell centres with the premaxillae.

The embryological evidence on which this mammalian vomer–reptilian anterior parasphenoid homology is based can be seriously questioned on several counts. First, the value of ossification centres as a guide to a bone's phylogenetic history is now known to be not always reliable. Secondly, if *Ornithorhynchus* is compared with *Emys* rather than with lizards, the relationship of the developing dumb-bell bone to the paraseptal cartilages is found to be similar to that of the reptilian premaxillae, while the vomers of *Emys* (which otherwise have normal reptilian relationships) ossify behind the cartilages just as does the vomer of the platypus. Thirdly, a recent re-examination by Green & Presley (1978; see also Presley & Steel, 1978) of the platypus embryonic material, previously described by Green (1930) and de

Beer & Fell (1937), suggests that the latter authors were wrong in their assertion that the dumb-bell bone does not ossify in continuity with the palatine processes of the premaxillae. As Green had earlier described, such a continuity does exist at an early stage of development but is lost during later stages. It seems likely, therefore, that the dumb-bell bone is, in reality, a detached portion of the premaxillae. Fourthly, bone has been described in a number of mammals (e.g. *Galeopithecus* and *Didelphis* – Parrington & Westoll, 1940; man – Reinbach, 1952; *Ornithorhynchus* – Green & Presley, 1978; *Didelphis* – Presley & Steel, 1978) which probably represents the remnant of the reptilian anterior parasphenoid. As Green & Presley and Presley & Steel have demonstrated in their study of the platypus and *Didelphis*, this remnant of the anterior parasphenoid develops in the prochordal mesoderm, while the vomer ossifies, as would be expected from a true palatal element, in the tectoseptal extension of the mesoderm of the maxillary processes.

It seems reasonable, therefore, to attach greater weight to the palaeontological evidence and to accept that the mammalian vomer is homologous with the fused vomers of the cynodont skull. As Parrington & Westoll (1940) have pointed out, to do otherwise would involve assuming a complete reversal of the evolutionary trends seen in the palate of advanced therapsids with the premaxillary palatine processes becoming reduced again in size, and eventually lost, and being replaced by the vomers which would have to have retraced their previous evolutionary history from paired flat bones of the palate to fused median nasal septum.

As already described, as the vaulting of the palate increases, the vomers are transformed from plates of bone lying more or less horizontally between the internal nostrils to a single fused plate lying in the median plane between the air passages. By the time that a complete secondary palate has evolved, the vomer forms the posterior and inferior part of the nasal septum, connecting posterosuperiorly with the reduced pterygoid bones and anteroinferiorly with the dorsal surface of the fused palatine processes. In mammals the forward expansion of the cranial cavity has led to the anterior part of the cranial base (presphenoid bone) coming to lie above the posterior part of the nasal passages. As a result the posterosuperior connection of the vomer is with the presphenoid (no longer covered by the parasphenoid). The anterior and dorsal part of the nasal septum was composed originally of cartilage but, in many mammals, this has been replaced, at least in part, by the bone of the mesethmoid ossification.

Because of the vaulting of the primary palate the main parts of the maxillae and palatine bones come to lie more vertically in the side walls of the vault. In consequence, the maxillae and palatine bones have, by the cynodont stage, assumed an essentially mammalian position in the lateral walls of the nasal passage with their palatine processes forming its floor (i.e. the secondary palate). The roof of the nasal passage is formed initially by the vomers and the most medial parts of the palatine bones. In mammals the reduction of the anterior part of the primary palate has led to the roof of the nasal cavity being extended upwards to the ventral surfaces of the nasal bones and anterior extremity of the frontal bone. Only posteriorly is the roof still formed by parts of the palatine bones and vomer, now aided by the presphenoid. In its central region the roof is completed by the cribriform plate connecting the septal and labyrinthine ossifications of the ethmoid.

The pterygoids of the pelycosaur palate are large bones which, in the typical reptilian manner, form the medial part of the palate behind the internal nostrils, bear a lateral flange and quadrate process, and are separated from each other by an interpterygoid vacuity. In theriodonts there is a general tendency for the pterygoids to become reduced in size and to meet in the midline ventral to the anterior part of the parasphenoid and anterior to the basipterygoid processes, so closing the interpterygoid vacuity. Probably as part of the reconstruction to provide more space for the enlarging jaw musculature, the quadrate process of the pterygoid shared in this reduction and became moved inwards to attach to the cranial base lateral to the parasphenoid. The cranial base is thus converted to a very rigid structure. With the fusion of the maxillary and premaxillary palatine processes, the pterygoids have come to form the roof and side walls of the nasal passage behind the posterior border of the secondary palate (i.e. in the region of the soft palate). These changes are well advanced in cynodonts (Fig. 32*b*). The interpterygoid vacuity is completely closed and the pterygoids are firmly attached to the ventral surface of the braincase. In the more advanced cynodonts and ictidosaurs the quadrate process of the pterygoid is reduced or absent. The quadrate ramus of the epipterygoid is more persistent and may be found as a slender rod still reaching the quadrate in those forms where the quadrate process of the pterygoid has already been lost.

Although on these palaeontological grounds there seems little reason to doubt that the mammalian pterygoid is a direct derivative of the reptilian pterygoid, this simple relationship has been questioned since Gaupp's (1908) discovery of separate dorsal and ventral pterygoids in the monotreme *Tachyglossus aculeatus*. Subsequently, evidence has accumulated that in other monotremes, as in marsupials and placental mammals, two elements are involved in forming the pterygoid (see Presley & Steel (1978) who also summarise previous evidence on this point). In *Ornithorhynchus*, as in *Tachyglossus*, the blastemata ossify separately and never fuse but in marsupials and placentals the blastemata fuse either before or after ossification has begun. The dorsal blastema (or ossification centre) gives rise to the main plate of the pterygoid while the ventral blastema (centre) becomes the hamulus. In many cases, secondary cartilage appears in the ventral centre. Several authors (e.g. de Beer, 1937; see also Presley & Steel, 1978, for a summary of views) have equated the dorsal centre with the transverse process of the parasphenoid while the ventral centre is believed to represent the sole remnant of the reptilian pterygoid. The fact that the palatine branch of the facial nerve passes in the Vidian canal above the dorsal centre (i.e. in the same relationship that it bears to the transverse process of the parasphenoid) is taken as evidence supporting these homologies. However, as Parrington & Westoll (1940) pointed out, this relationship to the palatine nerve would also be expected from the reptilian pterygoid–mammalian pterygoid homology since the nerve, after passing ventral to the basipterygoid process, almost certainly passed dorsal to the pterygoids. They also pointed out that the parasphenoid of the mammal-like reptiles generally lacks the well-developed transverse processes seen in some modern reptiles and may not have covered the ventral aspect of the palatine nerve. Parrington & Westoll preferred, therefore, to regard the main or dorsal part of the mammalian pterygoid as being derived from the reptilian pterygoid as indicated by the palaeontological evidence.

The ectopterygoid is already a much reduced bone in the sphenacodonts, lying on the lateral flange of the pterygoid. The bone is further diminished in theriodonts and is usually held to have been lost altogether from the mammalian skull. However, Parrington & Westoll (1940) suggested that it may be represented by the ventral centre or hamulus of the mammalian pterygoid.

The Parrington & Westoll interpretation has recently received strong support from the study by Presley & Steel (1978) of embryological stages of a wide variety of monotrematous and ditrematous mammals. They found that both the dorsal and ventral pterygoid elements develop in the tectoseptal extension, suggesting, as would be expected of the pterygoid and ectopterygoid bones, an origin from mandibular arch mesoderm (the parasphenoid, by contrast, being formed by prochordal mesoderm), and further that these elements in their ontogeny arise in positions very similar to those that they occupy in the mammal-like reptiles.

In the mammalian skull each pterygoid bone consists of a basal plate attached above to the ventral surface of the basisphenoid and presphenoid and enclosing the Vidian canal. From the basal plate a vertical lamina descends which forms the lateral wall of the nasal cavity in the nasopharyngeal region. The arrangement is, therefore, closely similar to that seen in advanced theriodonts. The major difference is that the basal plates do not meet in the midline in mammals, as they do in theriodonts, so exposing the ventral surface of the basisphenoid (the parasphenoid no longer covering this surface – Fig. 36*b*). This difference has been attributed to the reduced need for support resulting from the conversion of the large reptilian pterygoid muscle (part of the adductor internus) to the small tensor tympani and tensor palati of mammals (but see below). In most placental mammals the pterygoid element is much reduced, being replaced anteriorly by the perpendicular plate of the palatine.

The fate of the remaining major palatal element, the epipterygoid, has been dealt with in the section dealing with the side wall of the braincase.

LOWER JAW AND JAW ARTICULATION

The lower jaw of therocephalians shows several significant modifications from that of the sphenacodonts (Fig. 29). In the early pristerognathid forms the dentary is relatively enlarged and bears a slender but prominent coronoid process (Fig. 37*a*), while the postdentary bones are somewhat reduced in size. The surangular projects posteriorly from the medial surface of the dentary. Beneath and lateral to the surangular, the posteroventral region of the jaw is formed by the angular which exhibits a well-developed reflected lamina. The jaw joint lies in approximately the same horizontal plane as the tooth row. The articulating surfaces of the quadrate and articular are transversely widened and orientated so that their medial extremities lie anterior to their lateral extremities. The posterior border of the concave facet on the articular is marked by a pronounced ridge, the dorsal process. The retroarticular process of the articular projects downwards as it does in sphenacodonts, and in more advanced mammal-like reptiles is often curved forwards near to its tip. The size of the process is rather variable from group to group and this has led to the suggestion (Watson, 1951) that it is not, in fact, a direct derivative of the original reptilian retroarticular process (which, Watson believed, had disappeared together with the attached depressor mandibulae) but rather a neomorph giving attachment to the posterior pterygoid muscle. Although the matter cannot be considered finally settled, there appears to be no obvious reason for doubting that the retroarticular process of the mammal-like reptiles is equivalent to the similarly named process of other reptiles and that it gave attachment in the usual reptilian manner to the depressor mandibulae muscle. Parrington (1955) has suggested that the downward projecting, recurved form of the retroarticular process in advanced mammal-like reptiles was associated with the need to make a wide gape in order to use the powerful dentition efficiently. The medial surface of the jaw in the early types is not well known so that the arrangement of the splenial, prearticular and coronoid bones is uncertain. In more advanced (whaitsiid) therocephalians the prearticular abuts against the medial surface of the angular and the unit so formed runs forwards to be held between splenial and dentary (Fig. 37*b*). The coronoid connects the anterior ends of the surangular and angular–prearticular.

In the cynodonts the evolutionary trends in the lower jaw are well advanced and progressive. The procynosuchids have a dentary still further enlarged and with a stronger developed coronoid process than that of the therocephalians; in the later forms the beginnings of an angular process projecting from the posteroventral border are present (Fig. 37*c*, *d*). The lateral surface of the coronoid process is hollowed out for muscle attachment. The postdentary bones in the early cynodonts are not greatly reduced. The surangular and angular are still large bones although the reflected lamina is smaller and projects less far posteriorly than the corresponding structure in therocepha-

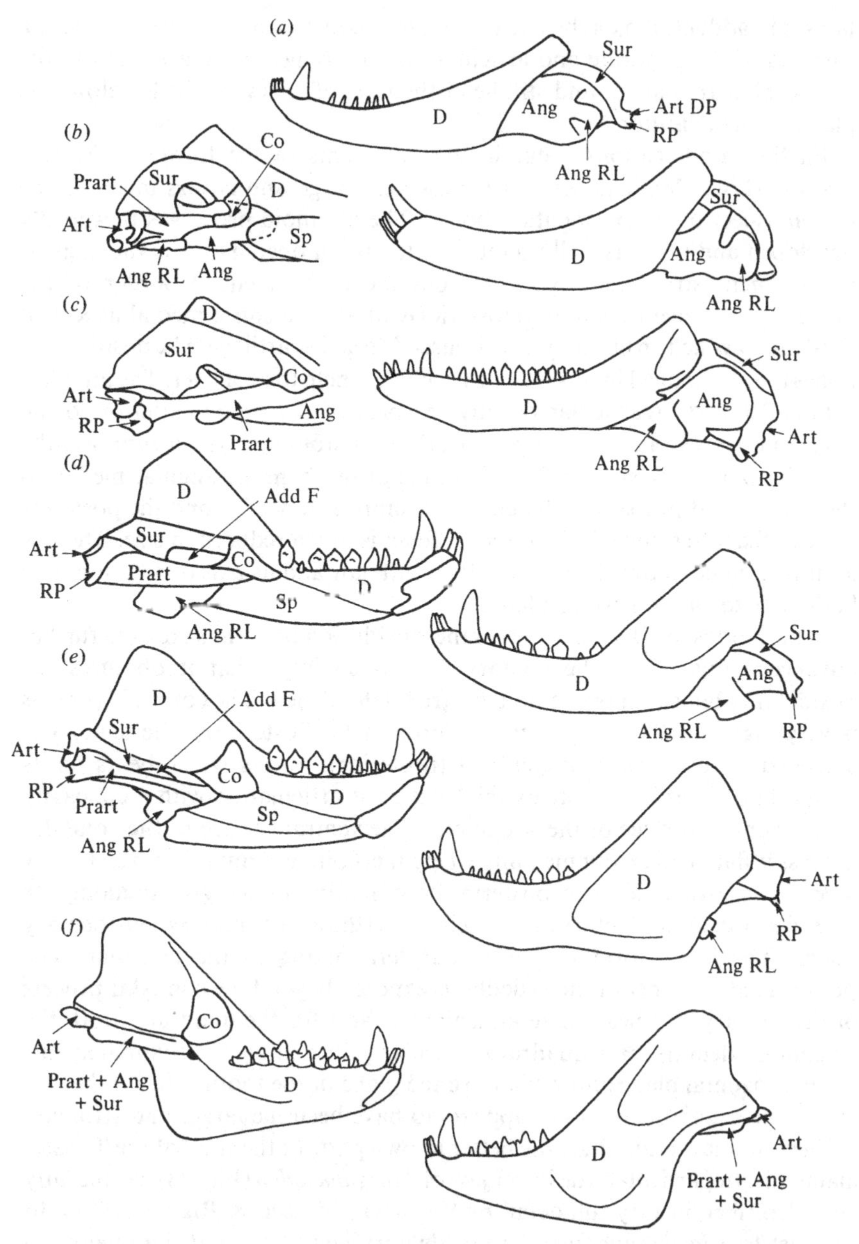

Fig. 37. The lower jaws of various therapsids: (*a*) pristerognathid therocephalian; (*b*) whaitsiid therocephalian; (*c*) *Procynosuchus*; (*d*) *Thrinaxodon*; (*e*) *Cynognathus*; (*f*) *Diarthrognathus*. Lateral views to right, medial views to left. Not to scale. For labelling see list of abbreviations.

lians. The adductor fossa shows a progressive reduction in size. The articulating surfaces of the quadrate and articular appear to have been arranged in the therocephalian manner and still lie at the level of, or even a little below, the plane of the dentition.

Further modifications occur in the cynodonts of the Lower to Middle Triassic (Fig. 37*e*). The dentary is now very large and covers much of the lateral aspect of the postdentary bones. The coronoid process is powerfully developed and projects well up into the temporal opening while the angle is a prominent structure projecting from the posteroventral border of the dentary. A groove runs anteroposteriorly along the entire medial aspect of the dentary which, in life, may have housed Meckel's cartilage. The postdentary bones are also lodged in this groove which they help to roof over. The angular, surangular and prearticular occupy the part of the groove inferior to the coronoid process and project posteriorly as a stout bar which supports the articular bone. The small reflected lamina arises from the angular medial to the dentary but probably extended for a short distance beyond the posterior limit of the latter bone. The adductor fossa is much reduced in size. The jaw joint is situated above the plane of the tooth jaw and its axis of rotation now lies closer to the transverse plane.

In the ictidosaur *Diarthrognathus* these evolutionary trends are even further advanced (Fig. 37*f*). The dentary is now so large that it obscures the postdentary bones almost completely from lateral view. The coronoid process is very high and the angle deep and prominent. Posteriorly, the dentary is extended to form a strong condylar (posterior) process which bears on its posterolateral surface a convex thickening for articulation with a concavity on the ventral surface of the squamous. The ventral surface of the condylar process is flattened except medially where it is hollowed out to form a shallow inverted groove which is a posterior continuation of the groove along the medial aspect of the dentary and which lodges the greatly reduced postdentary bones. The latter form a slender bar terminating at the articular. The posteromedial corner of the articular is exposed beyond the condylar process of the dentary and bears here an articular facet for the quadrate. Both the squamous–dentary and quadrate–articular joints lie in approximately the same horizontal plane and well above the plane of the tooth row. Of the two joints, the squamous–dentary appears to have been the larger and stronger.

The lower jaw is amongst the better known parts of the skull of the Triassic mammals. In particular, the lower jaw of *Morganucodon* (Fig. 38) has recently been described in very full detail by Kermack, Mussett & Rigney (1973). In contrast to *Diarthrognathus*, the postdentary jaw bones and the quadrate–articular joint are not greatly reduced compared to those of the advanced cynodonts, while the dentary has a rather more slender appearance, largely due to the relatively greater length of the tooth row. The squamous–dentary articulation appears to have been strongly built, the condylar process of the dentary being well buttressed, especially laterally. The articular surface of the

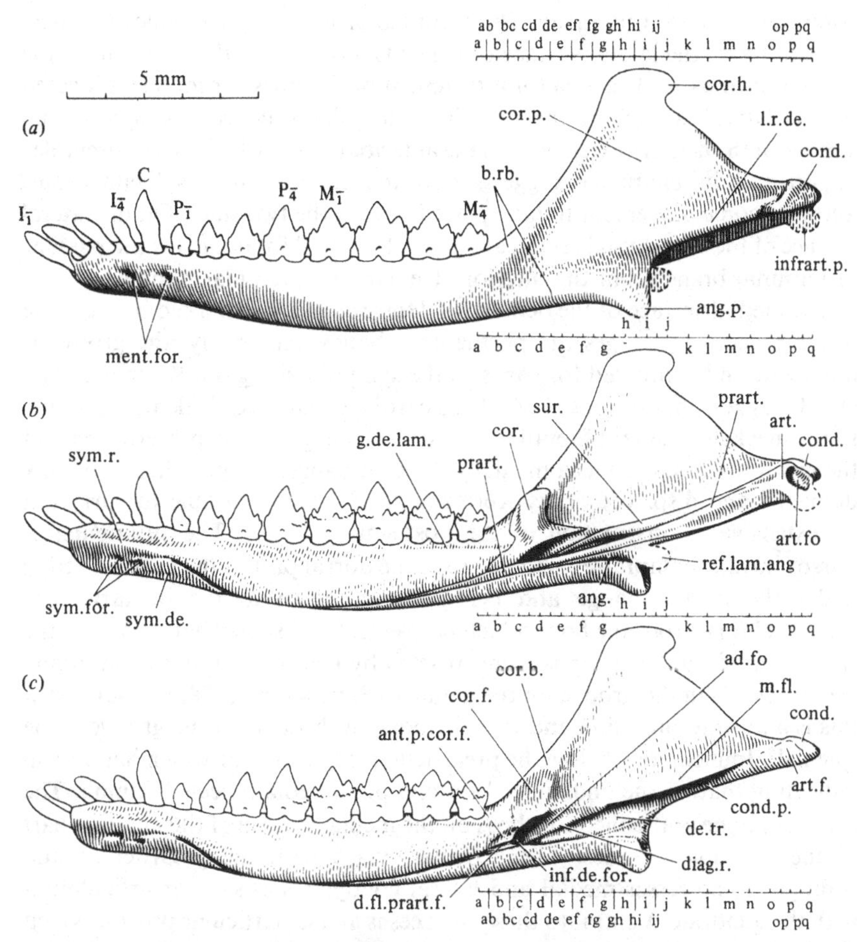

Fig. 38. Reconstruction of lower jaw of *Morganucodon*. (*a*) Left side, lateral view; (*b*) right side, medial view; (*c*) right side, medial view. Abbreviations: ad. fo. = adductor fossa; ang. = angular; ang. p. = angular process of dentary; ant. p. cor. f. = facet for anterior prong of coronoid; art. = articular; art. f. = facet for articular complex; art. fo. = articular fossa; b. rb. = bony rib; cond. = condyle; cond. p. = condylar process; cor. = coronoid; cor. b. = coronoid boss; cor. f. = facet for coronoid; cor. h. = coronoid hook; cor. p. = coronoid process; de. tr. = trough of dentary; d. fl. prart. f. = facet for dorsal flange of prearticular; diag. r. = diagonal ridge; g. de. lam. = groove for dental lamina; inf. de. for. = inferior dental foramen; infrart. p. = infra. articular process; l. re. de. = lateral ridge of dentary; ment. for. = mental foramen; m. fl. = median flange; prart. = prearticular; ref. lam. ang. = reflected lamina of angular; sur. = surangular; sym. de. = symphysis of dentary; sym. for. = symphysial foramen; sym. r. = symphysial ridge. (Reproduced with permission from Kermack, Mussett & Rigney, 1973.)

condyle faces posterodorsally. The coronoid process is a pronounced vertical plate of bone, the lateral surface of which is recessed for the insertion of the temporalis muscle. The area for attachment of the masseter muscle is located on the lateral surface of the prominent angular process. The symphysial surface of the dentary extends from the anterior tip of the bone to the premolar region and its character suggests that the connection was ligamentous, allowing some movement to take place between the two sides. On the lateral surface of the symphysial region are multiple mental foramina for the passage of terminal branches of the inferior alveolar neurovascular bundle.

The medial surface of the dentary of *Morganucodon* is marked by a groove which posteriorly lodges the postdentary bones. Anteriorly, the groove is narrowed and continued forwards to the symphysial region. Kermack *et al.* (1973) suggest that it housed Meckel's cartilage and is equivalent, therefore, to the meckelian canal of reptiles. The dorsal margin of the posterior part of the groove is produced as a medially directed flange. Despite their generally damaged and displaced state, Kermack *et al.* have been able to carry out an extensive reconstruction of the postdentary bones. The surangular is dorsoventrally compressed and occupies the dorsal part of the groove, fitting under the medial flange and being firmly joined to the dentary. The prearticular is a rod-like bone which lies ventral and somewhat medial to the surangular; its anterior tip is reconstructed by Kermack *et al.* as continuing well forwards in the groove on the dentary (Parrington (1978) has suggested this is a misinterpretation and that the bone in this part of the groove is the splenial). Immediately below the prearticular is the angular which has a form indicating that, in the undamaged state, it possessed a reflected lamina. The coronoid bone is attached to a boss on the medial surface of the anterior part of the coronoid process. The articular, fused with the prearticular and surangular, possesses medial and lateral condylar surfaces for articulation with the quadrate. Ventral to these surfaces is an infraarticular process which Kermack *et al.* homologise with the manubrium of the malleus of later mammals. The surangular appears to form about one-half of the lateral articular facet.

The lower jaw of the dryolestids (pantotheres) has also been described in detail (Krebs, 1971). In early forms (from the Jurassic) the dentary is of full mammalian proportions with a well-developed condyle, coronoid process and angle. Its medial surface exhibits markings which probably accommodated much reduced splenial and coronoid bones. In later dryolestids (from the Lower Cretaceous) the splenial has disappeared while the coronoid appears to have fused during ontogeny with the dentary in the region of the coronoid process.

Although there is no intention of implying that this assemblage of early mammals represents anything near to an evolutionary sequence, it does illustrate that the transition from the reptilian structure of the lower jaw to that typical of mammals was not clear cut and that reptilian features, such

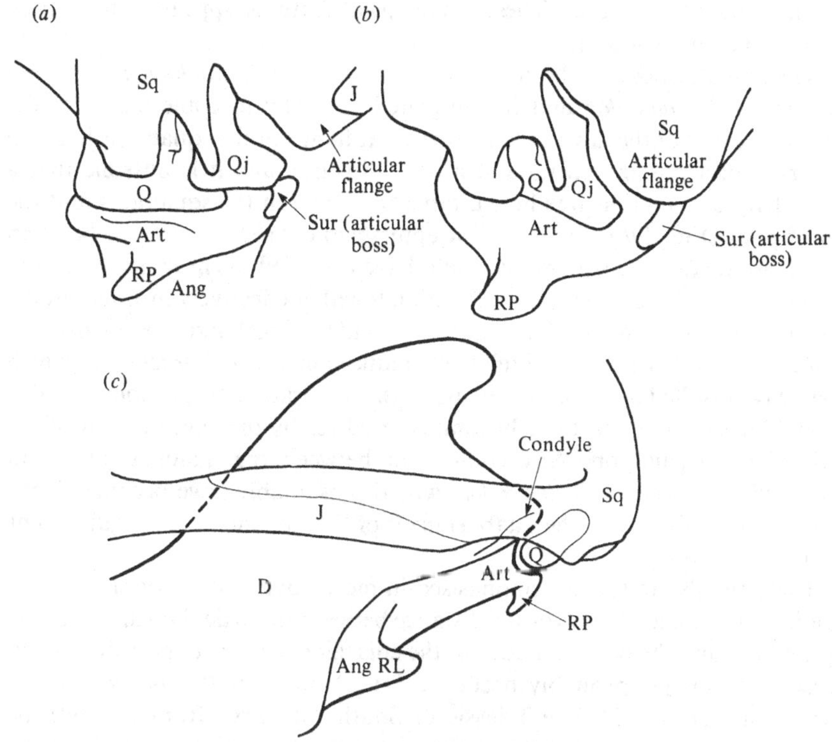

Fig. 39. Changes in jaw articulation in cynodonts. (*a*) Posterior view of right side in *Thrinaxodon*; (*b*) posterior view of right side in *Trirachodon*; (*c*) lateral view of left side in *Diarthrognathus*. Not to scale. For labelling see list of abbreviations. (Based on Crompton, 1972.)

as a quadrate–articular joint and the presence of postdentary bones, were retained in animals which by other criteria were mammals.

The changes that took place in the jaw articulation have recently been analysed in detail by Crompton (1972) in the light of new observations on the cynodont jaw. His conclusions indicate that these changes may have been more complex and may have varied more from one cynodont line to another than previously supposed. In *Thrinaxodon* the quadratojugal and surangular enter into the quadrate–articular joint. Crompton suggests that there were, in addition, ligamentous connections between an articular boss on the posterolateral surface of the surangular, on the one hand, and the articular flange of the squamous and the lateral tip of the ventral end of the quadratojugal, on the other, which may have helped to strengthen the jaw joint against the increased power of the adductor musculature (Fig. 39*a*) (although, as Kemp (1972*a*) points out, the orientation of such a ligament seems not altogether appropriate to provide maximum resistance against the upward and backward forces which were probably produced by the contracting musculature). These osteological features appear to have been neomorphic in cynodonts.

The articular boss on the surangular is enlarged in *Trirachodon* compared to that in *Thrinaxodon* and has migrated ventrally to come closer to the transverse axis of the jaw joint. The articular flange of the squamous has also enlarged in a posteroventral and medial direction and it is possible that a subsidiary articulation was here established between the squamous and the surangular (Fig. 39*b*). The loose (streptostylic) connection present between the quadrate and the squamous would have provided the mobility necessary to have allowed the two joints to function together effectively throughout the range of jaw movements. The quadrate, quadrotojugal and the postdentary bones of the lower jaw are all relatively reduced in size in *Trirachodon*, which may have resulted in some weakening of the quadrate–articular joint but this would have been at least partly compensated for by the development of the subsidiary articulation. Any connection between the quadrotojugal and surangular, as seen in *Thrinaxodon*, would presumably have become mechanically redundant with the establishment of the squamous–surangular joint and have been lost.

These trends are further emphasised in more advanced cynodonts. There is additionally a tendency for the distance between the articular surface of the surangular and the posterior edge of the enlarging dentary to be reduced and in some forms (as probably occurred, for example, in *Probainognathus*, a cynodont from the Middle Triassic of South America – Romer, 1969) the dentary may have invaded the squamous–surangular joint. A continuation of the phylogenetic enlargement of the dentary could clearly have led to this bone eventually monopolising the articulation with the squamous to the exclusion of the surangular and the establishment of the completed squamous–dentary joint such as is found in the ictidosaur *Diarthrognathus* (Fig. 39*c*), although here, Crompton (1972) suggests, the new joint may have been established lateral to and separate from the squamous–surangular joint rather than having been formed by invasion of the latter joint by the dentary. An unexpected finding in *Morganucodon* is the relatively unreduced nature of the accessory jaw bones and quadrate–articular joint as compared with advanced cynodonts and ictidosaurs. The reduction of these structures would seem to provide yet another example of parallel evolution towards the mammalian condition in that the advanced mammal-like reptiles, the cynodonts and ictidosaurs, show changes in the lower jaw which did not occur until a much later stage in the mammalian line.

The increase in size of the temporal opening, the bowing of the zygomatic arch, the reduction in size of the quadrate ramus of the pterygoid and the changes just recounted in the lower jaw were all undoubtedly related to an increase in size and differentiation of the adductor jaw musculature. Although direct observation of muscles is impossible in extinct forms, inferences can be drawn from the size and position of muscle markings as to how the muscles

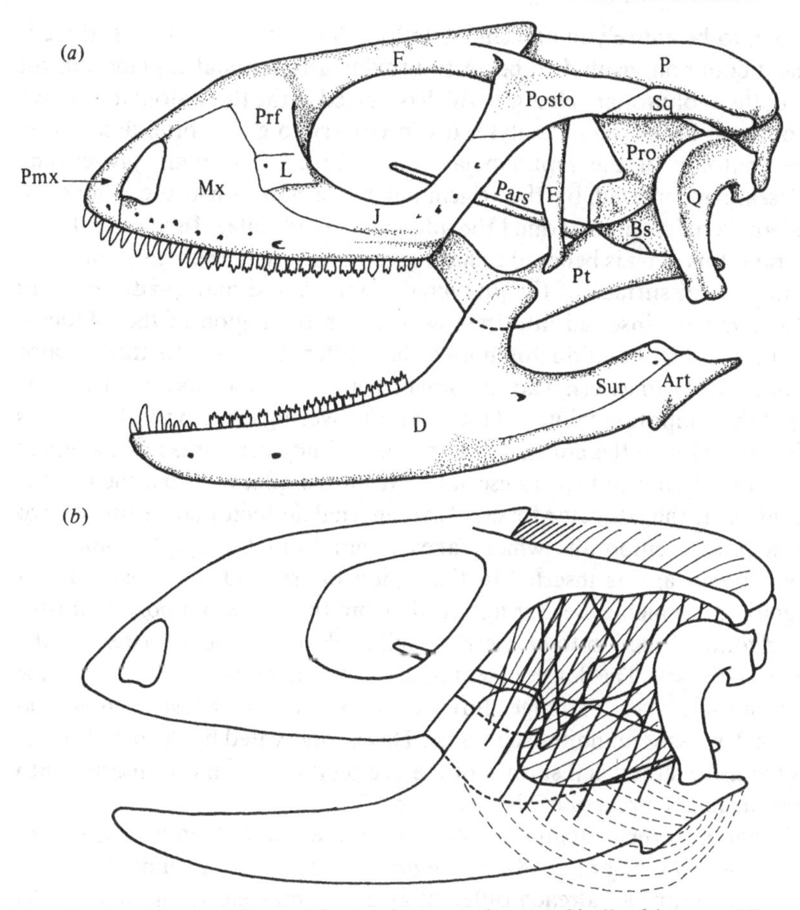

Fig. 40. To show reptilian adductor jaw muscles. (*a*) Skull of iguana (*Ctenosaura pectinata*) with lower jaw displaced downwards to allow pterygoid region to be seen. (*b*) Outline of skull with adductor musculature superimposed. Superficial external adductor (widely spaced continuous lines) arises from upper temporal arch (the lower arch has been lost) and is inserted into lateral aspect of lower jaw; deep external adductor (continuous fine lines) runs from skull roof to superior margin of lower jaw; pseudotemporalis part of internal adductor (closely spaced lines) arises from side of braincase anterior to deep external adductor and is inserted into coronoid process (this insertion is not shown); pterygoid part of internal adductor (broken lines) arises from pterygoid and is inserted into ventral part of posterior margin of lower jaw. For labelling see list of abbreviations. (After Bellairs, 1969.)

were arranged in life. Several authors (e.g. Adams, 1919; Watson, 1948; Parrington, 1955; Crompton, 1963*a*, *b*; Crompton & Hotton, 1967; Barghusen, 1968, 1972; Kemp, 1972*b*, 1979; Kermack *et al.*, 1973) have attempted to analyse from such indirect evidence the arrangement and function of the adductor jaw muscles of the mammal-like reptiles. Obviously such analyses contain an element of speculation and cannot be expected, therefore, to be entirely in agreement with each other. Nevertheless, there is sufficient common ground to begin to provide a functional explanation for many of the evolutionary changes which occurred in the theriodont lower jaw.

In order to follow these analyses it is necessary to give a brief description of the anatomy of the reptilian jaw musculature. The main jaw-closing muscles are innervated by the trigeminal nerve and constitute a mass of considerable magnitude, termed the adductor mandibulae. In ancestral land vertebrates this mass is believed to have taken origin in the subtemporal fossa from the lateral surface of the pterygoid, epipterygoid and quadrate bones and to have been inserted into the lower jaw in the region of the adductor fossa. In reptiles the adductor muscle has differentiated into three major components: (1) an external adductor which arises over a wide area from the bones of the temporal region and membranes covering the temporal openings and is inserted into the coronoid eminence and adjacent areas of the lower jaw; (2) a posterior adductor descending from the quadrate to attach to the lower jaw near the adductor fossa; (3) an internal adductor consisting of two parts, a pseudotemporalis which takes origin from the epipterygoid and adjacent bones and is inserted in the region of the adductor fossa, and a pterygoid part which arises far anteriorly from the pterygoid bone and runs in a horizontal direction to be inserted well back on the medial aspect of the lower jaw (Fig. 40). The development of the temporal openings provides space for this muscle mass to bulge durings its contraction. Most reptiles also possess a depressor mandibulae muscle. This is innervated by the facial nerve, arises from the superficial surface of the occipital region and is inserted into the retroarticular process of the articular.

Although in earlier attempts to reconstruct the jaw adductor musculature of *Dimetrodon* (Watson, 1948; Parrington, 1955) it was assumed that the external adductor had already differentiated into masseteric and temporalis components, more recent analyses (especially that of Barghusen, 1968) based on the greater information now available about the jaws of the mammal-like reptiles suggest that this was not so. It seems more likely that the external adductor was still a single muscle arising from the roof, posterior and lateral walls of the temporal fossa and being inserted into the dorsal margin of the coronoid and surangular and the bone immediately medial and possibly lateral to this dorsal margin. The insertion into the coronoid eminence may have been aponeurotic as is the case in some living reptiles.

In the procynosuchids an adductor fossa, clearly for muscle attachment, is present on the dorsolateral surface of the dentary (Fig. 37*c*). Barghusen (1968) suggests that the muscle which inserted here was derived from that part of the external adductor mass which was originally inserted on to the lateral surface of the aponeurosis attached to the coronoid process. The bowing out of the zygomatic arch, found in all cynodonts, would have allowed room for this muscle to have passed down between the zygomatic arch and the coronoid process to its insertion on the lateral surface of the process. Its origin was possibly from aponeurosis roofing in the temporal opening. In later cynodonts the muscular fossa is enlarged in a ventral direction, eventually extending over most of the lateral surface of the posterior part of the dentary, including the newly formed angular process (Fig. 37*d*, *e*) and the zygomatic arch becomes increasingly bowed outwards and develops markings suggestive of muscle attachment. Barghusen argues that these osteological features indicate a progressive enlargement of the lateral part of the external adductor muscle with a migration of its origin on to the zygomatic arch and of its insertion on to the more ventral region of the dentary. Such a muscle closely resembles and is almost certainly the antecedent of the mammalian masseter muscle (probably both deep and superficial parts). Other views of the origin of the masseter have been put forward (notably by Crompton, 1963*a*, *b*) but their proponents now seem generally to accept Barghusen's ideas (see, for example, Crompton & Hotton, 1967). Kemp (1972*b*), however, has more recently suggested that the masseter muscle was derived, not from the superficial part of the external adductor as postulated by Barghusen, but from a zygomatico-angularis muscle which he has described in whaitsiids as running from the inner surface of the squamous (i.e. posterior part of zygomatic arch) to the lateral surface of the body of the angular medial to the reflected lamina. He has suggested that the masseter muscle of *Procyno-suchus* was still essentially of this form and that the masseter of the more advanced cynodonts was derived by anterior extension of the origin and progressive transfer of its insertion on to the posteriorly enlarging dentary (Kemp, 1979). Whichever of these views is correct, it now seems well established that a sizable masseter muscle was first present in cynodonts, that it enlarged rapidly during the evolution of that group and in more advanced forms took origin from the zygomatic arch (especially its anterior part) and was inserted into the lateral surface of the angular process of the dentary.

There has been less dispute about the temporalis muscle. It seems evident that it represents that part of the external adductor remaining after the differentiation and separation of the masseter. Little rearrangement of the areas of origin and insertion and of fibre orientation would be required to produce the typical mammalian temporalis from such a source.

If the masseter muscle is a new development in cynodonts, the earlier view (e.g. Watson, 1948) that this muscle was inserted into the lateral surface of the reflected lamina of the angular in sphenacodonts and that the progressive diminution of the lamina seen in the sequence from pelycosaurs through therocephalians to cynodonts was consequent upon the insertion of the muscle being transferred forwards on to the lateral surface of the dentary must be incorrect. An alternative functional reason must then be found for the existence of the reflected lamina and for its phylogenetic shrinkage. Barghusen (1968) has suggested that the lamina may have served for the attachment of a posteroventrally directed branchiomandibularis muscle (such a muscle serves in lizards to depress the hyoid apparatus) and possibly also for an intermandibularis muscle. The reduction in size of the former and migration on to the dentary of the latter could account for the diminution of the reflected lamina. Kemp (1972*b*) concluded from a detailed study of the angular bone that it is more likely that the reflected lamina served to allow two muscles, whose fibres ran in different directions, to be inserted close together on the lateral side of the lower jaw. One muscle, he suggests, was the zygomatico-angularis and derived masseter (see above) while the other was the anterior pterygoid which probably took origin from the posterior surface of the lateral flange of the pterygoid bone, then ran posteroventrally to turn laterally around the ventral edge of the body of the angular and be inserted on the medial face of the reflected lamina. The forward migration of these insertions from the angular to the dentary, as postulated by Kemp, would account for the reduction of the reflected lamina seen in advanced therapsids. The possibility that the reflected lamina functioned as part of the auditory apparatus and not as the site of muscle attachment has recently been put forward by Allin (1975; this is discussed further below) which, if true, would make it unlikely that the reduction of the lamina was secondary to changes in the jaw muscles.

The attachments of the pterygoid musculature have been the subject of much speculation. In *Dimetrodon* Watson (1948) reconstructed an anterior pterygoid muscle (probably derived from the pterygoid part of internal adductor) as arising from the dorsal surface of the pterygoid and other palatine bones and running more or less horizontally backwards on the ventral surface of the flange produced by the prearticular to be inserted into the angular, medial to its reflected lamina, and the articular (the arrangement of the bones allowing the muscle to follow this path without major change of direction; see Fig. 29). A posterior pterygoid muscle (probably derived from the pseudotemporalis part of the internal adductor) may not have been clearly differentiated from the temporalis, although Watson described it as a separate muscle taking origin from the lateral surface of the posterior part of the pterygoid bone and being inserted into the medial aspect of the prearticular.

It has been widely, although not universally, held that both the lateral and medial pterygoid muscles of mammals are derived from the posterior pterygoid (see Parrington (1955) for a detailed discussion of this point). In this view, the anterior pterygoid is believed to have undergone a phylogenetic decrease in size, together with the postdentary bones, and to be represented in mammals by the tensor tympani, while the posterior pterygoid is thought to have shifted its insertion from the medial surface of the prearticular to the corresponding surface of the coronoid process and then split into two with the insertion of the lateral pterygoid migrating posteriorly and that of the medial pterygoid ventrally. Crompton (1963*b*), however, has postulated that the anterior relocation of the reflected lamina, the progressive thickening of the posteroventral corner of the dentary and the development of a depression on the posterior surface of the angle, seen at various stages in the sequence from procynosuchids to advanced cynodonts, are osteological indications that the anterior pterygoid progressively transferred its insertion from the angular to the dentary where it remains in mammals as the medial pterygoid. A slip of the anterior pterygoid, homologous with the mammalian tensor tympani, may have maintained its insertion into the articular. In this scheme, therefore, the lateral pterygoid alone is derived from the reptilian posterior pterygoid, the insertion of the muscle in advanced cynodonts having migrated, presumably, from the coronoid to the condylar process.

The arrangement of the adductor jaw musculature which is believed to have obtained in stem reptiles indicates that the lower jaw probably functioned during elevation as an approximation to a simple class III lever with the fulcrum at the jaw joint, and the muscular force acting between the fulcrum and the dentition. In such an arrangement with the upper and lower teeth in contact, the upward force of the contracting adductor muscle will be resisted more or less equally at the jaw joint and at the dentition. The jaw joint is thus subjected to heavy stressing. Clearly a more efficient arrangement would be for the force to be resisted in greater proportion by the dentition. A common aim of most functional analyses that have been made of the jaw muscles of the mammal-like reptiles has been to explain how such an increase in efficiency was achieved. It is generally assumed that there could have been no other way in which the jaw muscles increased so greatly in power (as the osteological evidence clearly indicates they did) while the jaw joint, despite the formation of a subsidiary articulation, underwent no corresponding strengthening and may even have been structurally weakened by the reduction of the accessory jaw bones. The lack of agreement about the arrangement of the jaw muscles of the mammal-like reptiles naturally precludes any great measure of accordance in the functional analyses but a number of common general conclusions can be drawn.

First, as pointed out by Parrington (1934), the forward migration involved in the transference of muscle insertions from the accessory jaw bones to the dentary would, of itself, help to apportion more of the thrust to the dentition. Secondly, extension of the tooth rows posteriorly would have the same effect. Thirdly, the tendency of some of the muscles to develop an increasingly oblique orientation would also tend to reduce the proportion of the force acting on the joint. In particular, the obliquity of the temporalis muscle, giving this muscle a posterior as well as a vertical component of pull, would have resulted in a relative decrease in the vertical force acting at the jaw joint. The development of a high coronoid process, as well as increasing the obliquity of the temporalis fibres, also helped to maintain a right angle between their average direction of pull and a line joining their insertion to the jaw joint so that the muscle retained its maximum efficiency in the position of jaw closure. The posterior force at the jaw joint produced by the posterior component of pull of the temporalis may have been offset by one or more of the other jaw muscles developing a compensating anterior component of pull. Despite the lack of agreement about the attachments of these muscles, it seems likely that both the masseter and the pterygoids may have functioned in this manner. As Crompton (1963*a*) has demonstrated, it is theoretically possible by such a distribution of muscular forces for virtually the entire counterbalancing thrust to be taken through the dentition with the jaw closed. Fourthly, it is improbable that such a complete cancelling out of forces at the jaw joint took place in even the most advanced theriodonts. Barghusen & Hopson (1970) and Crompton (1972) have argued that the still large residual forces were crucial factors in the development of the condylar process of the dentary, first to brace the postdentary bones and later to provide a subsidiary jaw articulation.

Kemp (1972*a*, *b*) has drawn attention to the general lateral direction in which the fibres of the jaw-closing muscles passed from their origins to their insertions in the whaitsiids and to the resulting tendency they must have possessed to displace the lower jaw medially during contraction. The oblique orientation of the articular surfaces of the jaw joint (that of the quadrate facing laterally as well as anteroventrally) probably served to prevent such displacement. The streptostylic attachment of the quadrate would have enabled the lower jaw to rotate despite the axes of the jaw joints of the two sides being inclined to each other. Essentially similar circumstances may have existed in procynosuchids (Kemp, 1979). The development of a large masseter muscle in advanced cynodonts would have added a considerable lateral component of pull which would have helped oppose the medially directed force produced by the remaining jaw-closing muscles and may have been an important factor in allowing the axis of the jaw joint to become more transversely orientated. Nevertheless, the quadrate appears to have retained its streptostylic mode of attachment and Kemp suggests that this may have allowed anteroposterior jaw movements to occur.

Kemp (1972*a*) has analysed the stresses induced in the whaitsiid lower jaw by the force of muscular contraction and has related the manner of jaw construction to the need to resist these stresses. The vertical pull of the muscles and the thrust taken through the dentition would have resulted in the lower jaw being compressed along its lower border and under tension along its upper border. The sutures in the lower part of the jaw tend to be butt joints which are well suited to resist compressive stress; those in the upper part of the jaw take the form of extensive overlaps where the large areas of collagenous connection would have provided a strong resistance to tension. Similar adaptive arrangements can be shown to have existed for the resistance of the stresses induced by the posterior and transverse components of muscle pull.

The reduced size of the postdentary bones in cynodonts would, presumably, have made the resistance of muscle-induced stresses even more critical. The vertical pull of the muscles and the reactions at joint and dentition would, as in whaitsiids, have resulted in the dorsal part of the dentary–postdentary contact being under tension and the ventral part under compression (Fig.

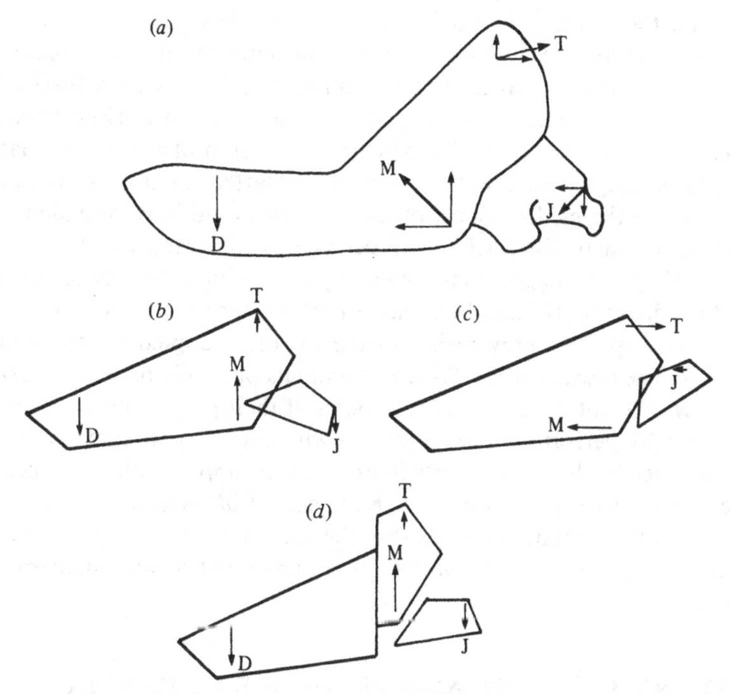

Fig. 41. To show principal forces acting on cynodont jaw: (*a*) resolved into vertical and horizontal components; (*b*) effect of vertical forces in producing a bending moment; (*c*) effect of horizontal forces in producing a bending moment; (*d*) effect of vertical forces in producing shear stresses. M = force due to masseter; T = force due to temporalis; D = resistance at dentition; J = resistance a jaw joint. (After Kemp, 1972*a*.)

41*b*). By contrast, the horizontal component of pull, increased in cynodonts due to the rearrangement of the adductor muscles, would have produced compression dorsally and tension ventrally (Fig. 41*c*), thus tending to cancel out the stresses produced by the vertical component of pull and leaving the region of dentary–postdentary contact only lightly stressed. The enlargement of the temporalis muscle and the development of a powerful masseter muscle in the cynodonts must have added significant shear stresses (Fig. 41*d*). Kemp suggests that the horizontal ridges found on the dentary of cynodonts were probably developed to strengthen the bone against these shear stresses. Weakness at the dentary–postdentary boundary would then be of no significance since collapse due to shear must obviously take place at two sections of the jaw simultaneously.

Kermack *et al.* (1973), in discussing their finding that the accessory jaw bones and quadrate–articular joint are relatively unreduced in *Morganucodon*, argue that, contrary to the general assumption, the arrangement of these structures in advanced therapsids represents a means of strengthening, not weakening, the lower jaw and its articulation. They point out that these animals were carnivores and that however efficiently the forces acting on the jaw joint were counterbalanced in biting on the cheek teeth, these forces must have been very great when struggling prey was being seized and killed between the incisor and canine teeth. The posterior extension of the dentary lateral to the postdentary bones and the fusion of the latter into a firm rod housed in a groove on the medial aspect of the dentary would have provided a jaw strongly resistant to the buckling forces produced during this type of jaw function. They also suggest that the changes in the morphology of the teeth and in the adductor jaw muscles indicate that late therapsids used their cheek teeth to slice up their prey rather in the manner adopted by mammalian carnivores. The insectivorous Triassic mammals probably used their teeth in a similar way to cut up the chitinous bodies of their prey. Such a scissor-like action would be performed most efficiently with a strong joint that allows little play and, since the lower teeth bite internal to the upper teeth, has a bearing contact lateral to the cutting edge. Kermack *et al.* believe that it was in response to these functional needs that the squamous–dentary joint evolved not only in the mammalian line but in several other advanced theriodont groups.

AUDITORY OSSICLES AND MIDDLE EAR CAVITY

Although comparative anatomical studies have established beyond doubt that the malleus and incus of the mammalian middle ear are derived from the articular (probably plus a contribution from the prearticular) and quadrate, respectively, of the reptilian lower jaw, precise knowledge of the sequence of events by which this transformation was brought about is precluded by the scanty fossil record of these small and delicate bones. Throughout the mammal-like reptiles the stapes remained the only ossicle in the sound conducting system of the middle ear. In the more advanced therapsids, however, several phylogenetic changes took place in the structure of the stapes and its relationships with the articular and quadrate, which appear to foreshadow the mammalian arrangement. From this basis a number of attempts has been made to infer the subsequent steps by which the articular and quadrate became freed from the lower jaw articulation and included in the middle ear cavity. The following account is based largely on the work of Westoll (1943*a*, 1944, 1945), Olson (1944), Parrington (1955), Hopson (1966) and Allin (1975).

In the majority of therapsids the stapes is perforate (for the stapedial artery), has a substantial articulation with the quadrate by means of a bony quadrate process, and possesses the base of what appears to have been a cartilaginous extrastapes. The hyoid process is missing as is the dorsal process in some forms. In *Thrinaxodon*, for example, the stapes is large but lightly built and extends from the fenestra vestibuli to contact, through the quadrate process,

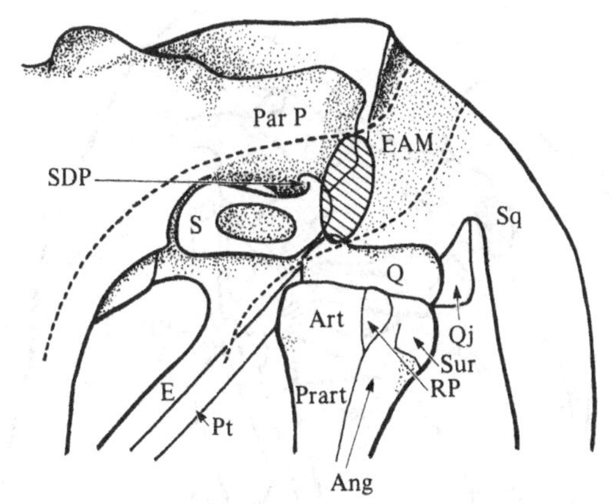

Fig. 42. Ventral view of right auditory region of an early cynodont such as *Thrinaxodon* to show possible position of tympanic membrane (hatched) and external acoustic meatus and tympanic cavity (both indicated by broken line). For labelling see list of abbreviations. (After Romer, 1956, Estes, 1961, and Hopson, 1966.)

the medial surface of the quadrate lying in its hollow in the squamous (Fig. 42). A dorsal process projects superiorly from its lateral part to meet the tip of the paroccipital process. It is usually assumed that a small extrastapes extended from the lateral end of the stapes to contact a tympanic membrane. The position of the membrane is not known for certain but it may well have been located in the notch in the squamous behind the quadrate, possibly attached to an anteroposteriorly orientated lip on the former bone. A sinuous groove on the posterolateral aspect of the squamous may indicate the site of the external acoustic meatus. The middle ear cavity probably extended forwards from the paroccipital process to the lateral flange of the periotic and quadrate processes of pterygoid and epipterygoid.

At some stage subsequent to the establishment of the squamous–dentary joint, the quadrate–articular articulation must have become redundant as a support for the lower jaw. As we have seen, this occurred after the mammalian grade had been attained, Triassic mammals such as *Morganucodon* still possessing a strongly built quadrate–articular joint. Since fossilised ossicles from this critical stage in the evolution of the middle ear are unknown, Hopson (1966) has used the structure of the middle ear in *Bienotherium* (a Late Triassic tritylodont) as a basis for inferring the possible steps by which the arrangement seen in *Thrinaxodon* might have been modified to that found in mammals, arguing that the tritylodonts were probably derived, like mammals, from a cynodont ancestor and may have undergone a considerable measure of parallel evolution with the mammalian line.

In *Bienotherium*, as in other late therapsids, the accessory jaw bones are reduced in size and occupy a groove on the medial side of the dentary. The reflected lamina of the angular is small and lies well forwards of the jaw joint.

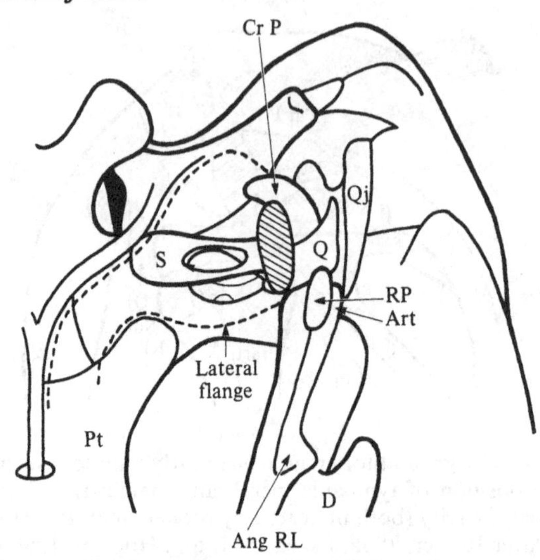

Fig. 43. Ventral view of right auditory region of *Bienotherium* to show Hopson's reconstruction of middle ear structures. Tympanic membrane hatched; limits of tympanic cavity indicated by broken line. For labelling see list of abbreviations. (Based on Hopson, 1966.)

In ventral view (Fig. 43) the middle ear region can be seen to have undergone several structural modifications from that of *Thrinaxodon*. The lateral part of the paroccipital process is expanded anteroposteriorly and bears two projections, the posterior of which may be homologous with the mammalian mastoid process while the anterior appears to be homologous with the crista parotica of monotremes and embryonic therian mammals. The posteromedial surface of the anterior projection bears a small process for attachment of the hyoid while the anterolateral surface probably provided the principal contact for the quadrate. The middle ear cavity appears to have been bounded laterally by the anterior projection, medially by the wings of the parasphenoid and anteriorly by the lateral flange of the periotic. The tympanic membrane was probably located beneath the anterior projection, behind the quadrate and in front of the hyoid. Hopson (1966) reconstructs the stapes as having contacted the medial surface of the quadrate below the anterior projection and as possessing a cartilaginous extrastapes extending to the tympanic membrane. The articular bone bears a curved retroarticular process to which was attached, presumably, the depressor mandibulae.

Hopson takes the arrangement and function of the ear ossicles in the monotreme *Tachyglossus* (Fig. 44*a*) to exemplify the ossicular mechanism of primitive mammals. In this animal the malleus has a movable articulation through its anterior process with the tympanic bone. The incus has a short process which extends posteriorly in a direction continuing the line of the

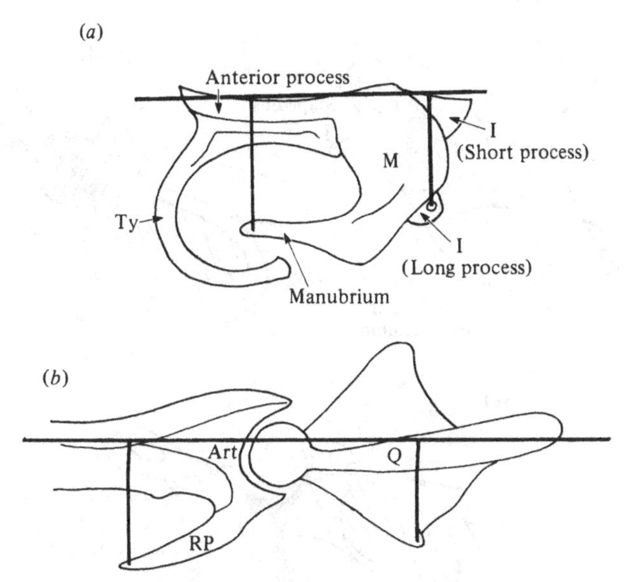

Fig. 44 To show derivation of malleus and incus according to Hopson. (*a*) Malleus, incus and tympanic of *Tachyglossus* in lateral view; (*b*) quadrate and articular of tritylodontid orientated as described in text to correspond with ossicular mechanism shown in (*a*). For labelling see list of abbreviations. (After Hopson, 1966.)

anterior process of the malleus to contact the crista parotica. The malleus and incus move as a unit about an axis passing through the anterior and short processes. Extending below this axis are the manubrium of the malleus, attached to the tympanic membrane, and the long process of the incus, attached to the distal extremity of the stapes. The manubrium forms the force lever arm activated by vibrations of the tympanic membrane, while the long process is the resistance lever arm of the system. Since the force lever arm is somewhat longer (as measured perpendicularly to the axis of rotation) than the resistance lever arm, the force exerted at the fenestra vestibuli is magnified (this magnification is in addition to that produced by the ear drum having a larger surface area than the fenestra vestibuli). The magnitude and velocity of the oscillations are, of course, correspondingly reduced (the increase in force and reduction in velocity are important factors in the impedance matching which is the principal function of the ossicles – see Chapter 6).

Hopson (1966) suggests that the arrangement of the quadrate and articular in *Bienotherium*, while related primarily to masticatory requirements, is pre-adapted to become part of the sound conducting mechanism of the middle ear. The articular with its curved retroarticular process bears a close similarity to the mammalian malleus with its recurved manubrium. In order for the articular to be transformed into a functional malleus it would be necessary

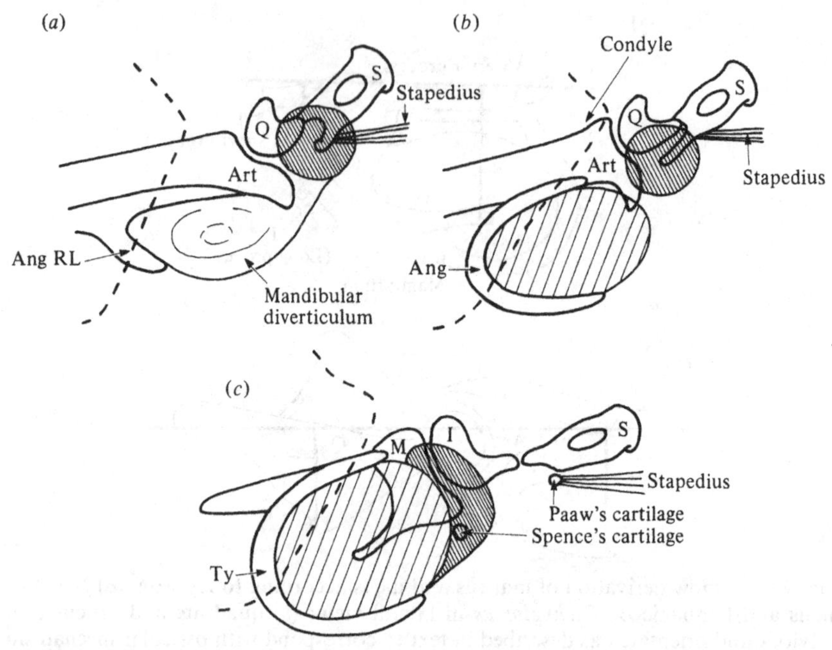

Fig. 45. Diagrammatic representation of stages suggested by Westoll (1945) in the evolution of the mammalian middle ear and drum (*c*) from the corresponding structures in the reptile (*a*). Reptilian tympanic membrane/pars flaccida shown by close hatching; mammalian tympanic membrane/pars tensa shown by widely spaced hatching. Outline of posterior part of dentary shown by broken line. For labelling see list of abbreviations.

only for a joint to be formed between the articular (together, possibly, with the fused prearticular) and the remaining accessory jaw bones such that the articular could rotate about an axis in the line of the postdentary rod. The tritylodont quadrate could similarly be readily converted into a functioning incus, all that would be required being a change in its orientation so that its long axis lies in the same line as the articular (Fig. 44*b*). The posteriorly directed flange of the tritylodont quadrate may have provided the contact with the stapes, in which case this flange was clearly preadapted to become the resistance lever arm of the ossicular mechanism.

The changes so far postulated would require an expansion of the tympanic membrane so that it could make contact with the manubrium (retroarticular process). It would also have been of advantage for the membrane to have lost its contact with the extrastapes as this would have 'short-circuited' the force-magnifying system provided by the addition of malleus and incus to the ossicular chain. Westoll (1943*a*, 1944, 1945 – see also Chapter 2) had earlier made the following suggestions as to how these modifications were achieved. He held that the mammalian tympanic membrane was a newly formed structure produced by a ventral mandibular diverticulum from the tympanic cavity meeting the skin in the notch between the body and the reflected lamina of the angular (Fig. 45). The mammalian drum thus came to be located lateral to the manubrium and away from the stapes and within the 'arms' formed by the two parts of the angular. With its detachment from the jaw and attachment to the braincase, the angular was transformed into the tympanic (ectotympanic) bone which forms an annulus supporting the drum of the mammalian ear. The reptilian tympanic membrane, originally located above and behind the angular region, underwent an anterior migration to meet the posterodorsal edge of the mammalian drum where it has persisted as the pars flaccida.

The principal difficulty with Westoll's suggestions is that they are inconsistent with the widely held view that the angular notch was occupied by the insertion of the pterygoid musculature, except possibly in the most advanced therapsids. In order to overcome this difficulty Hopson (1966) has proposed that the mammalian tympanic membrane was formed by a direct anteroventral enlargement of the reptilian drum. At the same time the reflected lamina of the angular underwent a posterior movement along with the angle of the dentary and thus came partly to surround the enlarging membrane. During the posterior migration of the angular the insertion of the pterygoid musculature was transferred to the angle of the dentary, leaving the angular free to provide attachment for the drum.

Subsequent to the disconnection of the tympanic membrane from the extrastapes, the lateral part of the columella auris became redundant and was lost, leaving just the bony stapes to provide the link between the incus and fenestra vestibuli. It is generally held (see, for example, Westoll, 1944) that the cartilages of Paaw and Spence represent the vestiges of the cartilaginous extrastapes. Paaw's cartilage is found in the tendon of the stapedial muscle in many therian mammals (Fig. 45). The stapedial muscle in therapsids was probably attached to the posterior surface of the proximal part of the extrastapes, in which case Paaw's cartilage would represent a vestige of this part of the extrastapedial structure. Spence's cartilage is found less frequently but when it does occur, it lies between the stapes and the hyoid cornu close to the chorda tympani. These relationships suggest that it is a remnant of the tympanic process of the extrastapes. Shute (1956), however, has denied these derivations on the basis that Paaw's and Spence's cartilages in ontogeny appear too late and display no connection with the stapes.

Westoll (1944) questioned the widely accepted view that the dermal part (goniale) of the anterior process of the malleus represents the reptilian prearticular, preferring the suggestion of Olson (1944) that the goniale and tympanic annulus together are homologous with the angular. He identified the prearticular with a dermal element which is occasionally found developing ventromedially to Meckel's cartilage while a similar element developing dorsomedially he equated with the surangular.

The derivation of the tensor tympani muscle has been disputed. Characteristically, the muscle is inserted into the malleus close to the root of the manubrium. The retroarticular process of the reptilian articular is the site of attachment of the depressor mandibulae. The different innervations of the depressor muscle (facial nerve) and tensor tympani (trigeminal nerve) rule out the possibility that these muscles are homologous (in fact, the reptilian depressor is believed to have been replaced by, or more probably converted into, the posterior belly of the mammalian digastric muscle – see Parrington, 1974; it may also have provided the source of the stapedius muscle which may already have been present in the theriodonts). Probably the most widely held view has been that the reptilian anterior pterygoid muscle represents the precursor of the tensor tympani (and also tensor palati), but as already described, Crompton (1963*a*) has suggested that the tensor represents only a small slip of the anterior pterygoid, the remainder having become the medial pterygoid muscle.

More recently, Kermack *et al.* (1973) have cast doubt on the view that the manubrium of the malleus is homologous with the reptilian retroarticular process. In their description of the lower jaw of *Cynognathus* they equate the retroarticular process with a boss on the surangular. They describe the downward projecting spur on the articular (which they term the infraarticular process) as giving insertion to part of the pterygoid adductor musculature rather than to the depressor mandibulae, agreeing in this respect with Watson (1951), who suggested that this process is neomorphic in advanced mammal-like reptiles, giving insertion to the posterior pterygoid. According to Kermack *et al.*, the manubrium of the malleus is homologous with the infraarticular process and the tensor tympani with the attached adductor muscle. However, these homologies have been strongly challenged by Parrington (1978), who prefers to regard the infraarticular process of Kermack *et al.* as equivalent to the retroarticular process of reptiles and the boss on the surangular as equivalent to the similarly located protuberance described on the cynodont surangular by Crompton (1972) as part of a secondary jaw articulation (see above).

A fundamentally different view of the structure of the cynodont middle ear from that traditionally adhered to has been advanced by Allin (1975) on the basis of examination of the auditory region of the skull in a wide selection of mammal-like reptiles and comparison with the equivalent cranial region in living reptiles and monotremes. He begins by emphasising that in pelycosaurs (as in captorhinomorphs) there is no definite osteological evidence (in the form of an otic notch) for the presence of a tympanic membrane; that the stapes, which is massive in pelycosaurs, while still large is much more lightly built in therapsids and has a direct bony attachment to the quadrate; and that the quadrate is progressively reduced in size in theriodonts and becomes increasingly more mobile. He postulates that the angular gap between the posterior border of the reflected lamina and the retroarticular process was occupied, in life, by a layer of dense fibrous tissue. As we have seen, the reflected lamina becomes progressively smaller, and the angular gap correspondingly wider in passing from early to late cynodonts. Allin rejects the view that the notch (his 'angular cleft') between the body of the angular medially and the reflected lamina and fibrous sheet laterally was occupied by muscle insertion as is widely assumed. He suggests instead that in all mammal-like reptiles except sphenacodonts (where it probably was the site of muscle insertion) it had become filled by an air-containing chamber (in this he follows Westoll (1944, 1945) although the latter described this chamber as developing only as a late stage in the evolution of the therapsid middle ear). This chamber could be either a diverticulum of the tympanic cavity, as proposed by Westoll, or of the pharynx. Allin also questions the osteological evidence for the presence of a conventional tympanic membrane in therapsids. In particular, he doubts that the groove on the posterolateral aspect of the cynodont squamous housed an external meatus, believing it to be uncharacteristically tortuous for such a purpose (he suggests that it may rather have been occupied by the depressor mandibulae muscle), and also points to the lack of definite evidence that the stapes possessed an extrastapedial extension contacting a postquadrate tympanic membrane.

Allin argues that the nature of the connection of the stapes with the quadrate is strongly suggestive that the latter element took part in the transmission of sound vibrations, probably relaying these from the lower jaw when this was in contact with the ground. Here he is in agreement with Tumarkin (1968) whose views on bone conducted hearing have already been discussed in Chapter 2. Allin departs from the latter author's views, however, in that he believes that the cynodonts were also efficiently equipped to detect air transmitted sound, the drum being provided, he suggests, by the reflected lamina, the fibrous tissue sheet in the angular gap and the associated soft tissues. The vibrations would then pass through the angular and other bones of the postdentary complex, the quadrate and then the stapes to reach the inner ear. Because of the relatively large mass of these components, such a mechanism would be most sensitive to low-frequency sound (see the account of impedance matching in Chapter 6). Only after the postdentary bones had become detached from the dentary, a stage not reached until the primitive mammals, could the angular become stabilised as a non-vibrating support for a tympanic membrane formed, presumably, from the fibrous tissue sheet in the angular gap and its associated coverings.

As Allin describes, his scheme would meet certain difficulties inherent in the more traditional accounts of the cynodont middle ear – in particular, the lack of real osteological evidence for the presence of a postquadrate tympanic membrane. It would also imply that the reduction of the postdentary bones, together with the associated enlargement of the dentary, and the change from a quadrate–articular to a squamous–dentary jaw joint, which are not altogether satisfactorily explained as part of a masticatory adaptation, progressed instead as part of an increase in auditory efficiency in which the range of sound sensitivity became extended into the higher frequencies.

SECTION III

FUNCTIONAL ADAPTATIONS OF THE SKULL IN MODERN EUTHERIANS

5

THE MASTICATORY APPARATUS

In Chapter 4 a description has been given of the sequence of events whereby the single-bone lower jaw and temporomandibular (squamous–dentary)[1] joint of mammals replaced the multiple-bone lower jaw and quadrate–articular joint of reptiles. Associated with these changes was the emergence of a heterodont dentition, with postcanine teeth having a tribosphenic cusp pattern, and a powerful and highly structured jaw musculature capable of complex synergistic activity. As a result of these structural modifications the jaws of mammals have become capable or being used in a functionally more efficient manner than was possible in the majority of mammal-like reptiles where it appears, from the morphology of the jaws and dentition, that jaw movement was effectively limited to a vertical snapping type of action. Although a form of chewing may have been present in some herbivorous groups of mammal-like reptiles, the use of the jaws in this type of movement is essentially a mammalian characteristic. It undoubtedly resulted in an improved efficiency of feeding which may well have been an essential factor in the survival of the very small creatures with high energy requirements (because of their big surface area to body volume ratio) which constituted the earliest mammals. In addition, the possession of a masticatory apparatus adapted for chewing conferred upon the mammals a potential for further evolution which has led to the diverse and remarkably effective feeding mechanisms encountered amongst the later members of the class.

It has been generally recognised (see, for example, Turnbull, 1970) that the eutherian mammals can be subdivided into four major groups – one generalised and three specialised – on the basis of the functional morphology of their jaw apparatus. In the generalised group (including insectivores, insectivorous bats and primates) this apparatus, both in its structure and function, has remained relatively unspecialised, probably approximating more closely than that of any other group to the condition that obtained in the earliest mammals. The three specialised groups include the ungulates, in which the jaws have become adapted to a herbivorous diet and in which the principal jaw action is of a milling or grinding nature; the rodents, in which gnawing is the predominant jaw movement; and the carnivores, in which the

[1] In order to comply with common usage, the terms 'dentary' and 'squamous–dentary' were used in Chapters 1 to 4. Again to comply with usual practice, these terms are replaced by 'mandible' and 'temporomandibular', respectively, when dealing with modern mammals.

jaws function in a powerful scissor-like manner. A number of other smaller groups exist, including the Edentata, Sirenia, Proboscidea and Cetacea, where a high degree of specialisation of the jaws has taken place along lines different from those in the three major specialised groups. Although the dental structures in these highly specialised groups show a number of interesting structural adaptations, there is at present so little information available about either the structure or the function of their masticatory apparatus that their inclusion would add little to the present discussion of the principles underlying the structural modifications of the mammalian jaws. These groups are not, therefore, considered further in the present chapter.

With the principal exception of the carnivores, mastication in mammals involves considerable excursive movements of the mandible as well as elevation and depression. This range of movement requires that the temporomandibular joints be rather loosely constructed, thus allowing the mandibular condyles to undergo translation (sliding) movements as well as rotation about a transverse axis. The temperomandibular joint is, in many respects, a typical synovial articulation. Its articular surfaces, formed by the inferior surface of the squamous and the condyle of the mandible, vary greatly in shape from one mammalian group to another, according to functional requirements but, in general, the condyle is convex and the squamosal surface reciprocally concave in at least one plane of space. The articulating bones are attached to each other by a fibrous capsule, the precise attachments, structure and ligamentous strengthening of which again vary according to functional needs. The temporomandibular joint is, however, unusual in the nature of the coverings of the articular surfaces. Both are lined by a layer of fibrocartilage, instead of the hyaline cartilage found in other synovial joints. The fibrocartilage is usually composed predominantly of fibrous tissue with only scattered islands of cartilage matrix. As described in Chapter 8, a layer of hyaline cartilage (the so-called condylar cartilage) is present beneath the fibrocartilage of the condyle, where it functions as an important site of endochondral ossification. A phylogenetic explanation of the peculiar nature of the articular coverings of the temporomandibular joint may be found in the fact that this joint is neomorphic in mammals, having been formed between surfaces originally covered by periosteum. Under the ensuing changed functional circumstances, the osteoprogenitor cells of the areas of periosteum involved in the joint are presumed to have changed their pathway of differentiation so that they produce chondroblasts instead of osteoblasts. The articular coverings have thus come to be formed by fibrocartilage, probably representing the original fibrous periosteum with the addition of a certain amount of cartilage matrix. There is some experimental evidence (discussed in Chapter 8) favouring this view.

In most mammals, the cavity of the temporomandibular joint is divided into two by an articular disc. The disc is composed predominantly of white fibrous tissue with rather less cartilage than is present in the discs of other joints (Rees, 1954). It gives attachment at its anterior edge to the lateral pterygoid muscle. The form of the disc varies, as will become apparent, from one mammalian group to another in a manner not always clearly correlated with the type of jaw movement. Despite intensive investigation, especially in man, the precise functional advantages that accrue from the presence of an articular disc are by no means clear. Four major functional roles are generally assigned to articular discs and menisci (see Barnett, Davies & MacConaill (1961) for detailed discussion). First, they undoubtedly serve to increase congruence of the articular surfaces over the range of joint movement and thus help increase the stability of the joint without appreciably limiting its mobility. Secondly, they probably help to protect the articular surfaces in that they distribute the forces generated within the joint during function over an effectively larger area than would otherwise be the case. Thirdly, discs and menisci may be important in ensuring lubrication of all parts of the moving surfaces, especially at the extremes of the range of jaw movement (MacConaill, 1950). Finally, the presence of a disc, in providing a double joint cavity, appears to facilitate a combination of more diverse types of movement than would be possible with a single cavity. The temporomandibular joint provides an excellent example of the last effect. Translation usually takes place in the subdivision of the articular cavity above the disc, the disc and condyle translating together across the squamosal articular surface. Simultaneously with the translation, the condyle can rotate against the undersurface of the disc in the inferior subdivision of the articular cavity. The components of sliding and rotation are built up into the complex functional movements of elevation, depression, protrusion, retrusion and lateral excursion of the mandible.

Despite the persuasiveness of these theoretical considerations, it has to be admitted that articular discs can be removed with frequently little or no functional impairment in either the short or the long term. In man, for example, the menisci of the knee joint and the disc of the jaw joint are surgically removed for traumatic or pathological conditions, often with no apparent reduction in joint stability or mobility and no major or inevitable long-term sequelae.

It is often assumed, following Gaupp (1912), that the temporomandibular articular disc represents a part of the lateral pterygoid muscle which became enclosed within the newly constructed temporomandibular joint during the evolution of the mammals from the mammal-like reptiles. It will be recalled from Chapter 4 that the posterior pterygoid part of the adductor jaw musculature, from which the lateral pterygoid may have been derived, probably inserted into the medial aspect of the prearticular. As the dentary approached the squamous to form the new jaw joint, it is possible that the distal part of the muscle, as it passed to the prearticular, became trapped between the approximating bone surfaces and subsequently incorporated within the new joint as the articular disc. Direct palaeontological evidence bearing upon this possibility is not, of course, available and the indirect ontogenetic evidence is conflicting. Harpman & Woollard (1938), for example, found a transient connection of muscle and tendon fibres passing from the lateral pterygoid to the malleus (which probably incorporates the prearticular) across the superior aspect of the condyle in the human embryo but not in the embryos of several non-human mammalian groups. Symons (1952) has described the lateral pterygoid in the human embryo as being continuous with the medial part of the articular disc and thence with the malleus.

The fact that the jaw joint in the majority of mammals is a rather loose structure allowing a wide range of translatory movements of the condyle is frequently taken as evidence that the joint cannot act as the fulcrum of a simple lever system during elevation of the mandible. Indeed, it has been suggested (see Chapter 4) that the joint had already become relieved of this function in the advanced mammal-like reptiles, where, as the jaw musculature increased in power and efficiency, its mode of action was modified so that the forces that it produced became borne in progressively greater proportion by the dentition, with consequent reduction of the reaction force at the joint. Whether or not this is correct, there is good evidence, which will be presented later in this chapter, that in many living mammalian groups such a redistribution of masticatory forces is present and there can be little doubt that with the principal exception of the typical carnivores it is the synergistic manner of action of the jaw muscles and, for movements in the intercuspal range, the shape of the occlusal surfaces of the teeth, rather than the fulcrum effect of the temporomandibular joint, which are the major determinators of the pattern of jaw movements.

In the following account, descriptions of the topographical anatomy of the jaw musculature are not included except where there is some functionally significant departure from the typical mammalian condition. The reader who requires further information is referred to the excellent works (each with an extensive bibliography) by Edgeworth (1935), Becht (1953), Schumacher (1961) and Turnbull (1970) dealing with the structure and function of the jaw muscles in a wide range of mammalian groups. Of the many earlier studies that by Toldt (1905) is outstanding.

One final general consideration to be borne in mind is that the degree of mobility at the symphysis varies over a wide range. The primitive condition was probably one in which considerable movement was possible. Amongst recent mammals the symphysis may, at one extreme, be so slack as to allow virtually independent movements of the two halves of the lower jaw while at the opposite extreme it may be immovable or even replaced by a bony union. In the former case, as will become apparent, the mobility may be utilised to a number of functionally different ends. In the latter case the two halves of the mandible make up, to all intents and purposes, a single unit articulating with the cranium at the two jaw joints as well as, in the intercuspal range, the occluding surfaces of the upper and lower teeth. The principal advantage of a non-movable symphysis appears to be that in combination with the unilateral chewing usual in mammals it allows the forces produced by the muscles on the balancing side (i.e. the side opposite to that on which chewing is taking place) to be applied more efficiently at the dentition on the working side. The forces produced by the muscles of the two sides are thus summated with a corresponding increase in the force of the bite.

MASTICATORY MOVEMENTS AND THE FUNCTIONAL ANATOMY OF THE MASTICATORY APPARATUS

METHODS OF RECORDING MASTICATORY MOVEMENTS

Early studies of mammalian patterns of masticatory movement were based mainly on inferences drawn from examination of dried skulls and from dissections of the craniomandibular apparatus, although a few included attempts to observe, with the unaided eye, the movements of the jaws during feeding in the living animal, a procedure fraught with difficulty because of the speed at which mastication proceeds and the obscuring effects of the soft tissues. The introduction of cinephotography (e.g. MacMillan, 1930) overcame the difficulty due to speed of movement in that each frame can be studied separately, thus allowing the analysis of jaw movement to be made at leisure. However, the difficulty arising from the presence of obscuring soft tissues remained and it was not until Butler in 1952 (see also 1973) introduced his technique for analysing wear facets on the occlusal surfaces of the teeth that a precise method of determining the pathways of jaw movement became available. A further advantage of wear facet analysis is that it can be applied to fossil as well as to living species. Its disadvantage, of course, is that it gives information about the relative movements of the upper and lower jaws for only the period during which the teeth are in occlusion or close approximation.

For a full analysis of the masticatory cycle clearly a method is needed which enables the complete range of jaw movements to be recorded without being obscured by the presence of soft tissues or the speed at which chewing takes place. The method that comes closest to meeting these desiderata is based on the technique of cineradiography which not only allows the bones to be visualised but also permits frame by frame analysis of the movements. As might be expected in view of its practical importance, the pattern of mastication has been most intensively studied in man and it is in this field that the technique of cineradiography has been principally developed (e.g. Klatsky, 1939, 1940; Berry & Hoffmann, 1956; Arderan & Kemp, 1960). In recent years, however, cineradiography has been adapted for use in animals (e.g Ardran, Kemp & Ride, 1958; Hiiemae & Ardran, 1968; see also Hiiemae, 1978, for a recent, excellent review of this work). Since resolution is reduced by superimposition of the teeth when they are in intercuspal range, cineradiography has, on occasion, been supplemented by wear facet analysis to provide a complete description of jaw movement throughout the whole of the chewing cycle.

An alternative, and less frequently used, technique for recording jaw movements is by means of a position recording system attached to the jaws either directly or by means of a facebow, the advantage of this method over cineradiography, so far as human subjects are concerned, being the avoidance of the dangers of irradiation. Such recording systems have been employed in man (Ahlgren & Öwell, 1970) and in the rhesus macaque monkey (Luschei & Goodwin, 1974).

Ideally, the techniques for tracing jaw movements should be supplemented by simultaneous recordings of the associated jaw muscle activity and of the movements of the mandibular condyle at the jaw joint. Practical difficulties have limited the number of studies that use such a combined approach but recently a number of investigations incorporating electromyographic recording of jaw muscle activity with radiographic determination of jaw movements has been reported in man and a few other mammals. These form an important component of the following descriptions.

TYPES OF MASTICATORY MOVEMENTS IN MAMMALS

Traditionally, three types of mandibular masticatory movement have been recognised in mammals: (1) *orthal* – vertical movement (i.e. elevation and depression) of the mandible; (2) *ectental* – transverse movement of the mandible; (3) *propalinal* – anteroposterior movement (protrusion and retrusion) of the mandible. These movements in various combinations make up the generalised, ungulate-grinding, rodent-gnawing, carnivore-shear and other, less common, mammalian masticatory patterns.

Detailed investigations of masticatory movements have been made for relatively few of the many thousands of mammalian species. Fortunately, such species as have been investigated include examples of most of the major masticatory patterns. A rather more mixed blessing is that three of the most intensively investigated species – man, the rat and the pig – have become adapted to a considerably more omnivorous type of diet than is usual for the groups to which they belong and, as Turnbull (1970) has emphasised, the group characteristics tend to become less pronounced in its omnivorous members.

MASTICATORY MOVEMENTS IN PRIMITIVE MAMMALS

In order to reconstruct the pattern of masticatory movement in the early mammals in which, presumably, the jaw apparatus was not highly specialised, Crompton and Hiiemae and their colleagues (Hiiemae & Jenkins, 1969;

Crompton & Hiiemae, 1970; Hiiemae & Crompton, 1971; Hiiemae, 1976; Crompton, Thexton, Parker & Hiiemae, 1977) have investigated various aspects of the structure and masticatory functions of the jaws in the American opossum (*Didelphis marsupialis*). In that it possesses several primitive features in the morphology of its jaws and dentition, including only slightly modified tribosphenic molars (dental formula = $I\frac{5}{4}\,C\frac{1}{1}$ cheek teeth $\frac{7}{7} \times 2$) and an unfused mandibular symphysis, *Didelphis* appears to provide an entirely acceptable analogue, so far as the jaw apparatus is concerned, of the hypothetical generalised condition of the early mammalian stock. Although marsupials are outside the general scope of the present work, these studies on *Didelphis* are of such fundamental importance to our understanding of mammalian mastication that they must be considered in some detail.

By using a combination of cineradiography and wear facet analysis, Crompton and Hiiemae (Crompton & Hiiemae, 1970; Hiiemae & Crompton, 1971) have analysed the chewing movements of *Didelphis*. These movements are essentially cyclical as in all mammals. The sequence of events within the cycles is modified according to whether the cycle is concerned with transferring food from the incisors to the postcanine teeth, with puncture-crushing (when the teeth do not intercuspate) or chewing (teeth do intercuspate). Since *Didelphis*, like the majority of mammals, is anisognathous (as originally coined, this term meant jaws of unequal width but in the present work it will be restricted, following modern usage, to describe specifically the situation where the upper dental arch is the wider of the two), mastication is restricted to one side at a time.

The chewing cycle in *Didelphis* consists of three strokes: (1) *closing* (*fast closing*) in which the lower teeth are moved rapidly upwards towards the upper teeth, this phase beginning with the lower teeth displaced laterally to the working side; (2) *power* (*slow closing*) in which the food is reduced by the movement of the lower teeth in close proximity to the upper teeth; (3) *opening* in which the lower teeth are moved downwards at first slowly (*slow opening*) and then more rapidly (*fast opening*). Maximum gape occurs at the end of fast opening and is used, by convention, as the starting point in describing the cycle. The chewing cycles of all other mammals that have been studied can be similarly divided into three strokes, although, in describing these, it is sometimes more convenient to deal with the opening stroke first.

When food is ingested by *Didelphis* it is first puncture-crushed to reduce its bulk and then chewed. Both the precise form and the number of puncture-crushing and chewing cycles vary according to the size and consistency of the ingested material. There is also variation within a sequence of cycles as the food becomes progressively reduced. As might be expected, the variation in relation to the nature of the food is greater in puncture-crushing than in chewing cycles.

Despite this variation, average data provide a useful summary of the main

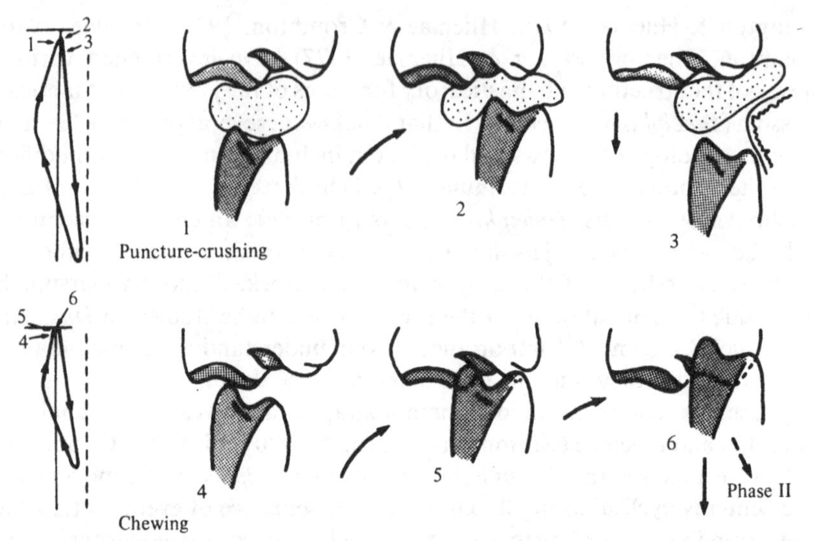

Fig. 46. Occlusal relationships of cheek teeth during puncture-crushing and chewing in *Didelphis*. Movement profiles (working side on animal's right) shown at left – horizontal bar indicates level of centric occlusion; broken vertical line represents median plane. Small arrows indicate positions on profile enlarged in occlusal drawings. Teeth shown in anterior view. 1 = tooth–food–tooth contact; 2 = minimum vertical dimension (arrow between 1 and 2 shows direction and angulation of movement of protoconid)); 3 = early in opening stroke (food being collected by tongue); 4 = beginning of power stroke; 5 = halfway through power stroke; 6 = centric occlusion (arrows between 4 and 5 and 5 and 6 show direction and angulation of movement of protoconid; broken arrow indicates course of movement in those mammals with a phase II in the power stroke). In 4–6 the food has been omitted for clarity. (Reproduced with permission from Hiiemae, 1976.)

features and indicate the essential similarity of the puncture-crushing and chewing cycles. Movement profiles of the two types of cycle in *Didelphis*, together with the occlusal relationships of the postcanine teeth during the power stroke and early part of the opening stroke, are shown in Fig. 46. In Fig. 47 average data for puncture-crushing and chewing are depicted graphically by plotting the angle of gape between the profiles of the upper and lower postcanine teeth against time. The puncture-crushing cycle lasts, on average, just over 300 ms while the chewing cycle has a duration of about 400 ms, the difference being largely due to the extended slow opening in chewing. Some 22 per cent of the cycle is taken up by the power stroke. The maximum gape during puncture-crushing tends to be bigger than in chewing. During the power stroke in both types of cycle the lower molars are moved relative to the upper molars in an upwards, anterior and medial direction so that a single cutting edge on a lower tooth works against one or possibly two cutting edges on the corresponding upper tooth (Fig. 46). For a short period in the power stroke (of the order of 17 ms in puncture-crushing and 50 ms

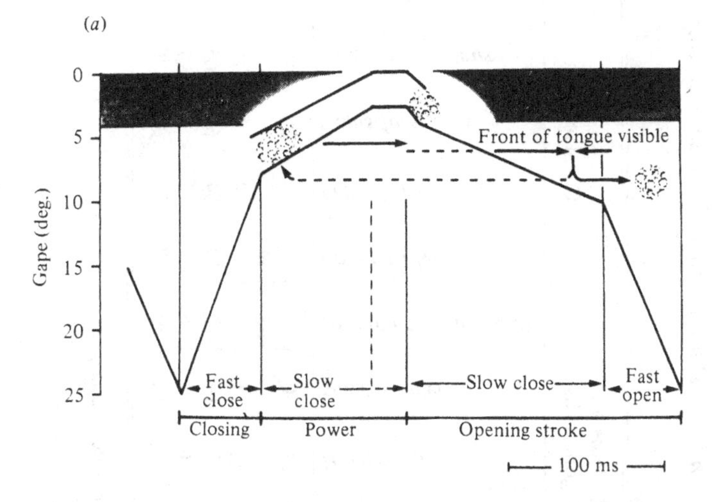

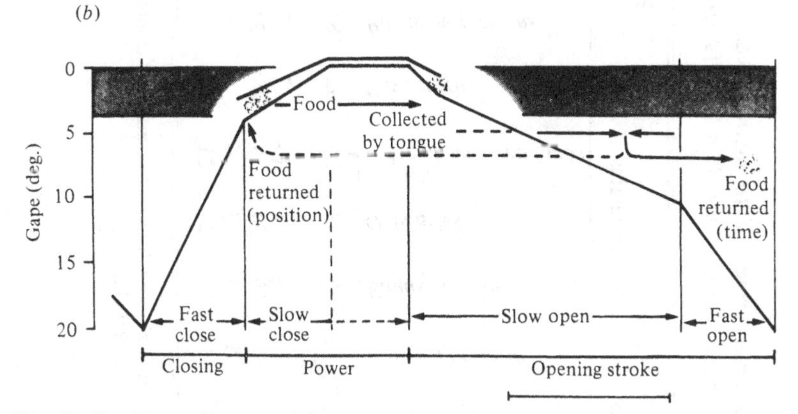

Fig. 47. Profiles and approximate durations of main stages of (*a*) puncture-crushing and (*b*) chewing cycles in *Didelphis*. See text for description. (Reproduced with permission from Hiiemae, 1976.)

in chewing), there is no vertical movement of the lower jaw but it seems likely that during this time the jaw continues to move anteromedially so that there is no truly isometric phase. Puncture-crushing involves chiefly the tips of the cusps whereas chewing involves the slopes of the cusps. In both types of cycle there is no transverse movement past centric occlusion (or its equivalent in puncture-crushing) – that is, there is no phase II (see below) in the power stroke.

During the slow opening phase the lower jaw may be progressively depressed or be depressed to about 7° then held, or even slighly elevated, before rapid opening begins. The latter profile seems to be especially related to the mastication of hard food. During this stage the tongue is used to manipulate the food within the mouth.

TABLE 2. *Percentage weights of jaw muscles (rounded to nearest whole number). Data derived from several sources, including Becht (1953), Davis (1955), Schumacher (1961) and Turnbull (1970)*

Muscle		Primates			Ungulates				Rodents and lagomorphs				Carnivores				
	Didelphis marsupialis	*Homo sapiens*	*Pongo pygmaeus*	*Macaca mulatta*	*Equus caballus*	*Ovis aries*	*Odocoileus virginianus*	*Sus scrofa*	*Rattus norvegicus*	*Hystrix* sp.	*Sciurus niger*	*Oryctolagus cuniculus*	*Panthera leo*	*Felis domestica*	*Canis familiaris*	*Ursus americanus*	*Tremarctos ornatus*
Masseter	25*				46	44	36	32	48†	61	50			26	20	16	11
		29	30	27								63	34				
Zygomaticomandibularis	9				9	8	10	17	7	11	11			9	6	12	15
Temporalis	57	45	51	57	15	24	29	28	33	17	19	12	59	54	65	65	65
Lateral pterygoid	2	14	6	5	4	3	4	6	4	6	7	5	< 1	< 1	< 1	1	1
Medial pterygoid	7	11	14	11	27	21	21	17	9	6	12	20	7	10	8	6	7

* Separation of temporalis and deep part of masseter is incomplete in *Didelphis*.
† Including maxillomandibular.

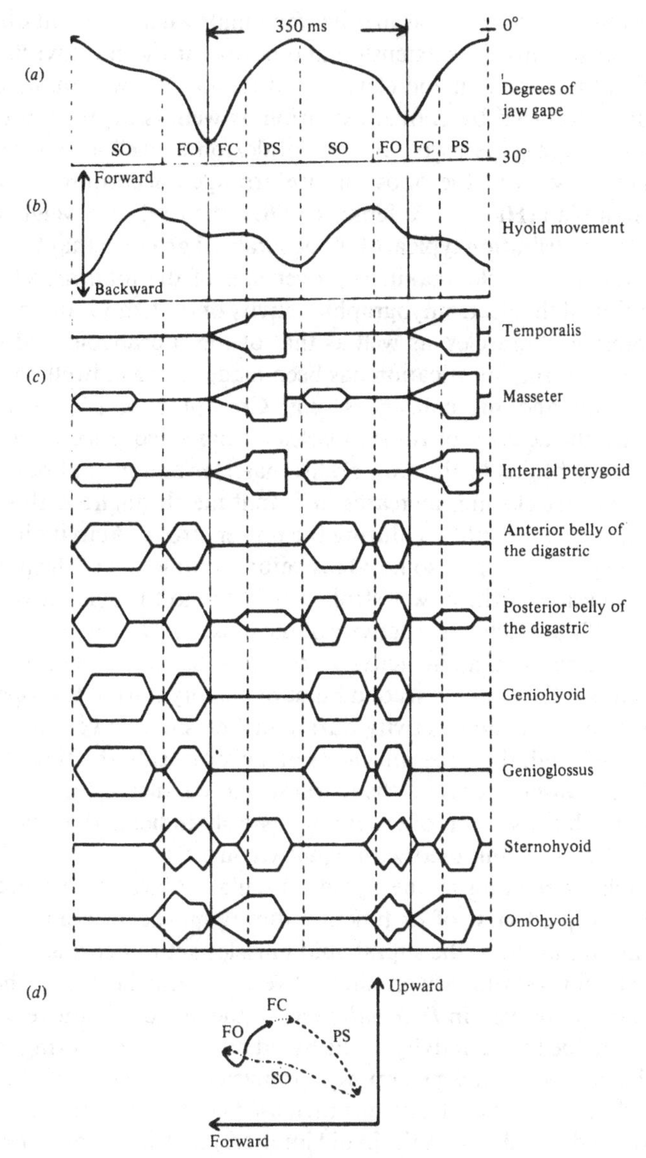

Fig. 48. Movements of jaw (*a*) and hyoid (*b*), and associated EMG activity in jaw and hyoid musculature (*c*) in *Didelphis* while chewing soft food. The path of movement of a point on the hyoid during chewing is shown in (*d*). FC = fast closing; PS = power stroke; SO = slow opening; FO = fast opening. In (*b*) movements of hyoid have been plotted on a plane parallel to the secondary palate. (Reproduced with permission from Crompton *et al.*, 1977.)

Mastication in *Didelphis*, as in other mammals where relevant observations have been made, involves extension and flexion at the craniovertebral joints in addition to depression and elevation of the lower jaw – an increase in the gape being produced by cranial extension as well as depression of the jaw, and a decrease in gape involving cranial flexion as well as jaw elevation.

The jaw muscles in the opossum are arranged according to the general mammalian plan (Hiiemae & Jenkins, 1969), their relative weights (Table 2) showing the distribution typical of the generalised group in that the temporalis is the dominant muscle, making up over half of the total muscle mass. An investigation of the electromyographic activity of the temporalis, masseter and medial pterygoid muscles, as well as that of the suprahyoid and infrahyoid musculature, during mastication has been made, using indwelling electrodes, by Crompton and his colleagues (e.g. Crompton *et al.*, 1977). Fig. 48 summarises the activity of these muscles during a sequence of chewing soft food. Activity begins in the temporalis, masseter and medial pterygoid with the onset of fast closing, increases in amplitude throughout this stage and reaches maximum amplitude during the power stroke. Activity in this group of muscles ceases abruptly some 50 ms before completion of the power stroke (the continuation of the power stroke being due to the delay involved in the expression of the mechanical activity of the muscle). During the phase of slow opening, electromyographic activity occurs in the suprahyoid muscles, then ceases, but is followed by a second burst of activity during fast opening. The infrahyoid muscles show activity during fast closing, power stroke and fast opening. Although the opossum chews on only one side at a time, the periods in which the musculatures on the working and balancing sides show activity are similar. This does not, of course, necessarily indicate that the muscles of the two sides are being used with equal vigour.

The findings relating to the hyoid musculature are at variance with the generally accepted view of its function during mastication and swallowing. According to this view, the suprahyoid muscles act to depress the lower jaw when the hyoid is stationary or, alternatively, to raise the hyoid when the jaw is stationary. However, in *Didelphis* at least, the hyoid is in movement during mastication, due to the activity of the hyoid musculature. During fast closing the hyoid moves slightly posteriorly, primarily as a result of the action of omohyoid; in the power stroke, the increased contraction of the sternohyoid and omohyoid muscles pulls the hyoid inferiorly as well as posteriorly; in slow opening, the hyoid moves sharply anterosuperiorly, due to the action of the suprahyoid muscles; in fast opening, all the hyoid musculature is in contraction and the hyoid moves posterosuperiorly. The base of the tongue moves with the hyoid. During closing and the power stroke, therefore, the base of the tongue is pulled backwards and downwards while in slow opening it is moved sharply upwards and forwards, facilitating the protrusion of the tongue which occurs during this stage and which is probably concerned with collecting and positioning the food crushed during the previous power stroke. The combined action of the hyoid muscles in fast opening serves to retract the hyoid somewhat and so allows the suprahyoids to produce rapid depression of the lower jaw (for further details regarding movements of the hyoid and tongue during feeding, the reader is referred to Crompton *et al.*, 1977; Hiiemae, 1978; Hiiemae, Thexton & Crompton, 1978).

Amongst eutherian mammals the morphology of the masticatory apparatus appears to have departed least from the primitive condition exemplified by *Didelphis* in the generalised group, comprised principally of the primates, insectivores and insectivorous bats. The primates are discussed in the next section as the major example of the generalised group. So far as the functional anatomy of the masticatory apparatus of the insectivores and insectivorous bats is concerned, information is sparse. For this reason, the analysis by Kallen & Grans (1972) of masticatory movements and associated electromyographic activity in the jaw muscles in the little brown bat (*Myotis lucifugus*) is of particular interest. As in *Didelphis*, the jaws are anisognathous, the symphysis possesses a considerable degree of flexibility and the jaw joint allows a wide range of condylar movement. Although the technique of cinephotography used by Kallen & Gans to record jaw movement does not allow such high resolution as cineradiography, the fact that mastication in *Myotis* is similar in essentials to that in *Didelphis* clearly emerges from their study. In both species, mastication takes place unilaterally, there appears to be the same division into puncture-crushing and chewing types of cycle, these cycles involve much the same sequence of movements, and the duration of the cycles is of the same order. Again, the general pattern of electromyographic activity in the jaw muscles during masticatory movements is similar in the two animals. This overall comparability, which, as we shall see, applies also to *Tupaia* and the primates, is consistent with the claim that the masticatory mechanisms in *Didelphis* and in the generalised group of eutherian mammals come closest, amongst living forms, to those that obtained in the early mammals.

MECHANICS OF MASTICATION AND THE ASSOCIATED STRUCTURAL MODIFICATIONS OF THE SKULL IN THE MAJOR EUTHERIAN MASTICATORY GROUPS

Primates (representing the generalised group)

The dentition and jaws of primates have tended to remain unspecialised, apart from the modifications of the anterior teeth seen in prosimians (see below), and most appear capable of enjoying a remarkable range of foodstuffs. The diet of nearly all prosimians consists chiefly of insects but can include small reptiles, birds and mammals, eggs, and also fruit and other herbivorous material. Monkeys are, in the main, herbivorous but will supplement their diet with grubs and insects as opportunity allows. Apes live principally on fruit, young shoots and leaves but the chimpanzee will also catch and eat small mammals and insects. In man, the primate tendency towards a varied diet reaches its maximum expression, practically anything capable of being swallowed and not immediately and fatally toxic being consumed. The most exclusively herbivorous primates are the Colobinae which possess specialised digestive organs, allowing them to thrive on a diet composed mainly of leaves.

It is appropriate to take the primates as our principal example of the generalised group because the masticatory apparatus in this order has, without doubt, been subjected to far more intensive investigation than that of any other mammalian group. The principal focus of interest in this context has been man himself on account of the clinical importance of the structure and function of his jaws and dentition. Considerable attention has also been devoted to this aspect of the anatomy of the great apes as well as of some of the better known fossil hominoids because of the relevance it has been thought to bear to the determination of phylogenetic relationships within the Hominoidea. The functional anatomy of the remaining primates has received considerably less attention but a number of recent studies has begun to fill this gap in our knowledge.

Although in many ways not representative of the primates generally, so many of our basic concepts of masticatory function have been worked out using the human subject that it is convenient to begin the discussion with our own species.

Man

The outstanding feature of the human masticatory apparatus is its relatively weak development (Fig. 49). The dental formula (permanent dentition = $I\frac{2}{2}C\frac{1}{1}P\frac{2}{2}M\frac{3}{3} \times 2$) is the same as that of other catarrhine primates, but the lateral upper incisors, lower first premolars and third molars are all tending to reduction, being congenitally absent in quite a high proportion of individuals (Lavelle & Moore, 1973). The jaws are small and this, together with the fact that the anterior basicranial angle is well below 180° (adult average is about 130°, see Chapter 8), results in the facial skeleton protruding but little in front of the prominent frontal part of the braincase. The teeth are of a correspondingly small size, the canines especially being reduced in comparison with those of most other primates. The jaw-closing muscles are also poorly developed and the jaw joints are rather weak structures which, because of the slackness of the capsule above the articular disc, allow a wide range of translatory movements of the mandibular condyles. The mandible is as wide, or even wider, in its tooth bearing part as the corresponding portion of the maxillary complex. Nonetheless, a degree of anisognathy is produced by the lingual inclination of the mandibular molars and premolars and the buccal inclination of the corresponding teeth in the maxillae. Chewing in man is, in consequence, unilateral as in most other mammals. The mandibular symphysis fuses at about the time of birth.

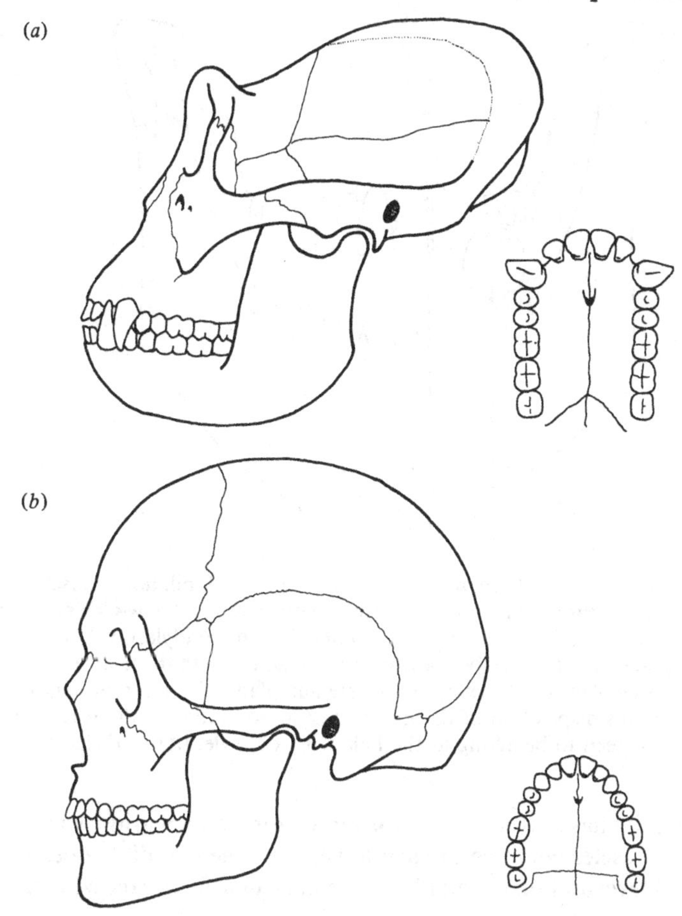

Fig. 49. Left lateral views of skulls and occlusal views of maxillary dental arcades of (*a*) male gorilla and (*b*) man. Not to scale.

Despite the very detailed knowledge of the anatomy of the human jaws, there are certain fundamental aspects about the way they function which are still controversial. Perhaps the most important of these is whether or not the temporomandibular joints are subject to a reaction force when the jaw-closing muscles contract. The early view was predominantly that the human mandible acts as a simple class III lever with the jaw joint providing the pivot. In such a system the relative magnitudes of the distances from the joint to the grinding teeth on the one hand, and to the calculated point of application of the effort force exerted by the jaw-closing musculature, on the other hand, indicate that only between one-third and two-thirds of this force would be effective at the teeth (the actual proportion decreasing from molars to premolars) and the remainder would be borne by the joint. The alternative view (strongly advocated by, for example, Wilson, 1920, 1921), based on the belief that the

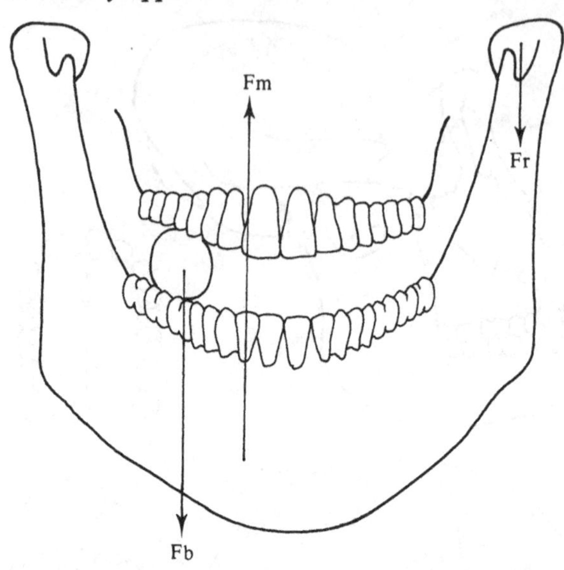

Fig. 50. Frontal view of human jaws to show effect of bilateral muscle action. Fm = resultant of forces exerted by bilateral contraction of jaw-closing muscles; Fb = force at molar teeth; Fr = reaction force at balancing condyle. On the assumption that working side muscles may be somewhat more active than those of the balancing side, Fm has been placed to working side of median plane. The fulcrum-making force resulting from this disposition of the muscle resultant relative to the force acting at the teeth can be seen to be acting at the balancing condyle. (After Hylander, 1975.)

human jaw joint has a structure inappropriate for stress-bearing, was that the jaw-closing muscles combine in such a way that the mandible does not act as a class III level and that virtually all the muscular force exerted is effective at the teeth.

Gysi (1921) provided a modification of the lever theory. He pointed out that the previous analyses had considered the lower jaw as though it were a unilateral structure. He suggested instead that the musculature is in action on both sides even though chewing is unilateral (this is now known to be the case, see below). The net effect of bilateral muscle action is not only to increase the force at the teeth but, through the rigidly fused symphysis, to reduce the reaction force at the working condyle. The balancing condyle becomes the effective fulcrum and bears the major part of the reaction force (Gysi provides a graphical and mathematical demonstration of the leverage system he proposes but the situation can be readily understood by reference to Fig. 50). In a simplified model of the human jaw Gysi showed that if, for example, a vertical force of 12 lbs is applied to each side (i.e. 24 lbs in all), 12 lbs of the combined force would be exerted at the first molar on the working side, just over 8.5 lbs at the balancing condyle and just under 3.5 lbs at the working condyle. Gysi countered the suggestion that the human temporomandibular joint, with its loose capsule and poorly fitting articular surfaces, is an inappropriate structure to act as a fulcrum of a lever system by pointing out that even the loosest articulation may be fixed at any stage of movement by an appropriate combination of muscle action and effectively converted to a fulcrum. In his view the reaction force at the joint does not represent wasted muscular effort; it is, in fact, necessary to convert the loosely structured joint into a stable fulcrum (and was, in consequence, termed by Gysi the 'fulcrum-making' force).

The controversy about the mechanics of the human jaws has continued since Gysi's publications. Recently, Hylander (1975) has reviewed the more modern work and has attempted to resolve the conflicting views. Probably the most frequently stated reason for doubting that the temporomandibular joint acts as a fulcrum of a lever system is the assumption that it is structurally incapable of bearing stress. This assumption is based principally on observations that the bone in the roof of the mandibular (glenoid) fossa is very thin, that the articular disc contains blood vessels, lymphatics and a synovial layer and that the mandibular neck is an apparently fragile structure – all features believed to be incompatible with the compressive stresses that would follow if the joint acted as a fulcrum. However, as Hylander points out, in the working position of the mandible the condyle articulates with the robust articular tubercle (not the floor of the mandibular fossa) from which it is separated by an avascular portion of the disc that lacks a synovial layer. Furthermore, Hylander demonstrates that the mandibular neck, despite its slender appearance, has a mechanical structure which confers upon it a quite adequate stress-bearing capability.

The second principal argument upon which the non-lever hypothesis is based is that, since the resultant of the forces produced by the jaw-closing muscles passes through the molar teeth, the distances from the jaw joint to the point of application of the muscle force, on the one hand, and to the main grinding teeth, on the other, are of identical magnitude and all the force is thus transmitted through these teeth with no reaction force at the joint. In fact, as Hylander demonstrates, most of the models from which this argument has been derived are anatomically incorrect, either in the positioning of the lines of action of the muscles or of the dental arcade. When the appropriate corrections are made the resultant of the forces passes behind the molar teeth, from which it follows that a proportion of the force is resisted at the joint.

Hylander concludes, in agreement with Gysi's earlier analysis, that the human jaw functions as a lever with the reaction force acting principally at the balancing condyle. He further points out that there is a number of ways in which this system could be made more efficient by increasing the ratio of the power arm to the load arm, as viewed in frontal projection (Fig. 50). Narrowing the dental arch would have this effect by reducing the load arm, while increasing bicondylar width would increase the ratio by adding an equal amount (in absolute terms) to both the power arm and load arm (in relative

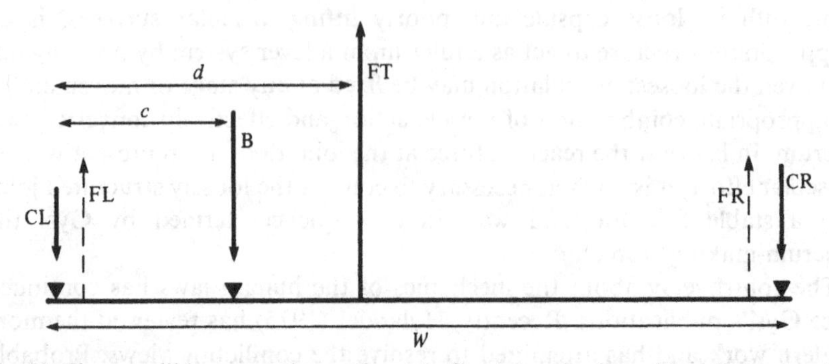

Fig. 51. The mandible as a stationary beam in frontal projection with unilateral molar biting. The left and right muscle forces (FL and FR) are replaced by a single force FT with a distance d determined by the ratio of FL to FR. The bite force B, muscle force FT, and the sum of the reaction forces at the working side and balancing side condyles (CL and CR, respectively) determined from the sagittal projection. W = bicondylar width; $CR = (FTd - Bc/W)$; $CL = (FT(W-d) - B(W-c)/W)$. During this phase of mastication the balancing side condyle is on the articular eminence and the working side condyle is in the mandibular fossa. Forces estimated from measurements on human cadavers, assuming equal muscle force applied on each side (figures expressed as percentages of bite force): total condylar reaction force = 78; CL = 15; CR = 63. (Reproduced with permission from Smith, 1978.)

terms, therefore, the power arm would gain more than the load arm). This may be the functional explanation of the large bicondylar dimension/dental arch width ratio encountered in, for example, the eskimo mandible where large chewing forces are undoubtedly generated.

More recently still, Smith (1978) has suggested that since the power stroke involves minimal vertical movement the mandible is better regarded, during this part of the chewing cycle, as a stationary beam rather than as a third class lever. This postulated beam has two supports (at the joint and at the bite point) in sagittal projection but three supports (at the working condyle, balancing condyle and bite point) in frontal projection. Analysis of the frontal projection (Fig. 51), using measurements taken from the human cadaver and assuming that the musculatures of the working and balancing sides contract with about equal force (no actual quantitative data are available but the findings of electromyographic analyses suggest that this assumption is not unreasonable, see below), shows that the reaction force at the balancing condyle would be about four times greater than that at the working condyle during molar biting. Smith derives essentially similar conclusions for the chimpanzee (*Pan troglodytes*), spider monkey (*Ateles belzebuth*) and macaque monkey (*Macaca* sp.).

The observation that the transverse axis about which the mandible rotates during depression and elevation passes through or close to the mandibular foramina has been frequently cited as evidence that the condyle cannot be acting as the fulcrum of a lever system (e.g. Tattersall, 1977). However, as just emphasised, powerful forces are produced by the masticatory muscles chiefly, if not entirely, during the power stroke of the chewing cycle. Such rotation about a transverse axis as takes place to produce the minimal amount of vertical movement involved in the power stroke probably does so about an axis passing through the condyles rather than through the mandibular foramina (see below). It appears, therefore, that the mandible undergoes a 'change of state' as the teeth come into intercuspal range. There is, thus, no inconsistency between the mandible acting as a class III lever or a stationary beam during the power stroke and rotating about an axis through the mandibular foramina during the opening and closing strokes.

Early attempts to analyse the human masticatory movements based on methods such as anatomical inference and graphic registration with face bows (see Schweitzer (1951) for a comprehensive review and bibliography of this work) have been largely superseded by investigations employing the more modern techniques of cinephotography and cineradiography. It appears from studies of the latter type (e.g. Ardran & Kemp, 1960; Ahlgren, 1966, 1976) that each individual possesses a characteristic chewing pattern that is established early in life (probably within a year or two of birth). There is a considerable amount of variation between these individual chewing patterns related, at least in part, to the prevalence in modern man of various types of malocclusion. Within the individual pattern, variation occurs from one cycle to the next and with the type of food being chewed.

Typically, in the presence of a normal occlusion (i.e. where the mesiolingual cusp of the upper molar engages the central fossa of the corresponding lower molar and where there are no gross malalignments elsewhere in the dental arcade), the mandible is lowered during the opening stroke from the rest position (in which the occlusal surface of the upper and lower teeth are separated by a gap of some 2 to 3 mm) and tends to deviate towards the working side, although in some cases there may be an initial deviation towards the balancing side. During the closing stroke a marked deviation towards the working side occurs in the early phase but eventually the mandible starts to move back somewhat towards the midline. The intercuspal (power) stroke includes, in the majority of subjects studied, a gliding contact between the occlusal surfaces of the upper and lower teeth on the working side during which the mandible is returned to its centric position. However, a particularly large degree of variation was found in this stage, both in the extent of the contact glide and in its angle. On average, gliding takes place over about 1 mm with the lower teeth moving in a superomedial and slightly anterior direction. Mills' (1955, 1963, 1978) analysis of the wear facets on human teeth indicates that the movement of the lower teeth from first contact to centric occlusion (which he terms the buccal phase and which is equivalent to phase I of Hiiemae and Kay in lower primates – see below) is followed by a shorter movement (the lingual phase = phase II of Hiiemae and Kay) during which the lower teeth continue to move medially but inferiorly rather than superiorly. During the buccal phase the distobuccal cusp of the lower molar slides in the groove between the two buccal cusps of the corresponding upper molar. In the lingual phase the distobuccal cusp of the lower molar slides against the buccal slope of the mesiolingual cusp of the upper molar. According to Ahlgren (1966) the total duration of the chewing cycle is between 500 and 750 ms (depending on the type of food) of which the closing and power strokes occupy just 50 per cent. Hiiemae (1978) gives a total duration averaging some 850 ms with the power stroke taking up 28 per cent of this time.

The fact that the degree of anisognathy present in the human jaws is small permits the teeth on the balancing side to come into occlusion during the power stroke, a condition termed balanced occlusion. Mills (1955) has suggested that a balanced occlusion is the condition typical for primates but the recent study by Kay & Hiiemae (1974, discussed further below) has indicated that in some primate species at least any contact between the teeth on the balancing side is of a transitory nature only.

Biting (incision) is a relatively simple movement in which the mandible is first moved forwards and depressed at the same time. It is then elevated so that the incisors meet edge to edge or with the edges of the lower incisors just behind those of the uppers. Severance of tough food is achieved by sliding the edges of the lower incisors upwards against the posterior surfaces of the upper incisors.

The human jaw joint possesses a squamosal articular surface that is far more extensive anteroposteriorly than is the condylar articular surface. Since the capsule above the articular disc is slack, the condyle, together with the disc, can make considerable anteroposterior translatory movements. The presence on the anterior part of the squamosal surface of a prominent articular tubercle results in the condyle moving in an inferior as well as an anterior direction as it translates forwards.

The movements of the condyles during mastication have been widely investigated using techniques such as extra-oral tracing (e.g. Campion, 1905) and more recently by radiological examination (e.g. Riesner, 1938; Thompson, 1946; Ardran & Kemp, 1960; Lindblom, 1960). Over most of the range of sagittal depression and elevation of the mandible, there is a combination of rotation (about a transverse axis) and translation of the condyles. The resulting overall mandibular movement is rotation about a transverse axis passing through, or close to, the mandibular foramina (Moss, 1959). The translatory component involves the condyle moving over a path averaging some 10 mm in length, with an inclination (because of the presence of the articular tubercle), relative to a horizontal line from the wing of the nose to the external acoustic meatus, of about 30°. That part of opening and closing, from occlusion to the rest position and vice versa, has been generally considered to involve a pure rotatory movement of the condyles, but the amount of movement is so small that any translatory condylar movement would be very difficult to record. Protrusion and retrusion involve bilateral translatory movements of the condyles while lateral swing is brought about by unilateral condylar translation on the side opposite to that towards which the mandible is swinging. The movements of the ipsilateral condyle during lateral swing are less certain, probably because considerable individual variation occurs, but it appears that, in many instances, the vertical axis about which the mandible turns lies posterior to this condyle. The ipsilateral condyle thus undergoes a small lateral (Bennett) movement.

In an attempt to improve the accuracy of determination of jaw movements during actual chewing movements, Gibbs, Messerman, Reswick & Derda (1971) have developed a technique for recording these movements by means of incremental transducers mounted between a maxillary reference bow and a face bow attached to the mandible. Information thus recorded is used to activate a reproducer mechanism which duplicates jaw movement in a cast replica of the subject's jaws. By using this technique they were able to show that, during the opening stroke of a typical chewing cycle, both condyles translate downwards and forwards. Early in the closing stroke, as the mandible moves towards the working side, the working condyle moves upwards, posteriorly and somewhat laterally (thus apparently confirming the existence of the Bennett movement). During the remainder of the closing stroke the working condyle remains nearly stationary and then, in the power stroke, moves medially to its closed position while the balancing condyle moves posterosuperiorly to return to its closed position.

The functional significance of the articular tubercle is still debated. The traditional view has been that it is related to the presence of overbite and overjet in the human dentition – the presence of the tubercle causing the condyles to move inferiorly during protrusion, so depressing the mandible sufficiently for the lower incisors to clear the overlapping upper incisors. In this way the incisors can be brought together edge to edge without interference from occluding posterior teeth. The tubercle and overbite are thus seen as functional correlates which allow the incisors to be used independently of the chewing teeth. Undoubtedly, the functional separation of incision and chewing is important and in some mammalian groups, as will be described, special morphological adaptations of the jaws are present to ensure that this occurs. However, doubt is thrown upon the role of the human articular tubercle in this context by the observation that overbite and overjet are consistent features only of modern man while a tubercle is found in human skulls of the greatest antiquity, including those of extinct forms, as well as in other hominoids in all of which the upper and lower incisors commonly meet edge to edge with the mandible in normal occlusion. An alternative possibility is that the presence of the tubercle is related to the relatively wide lateral excursions that the human mandible makes during mastication. This view has received support during recent years from Mills (1955, 1963, 1978) who has pointed out that since the molar and premolar teeth are tilted somewhat, such that the occlusal plane on each side slopes downwards and medially, the mandible must be depressed on the balancing side to prevent the buccal cusps of the lower molars fouling the more inferiorly projecting lingual cusps of the corresponding upper molars.

Most modern studies of the activity of the human jaw muscles during mandibular movement have used the technique of electromyography. One of the earliest, and still one of the best, of these studies is that by Moyers (1950). His principal conclusions may be summarised as follows: (1) Mandibular depression involves contraction of the lateral pterygoid muscle accompanied, in the second half of the movement, by contraction of the digastric, the activity in both muscles being bilateral. (2) Elevation of the mandible is brought about by the coordinated action of the temporalis, masseter and medial pterygoids of both sides. (3) Protrusion involves the simultaneous bilateral contraction of the lateral and medial pterygoids. (4) Retrusion is effected by the bilateral contraction of the middle and posterior fibres of temporalis. (5) Lateral movements are produced by contraction of the lateral and medial pterygoids of the side opposite to that towards which movement is occurring accompanied by contraction of the posterior fibres of the ipsilateral temporalis.

Subsequent electromyographic studies have confirmed and extended these general conclusions. Ahlgren (1966) and Møller (1966), for example, have focussed attention on the muscle activity occurring during the actual chewing cycle. Their findings emphasise the individual variation in muscle activity which is present (as it is in the movements comprising the chewing cycle) especially in relation to abnormalities in the pattern of occlusion and to the type of food being chewed. During the opening stroke the lateral pterygoids and digastrics were found to be active. In the closing stroke the medial pterygoids were the first elevators to become active, the onset of activity in the masseter and temporalis muscles occurring some 30 to 40 ms later. When chewing was deliberately restricted to one side, it was found that the temporalis of the working side became active in the closing stroke before the corresponding muscle on the balancing side but that the reverse was true for the masseter and medial pterygoid muscles.

Illuminating though these electromyographic studies are, there is still much about the functioning of the jaw muscles which is uncertain. There are, for example, few, if any, quantitative data available relating to the amount of force generated by the muscles of the working and balancing sides during chewing, nor is it clear how the jaw muscles combine to produce rotation about a transverse axis through the condyles during movement between occlusion and the rest position and how this axis is transferred to the region of the mandibular foramina during the remaining parts of elevation and depression.

So far as the latter point is concerned, it has been suggested (e.g. Tattersall, 1977) that during the closing stroke the temporalis and masseter–medial pterygoid muscle blocks form a couple with a centre of rotation at the mandibular foramen. However, as Smith (1978) has pointed out, a couple is produced by equal and opposite, non-colinear, parallel forces acting in the same plane – criteria which the temporalis and masseter–medial pterygoid blocks fail to meet. In fact, these muscles on contraction produce a vertical resultant.

While analyses based on hypothetical models have proved of great value in interpreting the biomechanics of the jaws, the investigation of the function of the human masticatory muscles now appears to have reached a stage where further progress will depend upon direct investigation in the living subject rather than theoretical speculation.

Great apes

The orang-utan, chimpanzee and gorilla all contrast sharply with man in possessing a very much more massive facial skeleton, together with a correspondingly powerful jaw musculature and robust dentition (Fig. 49). The canine teeth are well developed, especially in the male where they are long and dagger-like but even in the female they are sufficiently prominent for the upper and lower teeth to interlock to some extent. The lower first premolar is a sectorial tooth against which the upper canine hones. The dental arches, seen in occlusal view, have a U shape, with the right and left lines of postcanine teeth being parallel, rather than following a catenary type of curve as they do in man. Because the anterior basicranial angle is less than 180°, the facial skeleton of the apes, despite its massive size, is tucked under the frontal part of the braincase rather than being located directly anterior to it as in the majority of non-anthropoid mammals. The mandibular symphysis fuses early in postnatal life.

The structure of the jaw joint is, in essence, similar to that in man, although generally more massive and with stronger capsule and ligaments. It has been widely assumed that one functionally important difference between the human and ape joints is the absence, or very weak development, in the latter of an articular tubercle. This belief appears to have stemmed from a description by Lubosch (1906) of the temporomandibular joint in just six orang-utan, three gorilla and two chimpanzee skulls. Humphreys (1921, 1932) suggested further that the presence of an articular tubercle is a characteristic of *Homo sapiens*, separating him not only from the apes but also from extinct hominids such as Neanderthal man and *Homo erectus*. The extensive biometrical study by Ashton & Zuckerman (1954) has shown quite clearly that this view is incorrect and that the basic architecture of the squamosal articular surface is similar in man and apes as well as in the fossil forms *Australopithecus*, *Homo erectus* and Neanderthal man.

Because of the presumed absence of an articular tubercle and the presence of interlocking canines, it was at one time widely held that transverse chewing movements are impossible in the ape and that mastication takes place by a chopping type of action. However, studies of the wear facets on pongid teeth (Zuckerman, 1954; Mills, 1955, 1963, 1978) indicate that the movements of the mandible in the intercuspal range are closely comparable in apes to those already described in man and that transverse movements, including buccal and lingual phases (phases I and II), are involved in the power stroke in all living hominoids. In fact, it can be readily demonstrated by examination of the ape skull that a considerable amount of transverse movement is possible, although limited to a rather narrower path than in man by the presence of the large canines. In the absence of cinephotographic, cineradiographic or electromyographic analyses (the intractable nature of the great apes proving, apparently, an effective deterrent to these types of investigation), it seems reasonable, therefore, to assume that biting and chewing in man and the apes are essentially similar and involve similar muscular activities.

There can be no doubt, however, that the power of mastication in the great apes far exceeds that in man as indicated by the much greater mass of the pongid jaw musculature in both absolute and relative terms. The very large size of the temporalis muscle is associated with the formation of bony crests on the braincase which increase the area available for the origin of the muscle. The adult form of these crests has been intensively studied in the belief that cresting patterns may be relevant to the interpretation of the relationships of living and extinct hominoids (e.g. Zuckerman, 1954; Robinson, 1954, 1958; Le Gros Clark, 1955). A particularly detailed account of the cranial crests has been provided by Ashton & Zuckerman (1956) which is of especial interest in that they examined the development, as well as the adult form, of the cranial muscle markings in man and apes and in several species of monkey.

Ashton & Zuckerman found that in all adult male gorillas, in the majority of adult male orang-utans and in a small proportion of adult male chimpanzees the areas of attachment of the temporalis muscles on the braincase reach almost to the midline. When this happens the areas of attachment are increased by the development of a sagittal crest (Fig. 49). In the gorilla the crest may project by as much as 5 cm. Sagittal crests were found far less frequently in the female great apes. They occur rarely in the gibbon of either sex. Amongst the monkeys a crest was found in all adult male baboons (*Papio* sp.) but occurs rarely in the male white-nosed monkey (*Cercopithecus nictitans nictitans*) and never in the woolly monkey (*Lagothrix* sp.). Sagittal cresting is always absent in the females of these groups.

The growth processes underlying the formation of the cranial crests appear to be similar in each of the primates where they have been studied. In each group the posterior fibres of the temporalis, as judged by the extension of their area of attachment, grow more rapidly than the middle and anterior fibres and consequently approximate the advancing edge of the attachment of the nuchal musculature to produce, in the majority of cases, a nuchal crest. The least distance between the right and left temporalis muscles in the very young primate (milk dentition only erupted) is usually located well forward on the braincase. Subsequently, as the posterior fibres of the muscle grow more rapidly than the anterior ones, the point of least separation moves posteriorly so that, if a sagittal crest is produced, it usually forms first in the region between bregma and lambda.

The small size of the temporalis muscle in man, both modern and in such fossil forms as Neanderthal man and *Homo erectus*, together with the relatively large size of the braincase, precludes the formation of sagittal and nuchal crests. Despite these differences the general arrangement of the temporal and nuchal muscle markings suggests that the morphogenetic processes underlying their development are fundamentally similar to those in other primates.

There can be little doubt, in view of the experimental studies of Washburn (1947, see Chapter 8), that the ridges and crests associated with muscle attachments develop in response to the pull of the attached muscles. The overall expansion of the braincase during growth appears to be influenced principally by the enlargement of the enclosed brain but the outer table of bone is, in addition, responsive to the action of the temporalis and nuchal muscles (in the terminology of Moss, Chapter 8, the outer table is the skeletal unit associated specifically with the periosteal functional matrix provided by the attached muscles).

As well as serving to increase the area of attachment of the nuchal and temporal muscles the cranial crests together with the frequently associated supraorbital torus contribute to the strength of the cranium, enabling it to resist the huge muscular forces produced by these muscles.

Non-hominoid primates

Although at first sight the remaining anthropoids (Old and New World monkeys) and prosimians may seem to constitute groups that are too disparate to treat together, it is in fact convenient to do so, first because some of the important studies to be considered involve comparisons of monkeys and prosimians and secondly because it emerges from these comparative studies that the structure and function of the masticatory apparatus in the two groups possess many striking similarities.

The facial skeleton of monkeys is far smaller than that of hominoids in absolute terms but in relative shape and size has generally much in common with the facial skeleton of apes, although quite wide species variation exists. As in the apes, the facial skeleton is located anteroinferiorly to the frontal part of the braincase in association with an anterior basicranial angle below 180° (Fig. 52). The dentition is arranged in a U-shaped curve and the canines, especially in the males, project prominently above the occlusal surface of the remainder of the dentition. The mandibular symphysis undergoes early fusion.

The principal differences between the Old and New World monkeys lie in the molar and premolar teeth. The molars of the New World monkeys are rather primitive in form, the uppers bearing three cusps and the lowers often

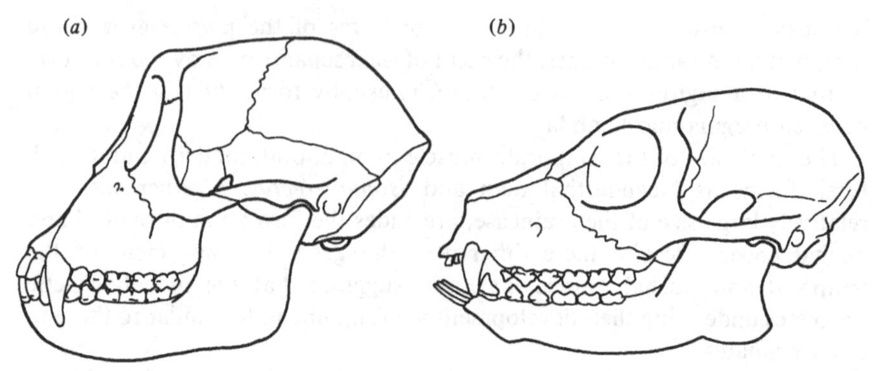

Fig. 52. Left lateral views of skulls of (*a*) *Macaca mulatta* and (*b*) *Lemur macaco*. Not to scale.

having a division into trigonid and talonid elements. In the Old World monkeys the molars are more specialised, both upper and lower teeth usually possessing four cusps with transverse ridges connecting corresponding buccal and lingual cusps (the lophodont condition). The New World monkeys have three premolars in each quadrant, one more than in the Old World group in which the dental formula is identical with that of hominoids.

The structure of the jaw joint in monkeys closely parallels that seen in apes and man. The articular tubercle is usually described as being low or absent but, in view of the inaccurate statements that have been made and accepted about the pongid articular tubercle, it would seem best to reserve judgement until comprehensive biometrical data are available. The jaw musculature of monkeys is relatively well developed and, as already described, sagittal and nuchal crests may be present in the males of some species.

The species variation found in the craniomandibular apparatus of monkeys is due to relatively small contrasts in size and shape rather than to fundamental differences in the architecture of the jaws. It seems likely that a proportion of the shape contrasts may itself be size-related, arising because an allometric relationship exists between variables. This has certainly been shown to be the case in the sexual dimorphism encountered in the shape of the primate skull (Wood, 1976) and there seems to be no reason why it should not also apply to species differences.

The most striking contrast between the craniomandibular apparatus of monkey and prosimian lies in the much greater specialisation found in the anterior teeth of the latter. In the majority of living lemurs (permanent dental formula $I\frac{2}{2}C\frac{1}{1}P\frac{3}{3}M\frac{3}{3} \times 2$), for instance, the upper incisors have tended towards reduction in size and are sometimes absent while the lower incisors have been formed into a dental comb used in grooming. In the most highly specialised dentition of all primates, that of *Daubentonia*, there is a single large gnawing incisor in each jaw, resembling the rodent incisor in being only partly covered by enamel and possessing a measure of continuous growth.

In addition to the differences in dentition, there are considerable structural differences in the jaws between prosimian and anthropoid primates. An unfused symphysis persists throughout life in the extant prosimians while in monkeys and hominoids the symphysis fuses soon after birth. The prosimian facial skeleton is typically elongated and situated anterior rather than anteroinferior to the braincase (Fig. 52). The anterior basicranial angle in the prosimian skull, in association with the more anterior location of the facial skeleton, is usually close to 180°. However, in some prosimians, for instance *Tarsius*, the facial skeleton is considerably reduced and the anterior basicranial angle is well below 180°, foreshadowing the condition typical of anthropoids.

In most fossil and all recent prosimians a postorbital bar is present but the orbital cavity is not separated from the temporal fossa by a bony partition, except in tarsiers, where a partial separation occurs. In anthropoid primates such a partition is present and usually complete except in New World monkeys, where a small gap of varying size may remain. The development of these postorbital skeletal structures has been related by Cartmill (1972) to the need to strengthen the orbits and thus protect their contents against distortion during contraction of the jaw muscles. Such strengthening may well have been crucial for the evolution of precise vision. An alternative, but not mutually exclusive, suggestion is that the postorbital bar and partition evolved in association with a tendency towards increasing incisal preparation of food. Cachel (1979) has shown, in a wide variety of primates, that there is a marked correlation between diet, the relative development of the anterior dentition and the weight ratio of the anterior temporalis (used in incision) and masseter muscles – both of the latter attributes being increased in species consuming diets of a frugivorous nature which require much incisal preparation. A postorbital partition could thus have evolved because of the extra bony area of origin it provided for the enlarging anterior component of the temporalis muscle.

Mills (1955, 1963, 1978) has found that the wear facets on the chewing teeth have a closely comparable distribution over a wide range of monkeys and prosimians (including *Tupaia* which is regarded by some authorities as being, in a number of respects, a living representative of the ancestral primate condition, although its precise taxonomic assignment is still debated) and closely resemble those encountered in the Hominoidea. From these observations he concludes that a common pattern of chewing, similar to that already described for hominoids, obtains throughout the primates and probably emerged at a very early stage of primate evolution (possibly having been already present in the earliest known primate *Purgatorius* from the Cretaceous). If true, this would indicate a quite remarkable retention of a relatively generalised mode of jaw usage in a large and, in many other respects, diverse group of mammals. The primates are well known, of course, for their tendency to retain generalised features, notably in the structure of their limbs.

The findings of Hiiemae and Kay (Hiiemae & Kay, 1972, 1973; Kay &

Hiiemae, 1974), in a study combining cineradiography and occlusal analysis of the jaw movements in the common tree shrew (*Tupaia glis*), the bushbaby (*Galago crassicaudatus*), the squirrel monkey (*Saimiri sciureus*) and the spider monkey (*Ateles belzebuth*), have provided broad confirmation of the conclusions based on wear facet analysis. The pattern of masticatory movements is virtually identical in *Tupaia*, the prosimian and the two monkeys and strikingly similar to that already described for *Didelphis*. In each case the food may first be pulped by puncture-crushing before chewing. The closing stroke of the chewing cycle in *Tupaia* (dental formula $I\frac{2}{3}C\frac{1}{1}P\frac{3}{3}M\frac{3}{3} \times 2$), for example, begins with the lower jaw at maximum gape and at, or close to, maximum lateral excursion towards the working side; the lower jaw is then elevated and moved somewhat medially so that the power stroke commences with the buccal surfaces of the upper and lower postcanine teeth of the working side in alignment. The mandible is then moved medially and superiorly. As the symphysis approaches the midline a slight anterior component of movement is added. A transitory contact may occur between the upper and lower teeth on the balancing side as centric occlusion is approached. During the power stroke the mandible of the working side turns about its long (anteroposterior) axis so that its lower border is everted, this being possible because of the presence of an unfused symphysis. The opening stroke begins with slow depression followed, in some cycles, by a pause, before rapid opening takes place with a lateral swing to the working side. In puncture-crushing, the power stroke involves a simpler movement than that seen in chewing, the mandible moving medially and slightly anteriorly.

Hiiemae and Kay found small variations between the species they studied in the duration of the chewing cycle and of its component strokes as well as in the precise pathways followed by the lower jaw but these seem relatively insignificant compared to the overall similarity. Even the presence of a fused symphysis in the two anthropoid species appears to have little effect, despite the fact that movement takes place at the unfused symphysis in *Tupaia* and *Galago* (the structural explanation of this finding is described below). One difference, of significance in the light of the findings of wear facet analysis, is that, while in *Tupaia* the power stroke is virtually complete at centric occlusion, in *Galago*, *Saimiri* and *Ateles*, as in the Hominoidea, it does not finish at centric occlusion but continues by a movement of the mandible towards the balancing side during which the mandible moves also somewhat inferiorly. The movement to centric corresponds with Mill's buccal phase and the continued movement from centric with his lingual phase. Hiiemae and Kay prefer to call these, respectively, phases I and II of the power stroke.

Hiiemae and Kay suggest that, despite the overall similarity in masticatory movements in the species they studied, certain evolutionary trends are apparent. One of these relates to the incisor teeth. In primitive mammals such as *Didelphis* the incisors appear to have little function in food separation or preparation and this is also the case for *Tupaia* and *Galago*, but in *Saimiri* and *Ateles*, as in other anthropoids, the spatulate incisors are used for biting all but the hardest foods. The fusion of the symphysis seen in the Anthropoidea may be related to this increased and more vigorous usage of the incisor teeth.

Another trend, Hiiemae and Kay suggest, has been for phase II of the power stroke to become prolonged. It will be recalled that this phase is completely lacking in *Didelphis*. In *Tupaia* phase II is very brief, the food being cut into small particles in phase I by the shearing blades produced by the attrition facets on the molar teeth. In *Galago*, *Saimiri* and *Ateles* this initial cutting is followed by a phase II in which further significant reduction of the food particles takes place by grinding. There has been a corresponding tendency for prolongation of the power stroke, this stroke occupying about 25 per cent of a chewing cycle which lasts some 240 ms in *Tupaia*, about 24 per cent of about 310 ms in *Galago*, about 26 per cent of some 360 ms in *Saimiri* and about 32 per cent of some 330 ms in *Ateles*. In order to improve the grinding ability of the teeth the advanced primates show a tendency for the wear facets on the molar teeth to form blade-ringed compression chambers rather than the shearing blades seen in more primitive mammals.

Information about the activity of the jaw musculature in monkeys has been provided by Luschei & Goodwin (1974) who carried out an electromyographic study of mastication, combined with position transducer recording of the associated mandibular movements, in *Macaca mulatta*. Although their description of mandibular movements is less detailed than that provided by Hiiemae and Kay, it appears that the pattern of movement in *Macaca* is similar to that in the various prosimian and anthropoid primates investigated by the latter authors. Muscle activity was found to occur on both sides but the amplitude of activity of the jaw-closing muscles on the working side exceeded that on the balancing side. The earliest activity during the closing stroke was recorded in the balancing masseter, medial pterygoid and lateral pterygoid muscles. The balancing temporalis and the muscles on the working side became active somewhat later. This distribution of electromyographic activity between balancing and working sides resembles that already described in man. The activity of the muscles of both sides ceased synchronously just as the teeth came into occlusion, the power stroke being completed presumably by contraction continuing in the muscles after electromyographic activity had ceased. A similar cessation of electrical activity during the power stroke has been noted in *Didelphis*. Activity in the macaque digastric muscle began immediately after cessation of activity in the jaw-closing muscles.

Luschei & Goodwin (1974) relate the pattern of distribution of muscle activity between the working and balancing sides to the timing of the transverse mandibular movements during the chewing cycle. The early activity of the balancing masseter and pterygoid muscles contributes the upward and transverse (towards the working side) component of the first part of the closing stroke, the transverse component being the result of the forward pull of the masseter and medial pterygoid (both of which have fibres running obliquely downwards and backwards from origin to insertion) in addition to that of the lateral pterygoid. Then, as the muscles of the working side come into action, they increase the upward force of the closing stroke as well as helping to return the mandible towards the midline.

Both the mandibular movements and the pattern of masticatory muscle activity were found to be remarkably constant from time to time within individuals and also between individuals. This constancy appears to be a common attribute of mastication in many mammalian groups, the most notable exception being man.

These studies of mastication in monkeys and prosimians do much to confirm the suggestion, noted earlier, that the pattern of jaw usage in the primates has undergone only modest change during this group's long evolutionary history. The retention of an unspecialised jaw structure and function, possibly from the earliest stages of mammalian evolution, has not unduly restricted the range of the primate diet, although it does appear to have limited the use of the jaws as food-gathering devices as compared to most other mammalian groups. Undoubtedly, the transformation of the primate forelimbs into skilful manipulative organs has had much to do with this lack of specialisation of the masticatory apparatus. An important effect of these trends is that the dominance which the jaws and their musculature possess in the skulls of most mammals is less marked in primates, especially in the anthropoids. This has allowed the facial skeleton to become relatively reduced in size leading to a better balancing of the head on the vertebral column, a trait which reaches its maximum expression in man and those varieties of extinct hominoid which had adopted an upright posture.

Functional significance of the structure of the primate symphysis

Fusion of the mandibular symphysis, beginning shortly after birth, is the typical condition in anthropoid primates but is lacking in extant prosimians (although present in some extinct groups). Undoubtedly, the significance of fusion at the symphysis is related to a need to strengthen and make rigid the anterior region of the lower jaw but the precise functional advantages that accrue from this arrangement are still debated. DuBrul & Sicher (1954) suggested that fusion occurs to resist the tendency towards separation produced at the symphysis by the medially directed force generated at the condyles by the contraction of the lateral pterygoids. This separating force is greater in anthropoids than in prosimians because the shortening of the cranium in the former group has brought about a posterior migration of the origins of the lateral pterygoids and hence an increase in the obliquity of their line of pull.

More recently, Hylander (1975) has proposed that a fixed symphysis is required to allow full use of the balancing muscles, pointing out that, in its absence, there will be a tendency for the force generated by the muscles on the balancing side to be dissipated by the balancing side of the mandible shearing past the working side at the symphysis rather than being transmitted across to the bite point on the working side.

Hylander's views gain support from the observations of Beecher (1977*a*, *b*) of the structure of the symphysis in a variety of modern prosimians (*Galago crassicaudatus*, *Lemur fulvus*, *Lemur macaco*, *Hapalemur griseus* and *Propithecus verreauxi*). Here, as in other mammals lacking full fusion at the symphysis, a pad of fibrocartilage is present in the anterosuperior part of the symphysis. The remaining symphysial tissue is composed largely of liagamentous fibres, many of which are arranged in a cruciate fashion (as seen in frontal view) while others run transversely. In *Galago carassicaudatus*, as in *Tupaia glis*, the mandible of the working side rotates about its long axis (lower border everting) during the power stroke. It may also move anteromedially relative to the mandible on the balancing side. As a result of these movements the tissues of the symphysis are subject to anteroposterior shear stresses as well as to compression superiorly and tension interiorly. There will be in addition, presumably, ventrodorsal shear stresses produced by the vertically acting force generated by the jaw-closing muscles of the balancing side. The arrangement of the tissues in the symphysis appears to be well suited to resist these various stresses – the pad of fibrocartilage resisting the superior compression, the transverse fibres resisting tension and anteroposterior shear and the cruciate fibres resisting ventrodorsal shear. Nonetheless, in spite of this apparent high degree of mechanical adaptation, the findings of the recent study by Hylander (1977) using in-vivo strain gauges to determine mandibular deformation, indicate that during unilateral biting the symphysis of *Galago* transmits only a small amount of force from the balancing side musculature to the working side. In *Lemur fulvus*, *Lemur macaco*, *Hapalemur griseus* and *Propithecus verreauxi* (all folivorous species) calcified and ossified areas are found in the symphysis, in association with the cruciate fibres, which may provide an improved resistance to ventrodorsal shear and thus greater ability to transmit forces from the balancing side musculature to the balancing side dentition. The completely fused symphysis of anthropoid primates could then be regarded as resulting from a continuation of this trend of structural modifications to resist ventrodorsal shear and hence to improve the transfer of muscular forces across the symphysis to allow full use of the balancing side musculature during chewing.

A further factor in the development of a fused symphysis may be the increased incisal activity of anthropoids, resulting in a need for strengthening of the symphysis to resist the consequent increase in mechanical stressing of this region (Kay & Hiiemae, 1974).

Ungulates

The structure of the jaw apparatus in ungulates, the major group of herbivorous mammals, is adapted to enable two separate and distinct masticatory movements to take place. Food is gathered by cropping grass, leaves or other vegetable matter between the upper and lower sets of incisor and canine teeth. The food is then transferred for chewing to the grinding battery of teeth made up of premolars and molars and whose number varies from one ungulate group to another. Characteristically, a long diastema separates the cropping from the grinding teeth and the jaws are anisognathous.

In the advanced artiodactyls, the ruminants, the upper set of anterior teeth is missing, being replaced by a dense pad of mucous membrane, an arrangement that reduces wear of the lower incisors and canines. The molar teeth are elongated anteroposteriorly and have high cusps that are crescent-shaped (the selenodont condition). The cheek teeth of the horse are also long and high but here the cusps are connected to each other by complicated ridges and the enamel of the crown is covered by a layer of cementum. In both cases the effect is to render the occlusal surfaces of the teeth continuously rough due to the different rates of wear of the various mineralised dental tissues.

Since cropping and chewing involve respectively and quite separately the incisor–canine and premolar–molar components of the dentition and since both activities involve the lower jaw making wide excursions in the horizontal plane, it is clear that the structure of the jaw joint must allow extensive translatory movements of the mandibular condyle. Typically, the ungulate condyle is wide transversely and rather short anteroposteriorly with a flattened articular surface. The squamosal articular surface is also flattened, the mandibular fossa being extremely shallow, and is usually bounded posteriorly by a postglenoid tubercle. The articular disc is thin and conforms closely to the shape of the condyle. The capsule is slack, especially between the squamosal articular surface and the disc, and shows no marked ligamentous thickenings. The mandible can be moved passively in all three planes of space, the disc and condyle moving together.

Despite this passive freedom of movement at the joint, the teeth follow fairly consistent paths of movement during cropping and chewing because of the presence of guiding ridges on the premolar and molar teeth. In the selenodont artiodactyls, for example, regional differences in the thickness of the enamel on the occlusal surfaces of the teeth result in transverse ridges developing with wear, the ridges on the upper teeth fitting into the valleys between the ridges on the lower teeth and vice versa. In the horse similar guiding ridges are developed during wear because of the presence on the occlusal surfaces of enamel, cementum and exposed dentine, each tissue having a different hardness and, therefore, a different rate of wear.

The masticatory movements in ungulates, apart from one or two isolated examples to be described below, have not been studied by the modern techniques of cineradiography and electromyography but attempts have been made to infer the nature of the jaw movements and the action of the jaw muscles from anatomical considerations. Smith & Savage (1959), for example, have used this method for the bushbuck (*Tragelaphus strepsiceros*). This animal, like other selenodont artiodactyls, is anisognathous, has prominent

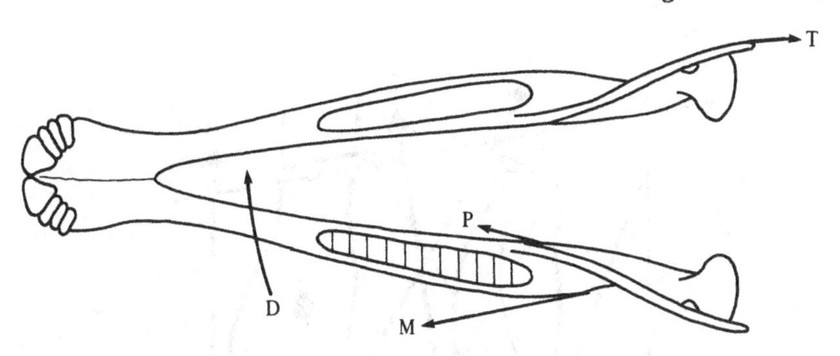

Fig. 53. Dorsal view of mandible of artiodactyl (bushbuck). Working side on left, molar teeth in occlusion indicated by hatching (the direction of the lines corresponding approximately with the orientation of the guiding ridges). Arrows M, P and T indicate lines of action of masseter, pterygoid and temporalis, respectively; arrow D indicates direction of jaw movement. (After Smith & Savage, 1959.)

transverse guiding ridges on the occlusal surfaces of the cheek teeth and an occlusal plane that slopes upwards and inwards (i.e. the occlusal surfaces of the lower teeth face upwards and outwards). It seems likely that, as a result of these anatomical constraints, chewing takes place unilaterally and involves a transverse power stroke in which the mandible moves medially, superiorly and a little anteriorly (Fig. 53).

Such a movement must involve translation at the jaw joints, probably the condyle on the working side moving forwards and that on the balancing side moving backwards, or at least being held in a posterior position. As can be seen from Fig. 53, the lines of action of the masseter and medial pterygoid muscles, as viewed in plan, are such that their contraction will produce anterior translation of the condyle of the same side while contraction of the opposite temporalis will move the condyle of that side posteriorly. The bushbuck, like other ungulates (Table 2), possesses (in relative terms) very large masseter and medial pterygoid muscles, moderately large lateral pterygoids but only small temporalis muscles, an arrangement which leaves little doubt that the muscular force producing anterior translation of the condyle on the working side is of far greater magnitude than the force moving the balancing condyle posteriorly. Smith & Savage (1959) describe this as a mechanically efficient arrangement in that only a small region of the mandible on the working side is stressed, whereas, if great muscle forces were produced by the balancing temporalis, stress would have to be carried forward along the balancing side, across the symphysis and back along the working side. However, Greaves (1978) has more recently proposed a model of artiodactyl jaw mechanics (see below) in which he suggests on apparently good grounds that, despite chewing being unilateral, bilateral jaw muscle activity is involved. As will be described, the few electromyographic studies that have been made of the jaw muscles in ungulates support the view that muscle

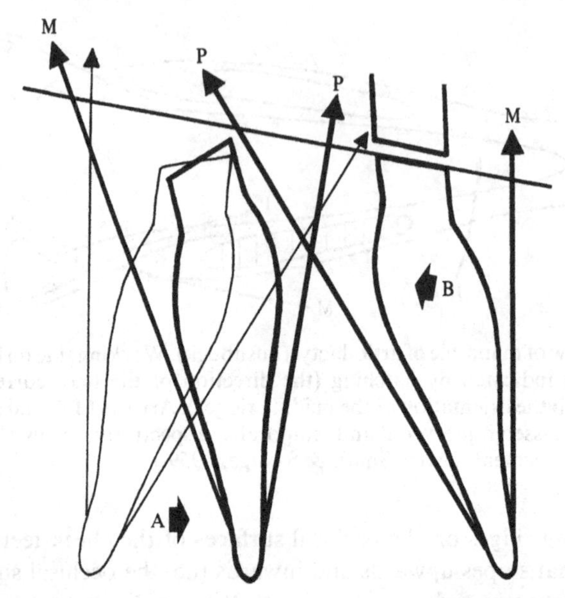

Fig. 54. Coronal section through artiodactyl lower jaw to illustrate Greaves' model of mandibular rotation. Working side on right. Mandible on left side rotates as indicated by arrow A, thus changing position of mandible and lines of action of muscles (before rotation indicated by thin lines; after rotation indicated by thick lines). Arrow B shows direction of mandibular movement on working side. M = line of action of masseter; P = line of action of medial pterygoid. (After Greaves, 1978.)

activity is bilateral. This does not, of course, negate the conclusion that the masseter and pterygoids are the principal movers in chewing. Greaves describes the presence of an unfused mandibular symphysis, such as is found in the majority of selenodont artiodactyls, as being an adaptation to bilateral muscle activity in that it permits the muscles of the working and balancing sides to cooperate in producing a unilateral transverse chewing stroke. With a fused symphysis the medial components of pull of the balancing and working medial pterygoids would be acting in opposition (Fig. 54). With an unfused symphysis, however, contraction of the balancing medial pterygoid will tend to rotate the mandible of that side about its anteroposterior axis so that the lower margin moves medially. In consequence, the line of action of the muscle, in the frontal plane, will be modified so that its medial component of pull will be reduced or even transformed into a slight lateral component. Thus, the medial component of the working medial pterygoid will now be unimpeded or even slightly enhanced and will add to the force moving the mandible of that side in a medial direction.

If Greaves' (1978) analysis is correct the contrast in evolutionary trends at the symphysis between the artiodactyls and the primates becomes explicable in terms of differences in jaw muscle action between the two groups. In artiodactyls, as we have seen, the persistence of an unfused symphysis may well be related to the action of the relatively large medial pterygoid muscle which, by virtue of its considerable medial component of pull, plays a major part in producing the transverse mandibular chewing movement. The trend in the primates, where the medial pterygoid is neither so large (Table 2) nor so medially directed (and, therefore, less active in producing transverse movements), has been for the symphysis to undergo fusion, the principal factor being here, apparently, the need to improve the transfer of vertically directed forces produced by the jaw-closing muscles of the balancing side to the dentition of the working side. In both cases the result is to increase the effectiveness of the power stroke by improving the transmission of the appropriately directed force from the balancing to the working side and thus increasing the efficiency of the bilateral muscle activity.

Smith & Savage (1959) describe the mechanism of chewing and the structural modification of the craniomandibular apparatus in the horse as being, in essence, similar to those just discussed for the bushbuck.

These authors also compare the mechanical implications of the orientation and relative sizes of the temporalis, on the one hand, and of the masseter and medial pterygoid muscles, on the other, in herbivores and carnivores. In the latter the temporalis is the dominant muscle (Table 2) and inserts into a prominent coronoid process. The condyle is situated relatively low down so that the moment arm of the temporalis is long while that of the masseter–medial pterygoid is short. By contrast, in herbivores the masseter–medial pterygoid block is relatively more powerful than the temporalis (Table 2) and, furthermore, because of the more superior location of the condyle, the moment arm of masseter–medial pterygoid is lengthened at the expense of that of the temporalis. In both cases, therefore, the principal adductor mass has its power further enhanced by elongation of its moment arm.

Scapino (1972) comes to essentially similar conclusions in his comparison of jaw muscle action in the horse with that in primitive mammals. He also illustrates the heightening of the condyle and the extension of the area of attachment of the masseter and medial pterygoid muscles in the sequence *Hyracotherium*, *Mesohippus*, *Merychippus*, *Pliohippus* and *Equus* and suggests that these changes, together with the development of a hypsodont dentition, were associated with the switch from a browsing to a grazing habit.

Detailed information of the masticatory movements and associated jaw muscle activity in ruminating ungulates has been provided by the study by de Vree & Gans (1973, 1975) of the pygmy goat (*Capra hircus*) in which the masticatory movements were recorded by cinephotography and the muscle activity by electromyography. As in other ruminants the grinding part of the dentition in the goat consists of a battery of molars and molariform premolars. Incisors occur in the lower jaw only and are separated from the premolars by a long diastema. As seen in frontal view the occlusal surfaces of the molar–premolar series have the characteristic ungulate slope (i.e. the surfaces of the upper teeth facing inferomedially). The grinding teeth are selenodont and possess transverse guiding ridges. The jaw joint consists of a condyle much widened transversely and a flattened squamosal articular surface. The articular disc is thin. The symphysis is unfused but bony struts projecting from the symphysial surface of each half of the lower jaw interlock and limit movement.

Food is taken by being grasped between the lips and then pulled back by retraction of the head. Stems and other tough material are broken off against the incisors by lifting the head. Chewing always proceeds unilaterally. During the opening stroke of a cycle in which relatively soft food (e.g. carrot) is being chewed, the mandible is depressed and swung over to the balancing side. The closing stroke of the next cycle thus begins with the lower jaw at, or close to, the most contralateral position with respect to the working side. From this position the mandible is moved dorsally in a smooth curve over to the working side. From here, the power stroke takes the mandible medially back towards the midline and also a little superiorly and anteriorly. The movements during the power stroke recorded cinephotographically in the goat thus correspond closely with those inferred from the shape of the occlusal surfaces of the teeth in the bushbuck. Considerable variation occurs in this pattern, both in the pathways followed by the mandible and in the time spent in each stroke, depending on the type of food being chewed. In general with a tough food (such as alfalfa) the power stroke is prolonged and the jaws remain incompletely closed for a longer period than with softer foods.

The electromyographic analysis that de Vree & Gans (1973, 1975) provide of the activity of the jaw muscles during the chewing cycle demonstrates that the muscles act in synergistic combination in which multiple units fire simultaneously. The actual timing of activity in each muscle varies with the type of food being chewed. In general, the working side temporalis and the balancing side posterior masseter are active in the closing phase. Since the temporalis has a retrusive and the posterior masseter a protrusive effect, the contraction of these two muscles, in addition to elevating the mandible, tends to swing it from its position towards the balancing side at the beginning of the closing stroke over to the working side. This action is aided by contraction of the balancing side lateral pterygoid. During the power stroke most of the adductors are active bilaterally, their combined effect being to move the lower jaw transversely towards the balancing side as well as to exert an upward force. This transverse movement is promoted by contraction of the working side protrusive (principally posterior masseter) and balancing side retrusive (temporalis and zygomaticomandibularis) muscles. The lateral pterygoid and digastric muscles of both sides become active towards the close of the power stroke and remain so during the opening stroke. The balancing side lateral pterygoid continues to contract in the closing stroke of the next chewing cycle adding, as already noted, a further balancing side protrusive component to the action of the working side temporalis and balancing side posterior masseter in producing transverse movement of the mandible during this stroke.

A rather similar study, employing cinephotography, cineradiography and electromyography, has been made by Herring & Scapino (1973), using the miniature domesticated pig (*Sus scrofa* dom.). This study is given particular interest by the fact that the pig is an omnivorous representative of the Artiodactyla and possesses several morphological features of the dentition, jaws and jaw musculature that are unusual amongst ungulates. First, so far as the dentition is concerned (the dental formula is $I\frac{3}{3}C\frac{1}{1}P\frac{4}{4}M\frac{3}{3} \times 2$): the molar teeth are bunodont (possessing numerous rounded cusps), the lower incisors are horizontally implanted so that their lingual surfaces receive the biting edge of the vertically implanted upper incisors, the canines form huge tusks in wild pigs although in domesticated varieties they are less extensively developed, and the dental arcades are isognathous or approximately so. Secondly, the symphysis is fused and the jaw joint allows a range of movement that is exceptionally wide, even amongst ungulates. Thirdly, the jaw musculature, although being fairly characteristic of ungulates in its relative weights (Table 2), is unusual in the extreme degree of pinnation found in its internal architecture.

Small particles of food are ingested with the aid of the lips. Larger items are first broken up by the incisors or (in the case of very tough food) by the premolars. Ingested food is then transferred by the tongue to the cheek teeth. The chewing cycle is of short duration lasting, on average, about a third of a second, some 20 or so cycles being required to reduce each lot of ingested food. The first few cycles may be of a relatively simple chopping nature equivalent to the puncture-crushing cycles described by Hiiemae and her colleagues, in *Didelphis*, *Tupaia* and primates. In the more complex full chewing cycles, transverse mandibular movements are involved. The opening stroke is generally in a vertical direction so that the chewing cycle begins with the mandible close to its midline position. During the closing stroke the mandible is deviated to one side or the other and then, in the subsequent power stroke, is moved transversely (and usually somewhat anteriorly) in the opposite direction back towards (or even beyond) the midline. In some cycles, however, the mandible may continue to move away from the midline during the first part of the power stroke only to reverse its direction later in this stage. Whatever its position at the end of the power stroke the mandible is usually returned close to its midline position in the first part of the opening stroke of the subsequent cycle. The power stroke occupies about 25 per cent of the total duration of the cycle.

The direction of transverse movement is most frequently reversed from one chewing cycle to the next, although occasionally a run of cycles may occur in which one direction is maintained. This regular alternation of direction is unusual amongst the mammalian species so far investigated but appears to be typical of the Suidae, having also been described in the warthog

(*Phacochoerus*) by Ewer (1958). Although the jaws of the pig are isognathous, the lateral deviation of the mandible during the closing stroke would seem to imply that maximal masticatory effective contact between the upper and lower teeth takes place only on the side towards which deviation occurs. Thus, as in anisognathous mammals there is a working and balancing side of the jaws, but these change over with each alternation of direction. It is not clear whether the food is gathered together into a single bolus which is transferred from side to side with each alternation or whether a certain amount of food is retained on each side of the mouth to be chewed only during alternate cycles.

The methods of observation used by Herring & Scapino (1973) do not allow precise temporal correlations to be established between the jaw movements and electromyographic activity in the jaw muscles. Their findings leave little doubt, however, that in the pig there is the same complex synergistic activity of the jaw muscles as that observed in all other mammals studied electromyographically, in which multiple muscles fire bilaterally to produce the continuity of vertical and transverse movements making up the chewing cycle.

A long diastema between the anterior and posterior sets of teeth is widespread amongst herbivorous or predominantly herbivorous mammals, being found in rodents and lagomorphs as well as in ungulates. In each of these groups the presence of the diastema appears to be functionally advantageous in that it provides a wide separation of the incisive and chewing movements. Ardran *et al.* (1958) have observed in ungulates and lagomorphs that the diastemal portion of the tongue is concerned predominantly with manipulating food before passing it backwards for chewing between the molars. They suggest, therefore, that the functional importance of the diastema is that it allows a quantity of herbage to be cropped and stored in the front of the mouth, to be then chewed at leisure with the head held up and the surroundings under observation for the presence of predators. Further, the presence of a long diastema allows the muzzle region of the skull to be reduced in width and height so that the cropping of small items of vegetation can be more conveniently carried out than if the face were short and blunt. The particular advantages of the presence of a diastema in gnawing animals are discussed in the section on rodents.

A mechanical explanation for the absence of teeth in the diastemal region has recently been advanced by Greaves (1978). His conclusions are based on the application of lever mechanics to a simplified model of the jaw mechanism in selenodont artiodactyls in which he assumes, following the analysis for the human mandible by Gysi (1921), that the jaw acts as a class III lever with the fulcrum being provided by the jaw joint on the balancing side. It can be seen from the foregoing discussion of this type of leverage system in primates that graded muscle activity on the working and balancing sides can, by its effect on the position of the muscle resultant, produce a similar force at each point within a circumscribed region of the jaw. In artiodactyls this region coincides with the part of the jaw occupied by the grinding teeth. Its anterior limit is determined by the point at which the effective force begins to decline because of reduced mechanical advantage. It is possible, therefore, that the reason why elongation of the jaws to produce a muzzle of suitable proportions for herbivorous feeding was not accompanied by a corresponding elongation of the molar–premolar tooth row was because of the poor mechanical advantage that the more anterior teeth of the row would have possessed. Instead, the teeth remained limited to the region of greater mechanical advantage with the result that a long diastema became opened up between the grinding and cropping teeth.

The posterior limit of the grinding tooth row is probably determined by the need to maintain equilibrium. According to Greave's model, the jaw, in horizontal plan, can be considered as a triangle supported against the cranium at the two jaw joints and at the area of contact between upper and lower teeth. In order for stability to be preserved the muscle resultant must be located within this triangle otherwise the mandible will tilt. If the posterior limit of the tooth row is too far back the triangle will not encompass the muscle resultant since, when the last tooth is taking the masticatory force, the resultant will be directed anterior to the side of the triangle running from the balancing jaw joint to the last tooth. Hence, the lower jaw will be tilted so that the jaw joint on the working side will tend to be dislocated.

It is apparent from Greave's model that the length of the grinding tooth region determined by these factors will be approximately equal to the distance from the jaw joint to the posterior limit of the tooth row. The camel provides the principal exception to this prediction in that it possesses a longer tooth row than would be expected. However, as Greaves points out, the right and left tooth rows in this animal converge anteriorly and this will act to increase the force at the anterior teeth by reducing the resistance lever arm (i.e. the distance from the fulcrum – the balancing jaw joint – to the point of application of the force), thus extending the area over which an unreduced mechanical advantage applies.

Clearly the correctness or otherwise of the model described by Greaves depends upon a number of assumptions which are as yet unproven and which, even if eventually vindicated, must considerably oversimplify the actual situation existing in the living animal. Even so, the model is valuable in that it has generated a number of testable hypotheses about the form and function of the artiodactyl jaw.

Rodents and lagomorphs

The characteristic specialisation of the Rodentia is the ability to gnaw, a type of incisive movement capable of reducing hard fibrous substances in which the separated material is not always ingested. Gnawing is carried out by means of upper and lower pairs of large, continuously growing incisors. The enamel (of specialised microstructure) is limited to the anterior surfaces of these teeth and is kept honed to a chisel-like edge by the self-sharpening action of gnawing. So far as both number of species and number of individuals are concerned, rodents are amongst the most successful of modern mammals. At present, more than 1600 rodent species are recognised. They make up a most varied group but all possess the ability to gnaw.

Gnawing is not, however, restricted to rodents. Amongst living mammals a similar type of jaw activity occurs in the lagomorphs and in the rare and unusual primate *Daubentonia*. Also the extinct Multituberculata possessed rodent-like incisors which they probably used for gnawing; the Multituberculata, like the rodents, appear to have been a highly successful group, and survived from the Early Jurassic to the Late Eocene. Clearly, therefore, gnawing has proved a most potent adaptation.

In addition to the specialisation of the incisors, gnawing is associated with wide-ranging structural modifications in the remainder of the dentition, in the skull and in the jaw musculature. In the rodents, second incisor and canine teeth are absent and the central incisors are separated from the cheek teeth by long diastemata. There are generally three or four molars in each quadrant of the jaws (where the number exceeds three, the first cheek tooth is, in fact, a premolar) but some species have less. These teeth are grouped together, without gaps between them, to make up a grinding battery. The occlusal surfaces are often elongated and folded in a complex manner so that enamel, dentine and cementum are exposed, differential rates of wear of these tissues producing ridges and grooves. In a number of forms the molars are hypsodont (e.g. some squirrels) or ever-growing (e.g. some heteromyids). As a result of these various structural modifications the molar batteries provide an apparatus efficient for chewing coarse fibrous material. This aspect of jaw usage has, without doubt, also been of great adaptive significance, many rodents, on account of their very small body size (and, therefore, relatively large surface area), requiring to masticate several times their own body weight of foodstuff daily.

Both gnawing and chewing in rodents involve predominantly anteroposterior (propalinal) movements of the mandible. The characteristic structural modifications of the rodent jaws and jaw musculature are all related to this type of movement. The jaw joint is of a loose nature, without pre- or postglenoid processes, allowing a wide range of translation. The jaw musculature is much modified so that the bulk of its fibres have a major anterior or posterior component of pull. The nature of these modifications varies in its detail to provide the basis on which the Rodentia is divided, classically, into its three suborders (Fig. 55).

The modifications of the jaw musculature involve principally the masseter and zygomaticomandibularis. In all three suborders the fibres of the superficial part of the masseter have a predominantly anteroposterior orientation, arising at or close to the base of the anterior root of the zygomatic arch and being inserted along the ventral margin of the angular process of the

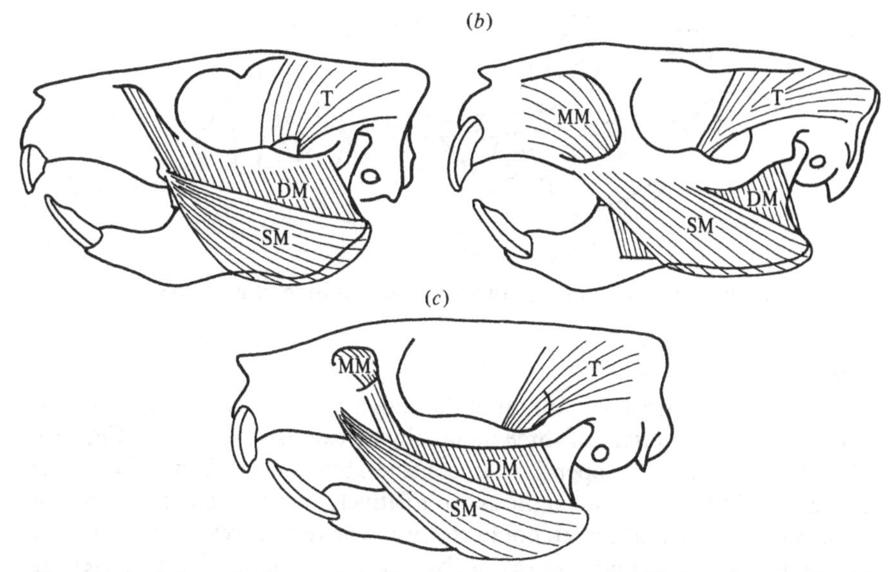

Fig. 55. Left lateral views of skulls of (*a*) sciuromorph; (*b*) hystricomorph; (*c*) myomorph. Abbreviations: SM = superficial masseter; DM = deep masseter; MM = maxillomandibular; T = temporalis (see text for description).

mandible. In the Sciuromorpha the origin of the deep masseter has been extended forwards, medial to that of the superficial part, on to the anterior face of the zygomatic arch and, from here, dorsal to the infraorbital foramen, to reach the bones of the snout. In the Hystricomorpha, on the other hand, it is the zygomaticomandibularis which has been involved in an anterior migration, the origin of its anterior part having spread from the medial surface of the zygomatic arch forwards through the enlarged infraorbital foramen on to the snout (the resulting detached portion of the zygomaticomandibularis being usually termed the maxillomandibular). The result in both cases has been to increase the anterior component of pull of the muscles. In the Myomorpha, the most successful of modern rodents, both the deep masseter and the zygomaticomandibularis have undergone corresponding modifications. The temporalis of modern rodents is small and its area of origin is restricted to the cranial wall posterior to the coronoid process. Wood (1965) has traced the probable evolutionary pathways by which these muscular modifications and the associated changes in the dentition were achieved and comes to the conclusion that once these pathways were channelled into a single major direction by the development of the gnawing habit there was much subsequent convergent and parallel evolution within the rodent order.

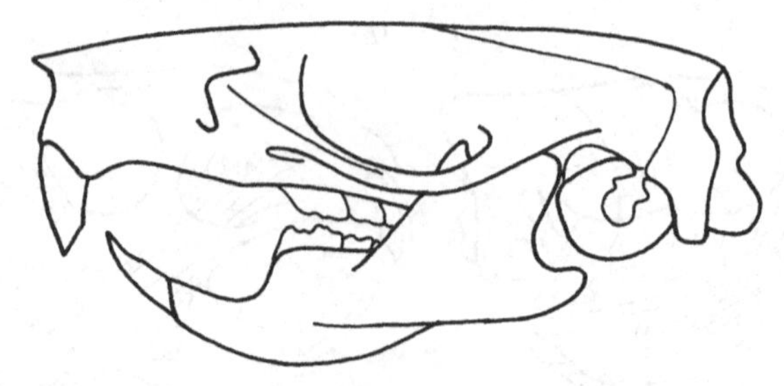

Fig. 56. Left lateral view of rat skull to show rest position of mandible.

Rat

The rat possesses in each quadrant of the jaws a continuously growing incisor separated from a battery of three molars by an extensive diastema (Fig. 56). The diastema in the upper jaw is considerably longer than that in the mandible. The rat is isognathous, a relatively uncommon condition in mammals generally, although not in rodents. The two halves of the mandible are joined in the symphysial region by a pad of fibrocartilage as well as ligaments which together allow a considerable range of movement. The jaw joint is elongated with its long axis lying in a sagittal plane. The squamosal articular surface is concave in transverse section but flat along its antero-posterior axis, although inclined somewhat anteroventrally. The condyle is about half as long and two-thirds as wide as the squamosal articular surface. The articular disc is thin centrally but thickened anteriorly and posteriorly. The capsule is slack and allows the condyle to slide to and fro over a distance of about 6 mm as well as to rotate a little about a vertical axis.

The jaw musculature of the rat displays the highly specialised structure typical of myomorphines. This departs from the characteristic mammalian arrangement in so many functionally important features that it will be helpful to include a rather fuller anatomical description than has been given of the jaw muscles of other mammals. The following account is based on Turnbull (1970), Hiiemae (1971*a*, *b*), Hiiemae & Houston (1971) and Weijs (1973) (Fig. 57).

The masseter muscle, as revealed by its relative weight (Table 2), is by far the most powerful of the jaw-closing muscles. In the overwhelming pre-eminence of one such muscle the rodents resemble the carnivores, although in the latter it is the temporalis which is predominant. The superficial part of the masseter arises by a tendon from the lateral surface of the maxilla and runs backwards and somewhat downwards to be inserted into the ventral border of the mandible as far posteriorly as the tip of the angular process. The deep masseter consists of more vertically orientated fibres arising from the zygomatic arch and being inserted into the lateral surface of the mandible over a wide area. The small maxillomandibular muscle arises from the medial wall of the infraorbital foramen, through which it passes posteriorly before

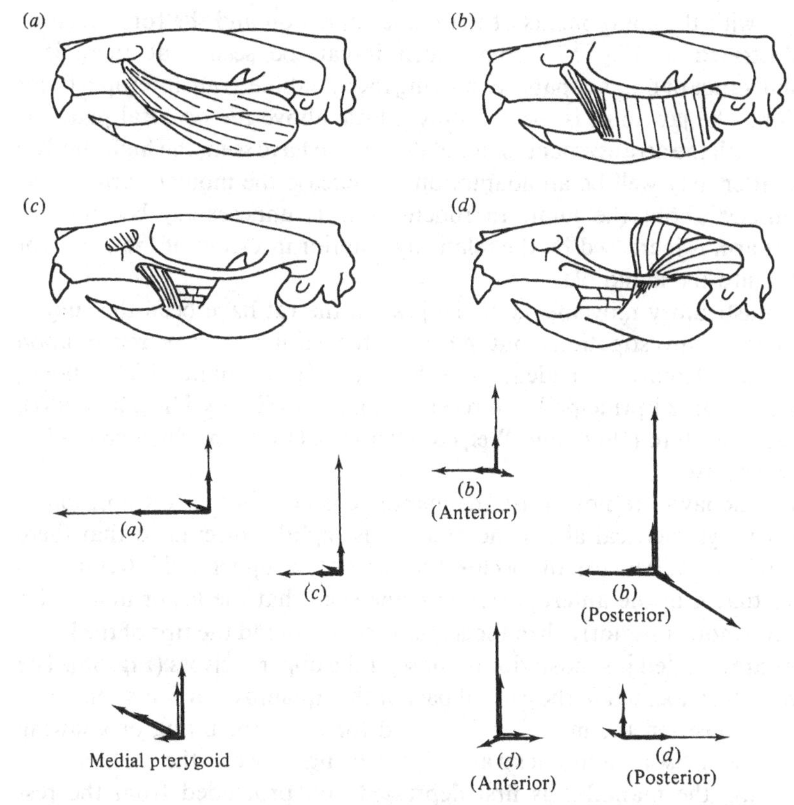

Fig. 57. Principal muscles of mastication in the rat. (*a*) Superficial masseter; (*b*) deep masseter; (*c*) maxillomandibular; (*d*) anterior and posterior temporalis. Below are given the components (thin lines; indicative of amplitude of movement) and force vectors (thick lines) of the muscles. (After Hiiemae, 1971*b*.)

descending vertically to be inserted aponeurotically into the mandible along the oblique line below the first two molar teeth.

The remainder of the zygomaticomandibularis muscle originates from the zygomatic arch and is inserted into the mandible with the maxillomandibular and also by a muscular insertion just below the sigmoid notch. The fibres of the anterior part of the temporalis run anteroventrally from the temporal ridge and lateral side of the cranium below the ridge to the coronoid process. The fibres of the posterior part of the temporalis arise from the posterior part of the temporal region and run anteriorly to the coronoid process. The medial pterygoid consists of fibres running posteroinferiorly as well as laterally from their origin on the pterygoid fossa to their insertion into the medial side of the mandibular ramus. The lateral pterygoid arises from the lateral surface of the lateral pterygoid plate and passes posterolaterally to be inserted into the condyle and articular disc. The principal lines of action of the muscles, together with the components of these lines of action and the force vectors, are illustrated in Fig. 57, from which it can be seen that there is a preponderance of fibres capable of moving the mandible in an anteroposterior direction. The jaw joint is located only a little above the occlusal plane by contrast with the arrangement in ungulates. As we have seen, the high condyle of the latter may well be an adaptation to increase the moment arm of the jaw muscles about the joint. In rodents this is unnecessary because the moment arm is increased by the relatively anterior insertion of the adductor muscles into the mandible.

The masticatory movements of the jaws in the rat have been the subject of numerous investigations but most of the earlier studies relied upon anatomical inference or inadequate technology. The account which follows, therefore, is based principally on recent detailed studies by Hiiemae (1967), Hiiemae & Ardran (1968) and Weijs & Dantuma (1975) in which cineradiography was used.

When the jaws are not in use the mandible is held in the rest position in which it is symmetrical about the midline, is slightly lowered so that there is a freeway space between the occlusal surfaces of the upper and lower molars and is situated in the anteroposterior plane such that the lower first molar lies a little more anteriorly then the upper first molar and the tips of the lower incisors are situated just posterior to those of the upper incisors (Fig. 56). The condyle is then located in the central part of the squamosal articular channel. From this position the mandible is moved forwards for biting or gnawing between the incisors, and backwards for chewing between the molars.

In biting, the mandible is first depressed and protruded from the rest position. The upper incisors are then pressed into the food and the lower incisors moved upwards until the tips of the two sets of teeth approximate. At the end of the bite the lower incisors move downwards and backwards, from which position the cycle can recommence. In gnawing, the lower incisors are moved forwards and upwards several times instead of in a single stroke. During incisive activity the condyle is in articulation with the anterior two-thirds of the squamosal articular surface.

Chewing is an entirely separate event from incision. It proceeds in a fairly constant manner, the principal variation being in the duration of the cycle with the physical consistency of the food. Chewing in the isognathous rat appears to be essentially bilateral. Food is collected by the tongue and cheeks and transferred to the region of the first molars. With completion of the last incisive cycle the mandible is transposed into the chewing position by a movement backwards through the rest position. The chewing cycle consists of three strokes: (1) closing, in which the mandible moves upwards and backwards; (2) power, in which the mandible translates forwards, grinding the food between the occluding surfaces of the upper and lower molars; and (3) opening, in which the mandible is depressed and then retracted. During these strokes the condyle is in articulation with the posterior two-thirds of the glenoid fossa. The total duration of the chewing cycle is about 190 ms of which some 28 per cent is occupied by the power stroke.

In essence, the cyclical chewing activity in the rat has the same sequence of strokes as that observed in other mammals, although the directions of movement in each stroke are predominantly propalinal rather than transverse as is the case in most non-rodent mammals that use their molars for grinding. Chewing in the rat does involve, however, slight transverse movements, the lower molars of both sides moving medially during the power stroke and laterally during retraction of the mandible. The range of movement in both cases is of the order of 0.4 mm at the third molar on each side. These transverse bilateral movements doubtless increase the efficiency of grinding and are made possible by the laxity of the symphysial joint which thus serves a rather different function from that in other mammalian groups, where its chief action appears to be to allow the exertion of maximum vertical forces at the occluding teeth of the working side. Chewing in the rat is not, therefore, exclusively propalinal nor are propalinal movements restricted to the chewing cycle, for, in addition to the relatively short movements of this type which occur during the cycle, there are the much longer propalinal translations by which the mandible is switched from the incising to the chewing position and vice versa.

In order to analyse the jaw muscle activity accompanying these movements, Weijs & Dantuma (1975) used a combination of cineradiography and electromyography. They quantified the degree of electromyographic activity by (1) dividing the chewing cycle into 16 phases, each of about 10 ms duration – phases 1 to 6 comprising closing, 7 to 11 the power stroke and 12 to 16 opening; (2) expressing the pulse numbers in each of the phases of the cycle as percentages of the total pulse numbers for the complete cycle to give a relative activity spectrum; and (3) expressing phase pulse numbers as percentages of the maximum pulse number observed in a sample of cycles to give the contraction percentage (the maxima corresponding to maximum muscular contraction or to a fixed percentage of maximum contraction).

As would be expected in view of the observation that the masticatory movements of the lower jaw are symmetrical about the median plane, electromyographic activity was found to be generally similar in its timing in the muscles of the two sides. In chewing, activity (as determined from the relative activity spectra) usually begins in the anterior deep part of the temporalis in phase 1, or slightly before in phases 15 or 16 of the preceding cycle, and reaches a maximum in phases 5 and 6. Activity begins in the remainder of the temporalis about 15 ms later and reaches a peak for the superficial anterior portion in phases 5 and 6 and for the posterior part in phases 7 and 8. The whole muscle is inactive during phases 11 to 15. The masseter becomes active about 20 ms after the anterior temporalis. In the superficial masseter, firing reaches a maximum in phases 6 to 9 while the deep masseter increases its level of activity a little in advance of the superficial portion of the muscle. Activity in all parts of the masseter falls rapidly in phases 9 and 10 and remains low during the remainder of the cycle. The activity pattern of the pterygoids is similar to that of the superficial masseter except that firing is sustained through to about phase 12, especially in the lateral pterygoid. All the hyoid muscles (both bellies of digastric, mylohyoid, geniohyoid and transverse mandibular) are quiescent during phases 2 to 4, fire at varying amplitudes in phases 5 to 9 and then fire strongly in phases 10 to 16.

In incising, the hyoid muscles show high activity in the opening phase, and the anterior translation of the condyles is accompanied by strong firing of the pterygoid muscles. During the incisive bite proper the suprahyoid muscles show phasic activity alternating with activity in the adductor musculature. The lateral pterygoid remains active during both suprahyoid and adductor periods. In the final phase the suprahyoid and lateral pterygoid become quiescent and the adductor musculature shows only incidental activity. The lower jaw returns to the closed position, presumably as a result of elastic recoil of the soft tissue.

In order to analyse further the muscular forces developed during mastication, Weijs & Dantuma (1975) estimated the relative magnitude of the force developed by each muscle by multiplying the physiological cross-section of the muscle by the appropriate contraction percentage. The working line of each muscle was next determined by taking the points of origin and insertion of the theoretical central fibre representing the average of all the fibres in the muscle. It was then possible to calculate the moments of the muscular forces so defined about the condylar contact area in the sagittal, horizontal and transverse planes by multiplying the force by the appropriate moment arm. Finally, the resultant of the moments of all the individual muscles was calculated for each phase. The whole procedure was repeated for eight jaw positions during mastication. (*a*) Chewing: start closing (phase 1), halfway closing (2, 3), start grinding (4, 5, 6), halfway grinding (7, 8, 9, 10), end grinding (11, 12, 13), halfway opening (14, 15), end opening (16). (*b*) Incision: bite position (incisors just contacting food). By means of this procedure Weijs and Dantuma were able to achieve a more dynamic description of mastication than can be obtained from the more usual practice of considering the muscle forces acting in just one jaw position.

During closing (phases 1 to 6) the resultant in the sagittal plane was found to be directed posterosuperiorly in phase 1, superiorly in phase 2 and anterosuperiorly in the remaining phases. It continues to be directed anterosuperiorly in each of the phases (7 to 11) of the power stroke and in the first phase (12) of opening. In phase 13 it is directed superiorly and in phases 14 to 16 posterosuperiorly. Allowing for the fact that the anterior slope of the squamosal articular surface is such that vertically directed forces will produce a posterior translation of the condyle, there is clearly good correlation between the direction of muscle force and direction of movement.

In the horizontal plane, the resultant is directed posteromedially in phase

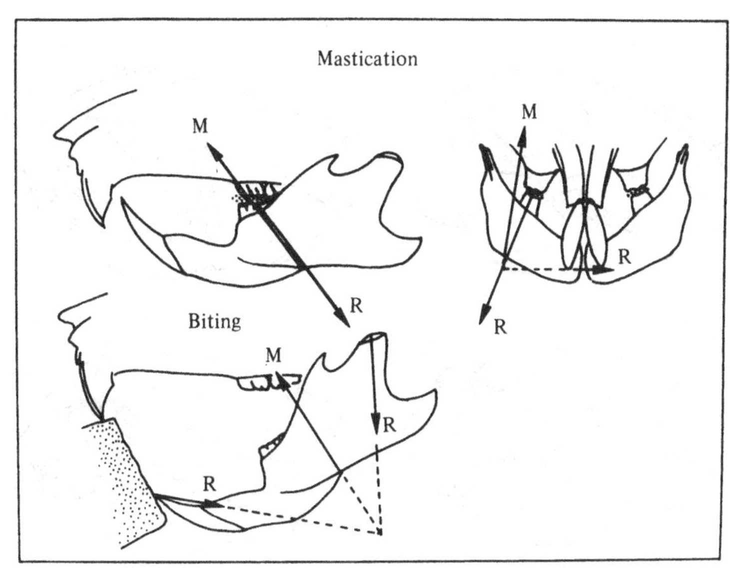

Fig. 58. Static equilibrium condition during molar and incisor food contact in the rat. During mastication the temporomandibular joint remains unloaded. In frontal view, molar reaction is perpendicular to the occlusal surface and the effect of the symphysial cruciate ligament is to balance the muscle resultant and molar reaction. During incisive action the joint becomes loaded and the biting force makes a small angle with the surface of the food pellet. M = muscle resultant; R = reaction. (Reproduced with permission from Weijs & Dantuma, 1975.)

1, anteromedially in phases 2 to 11, and medially in phases 12 to 16. The medial component is, in fact, very small in the closing phases, but, in the opening and power phases, it is sufficient to pull the condyles medially and so cause spreading of the incisors and the slight medial movement of the molar region already noted. In the transverse plane, the resultant is directed superomedially and so tends to turn the lower border of the mandible inward, a tendency resisted by the meeting of the two halves of the mandible at the symphysis.

The force exerted by the adductor musculature increases throughout phases 1 to 6 to reach its peak in phases 7 and 8. It then decreases rapidly in phases 9 to 11. During the power phases the mandible is subjected to four groups of forces: (1) the resultant of the muscular forces, (2) the reaction force in the molars, (3) the reaction force in the symphysis, and (4) the reaction force in the jaw joint. In sagittal projection the muscular resultant passes through the middle of the lower molar row which is, presumably, also the point of application of the reaction force in the teeth (Fig. 58). The jaw joint remains, therefore, unloaded. In transverse projection the reaction force at the molars is perpendicular to their occlusal surfaces, and the symphysial cruciate ligament provides a small balancing inward pull. Hence, the reaction force at both the jaw joint and symphysis is relatively small during chewing.

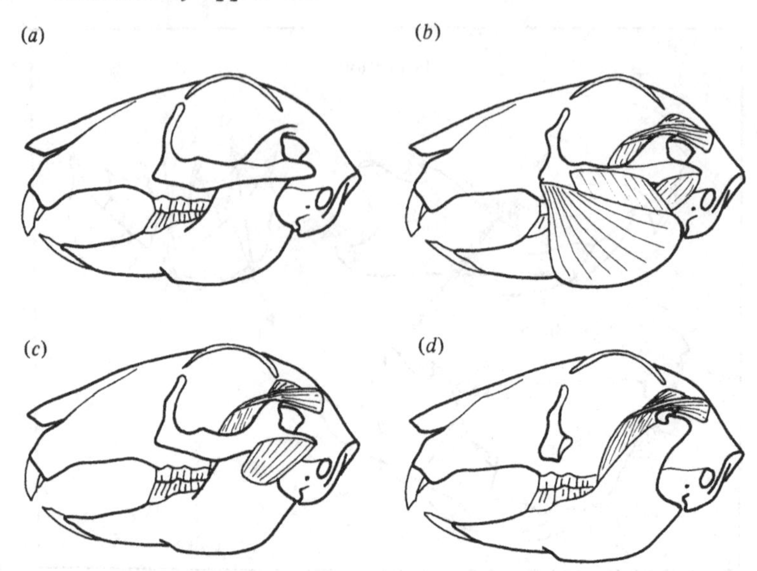

Fig. 59. To show masseter and temporalis muscle in the rabbit. (*a*) Left lateral view of the skull; (*b*) superficial masseter and temporalis; (*c*) deep part of masseter; (*d*) temporalis muscle (zygomatic arch removed to show part of muscle arising from posterior orbital wall).

During incising the mandible is depressed and protruded by the lateral pterygoid and suprahyoid musculature. In this position the resultant of the muscular forces is directed rather more vertically than during chewing because of the relatively more vertical orientation of the working lines. The horizontal component of the muscle resultant is resisted by the incisors and its vertical component by the articular surfaces of the jaw joint (Fig. 58).

Rabbit

Despite an overall similarity between the craniomandibular apparatus of the rabbit and that of rodents, several differences are present in which the lagomorph comes closer to the ungulate condition. First, the maxillary dentition of the rabbit includes a small lateral incisor situated just behind the large, continuously growing central incisor. Secondly, the masticatory muscles lack the characteristic rodent specialisations, although they are still clearly adapted to provide forces acting in a predominantly anteroposterior direction (Fig. 59). Thirdly, the jaw joint lies well above the occlusal plane, so increasing the moment arm of the jaw muscles about the joint and thus offsetting the poorer mechanical advantage that they enjoy, as compared with the corresponding rodent muscles, because of their more proximal attachments to the mandible. Fourthly, the incisors are separated from the cheek teeth by long diastemata which are of approximately equal length in the upper and lower jaws. Nonetheless, the possession of the ability to gnaw makes it more instructive to deal with the lagomorphs in conjunction with the rodents rather than with the ungulates.

The masticatory movements in the rabbit have been investigated by Ardran *et al.* (1958), using cinephotography and cineradiography, an investigation which has largely superseded earlier studies based on anatomical observation alone. The rest position in the rabbit is similar, in essence, to that seen in the rat but, because of the equal length of the upper and lower diastemata in the rabbit, the tips of the lower incisors are held closer to those of the upper incisors than occurs in the rodent. During biting the mandible is lowered and then protruded from the rest position so that on closing the upper and lower incisors meet edge to edge. This movement is usually sufficient to sever soft food but with tougher material the lower incisors are raised further so that they slide upwards against the posterior edge of the upper incisors.

Earlier descriptions of chewing in the rabbit varied in the direction of movement attributed to the lower teeth during the power stroke but it is now clear from the study of Ardran *et al.* (1958) that the principal movement during this phase is in a lateral to medial direction. The rabbit is anisognathous and chewing is, in consequence, unilateral. At the beginning of closing the mandible is symmetrically situated about the median plane. As closing takes place the jaw moves outwards towards the working side until, as the cheek teeth come into occlusion, the buccal edges of the lower teeth lie immediately below those of their upper opponents. During the power stroke the lower molars move medially across the uppers, thus grinding and crushing the food. Once the mandible has reached its median position the opening phase begins, the jaw being depressed to the open position without lateral divergence. As in rodents, the chewing cycle takes place very rapidly; in chewing grass, for example, the rate is in excess of 300 cycles per minute.

Unfortunately, there appear to be no published electromyographic analyses of masticatory muscle function in the rabbit.

The common feature of jaw usage in rodents and lagomorphs is, of course, the importance of propalinal movements in one or more phases of mastication. However, as already emphasised, it is as yet impossible to generalise about chewing movements in rodents. At least some members of the Hystricomorpha and Sciuromorpha resemble lagomorphs in being anisognathous and chewing unilaterally, probably by ectental rather than propalinal movements. Weijs & Dantuma (1975) suggest that the bilateral propalinal mode of chewing seen in the rat may well represent the most efficient usage for chewing of a musculature already adapted to a protracting function as part of the gnawing mechanism. If true, this would indicate that the craniomandibular apparatus of the rat, and possibly of other myomorphine rodents, is in evolutionary terms more advanced than that of rodents chewing by unilateral ectental movements.

As in ungulates, the incisors and cheek teeth are separated by a wide diastema in both rodents and lagomorphs. The functional significance of this

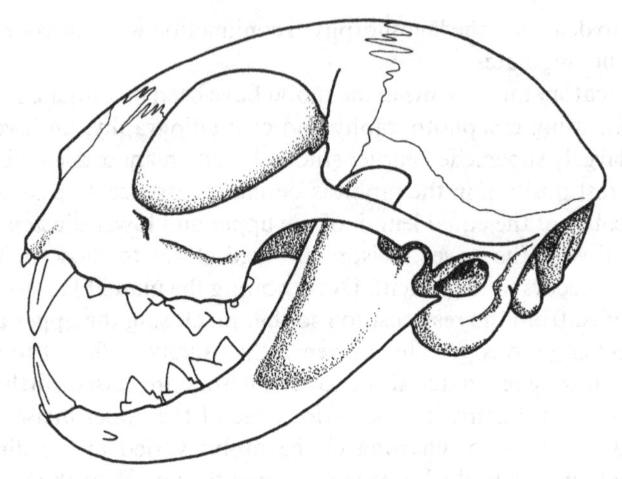

Fig. 60. Left lateral view of cat skull.

gap in the dentition in ungulates has already been discussed but whether or not the mechanical considerations, particularly the fulcrum nature of the jaw joint, assumed to apply in that group also apply in the gnawing mammals is doubtful. In the latter, diastemata may be more especially advantageous in that, by facilitating the functional separation of incisal activity and chewing, they allow gnawing to proceed without the reduced material being necessarily ingested. A marked discrepancy in the size of the maxillary and mandibular diastemata, such as occurs in the rat, enhances this functional separation.

Carnivores

Although the jaws and dentition of the typical carnivore are clearly adapted for the killing and eating of prey, by no means all members of the group are exclusively carnivorous. Amongst terrestrial mammals the Felidae (permanent dental formula $I\frac{3}{3}C\frac{1}{1}P\frac{3}{2}M\frac{1}{1} \times 2$) are the most highly specialised for the predatory way of life. As in fissipeds generally, the canines are large and dagger-like and the most posterior upper premolar and the lower molar (the carnassial teeth) are blade-like for the slicing of flesh and bone (Fig. 60). In the Felidae the premolars immediately anterior to the carnassial teeth are also blade-like in nature, adding further slicing elements to the dentition. The upper molar lies transversely and is functionless. The corresponding tooth in the Hyaenidae ($I\frac{3}{3}C\frac{1}{1}P\frac{4}{3}M\frac{1}{1} \times 2$) also lies transversely but is rendered functional by occluding with a small talonid element of the lower molar. There is also an articulation between a talonid element on the third lower premolar and the lingual cusp of the upper carnassial. These occlusal relationships provide a 'stop' which serves to protect the tooth supporting structures from the very heavy forces developed between the carnassial teeth when slicing bone.

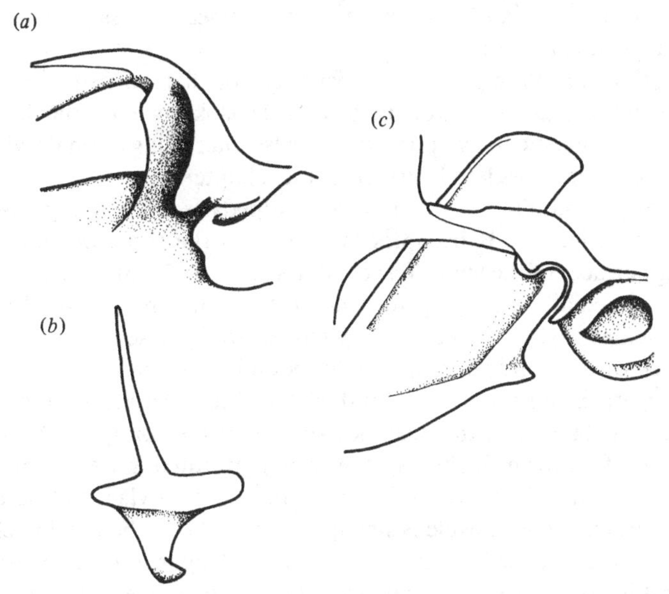

Fig. 61. Close-up views of left jaw joint of cat skull. (*a*) Ventral view of squamosal articular surface; (*b*) posterior and slightly dorsal view of mandibular condyle; (*c*) lateral view. Not to scale.

The dentition of the Canidae ($I\frac{3}{3}C\frac{1}{1}P\frac{4}{4}M\frac{2}{3} \times 2$) is rather less specialised than that of the Aeluroidea but is still clearly adapted for a predominantly, if not exclusively, meat diet. The canines are prominent and the carnassials are powerfully developed. The lower carnassial bears a talonid which articulates with the upper first molar. The third lower molar is a peg-like tooth with no opponent in the upper jaw. The mustelid carnivores (permanent dental formula typically $I\frac{3}{3}C\frac{1}{1}P\frac{4}{4}M\frac{1}{2} \times 2$) are a rather varied group so far as diet is concerned. The weasels are predominantly carnivorous while the badgers consume a wide-ranging diet including vegetable matter. Both have typically carnivorous dental features. The otters are piscivorous, possessing rather slender curved canine and cheek teeth for the retention of their slippery prey.

The Ursidae ($I\frac{3}{3}C\frac{1}{1}P\frac{4}{4}M\frac{2}{3} \times 2$) and Procyonidae ($I\frac{3}{3}C\frac{1}{1}P\frac{4}{4}M\frac{2}{2} \times 2$) provide the principal exceptions to the predominantly meat-eating nature of the carnivores. Most bears are to some degree omnivorous while at least one member of this group, *Tremarctos ornatus*, is largely herbivorous. Members of the Procyonidae also tend to be omnivorous apart from the pandas, which are highly specialised, exclusively herbivorous feeders. The nature of the jaw mechanisms in the bears and pandas is of particular interest in that they represent secondary readaptations of mechanisms that were presumably at one time adapted to more typical carnivore feeding habits. In both of these groups the molars have tended to become broadened with numerous cusps, the carnassial teeth have lost their blade-like form and the canines are less prominent than in more typical carnivores.

The teeth of the Pinnipedia are specialised for diets of fish (the seals) or shellfish (the walruses). Unfortunately, little is known of the functional anatomy of the masticatory apparatus in these interesting animals which will not, therefore, be considered further in this chapter.

The temporomandibular joint of all fissipeds is built along essentially similar lines (Fig. 61). It occupies more or less the same plane as do the occluding surfaces of the teeth. The condyles are cylindrical in form with their long (lateromedial) axes lying in a common transverse plane. The squamosal articular surface is reciprocally curved to fit closely the corresponding surface of the condyle. It is usually bounded anteriorly by a raised lip and posteriorly by a prominent curved postglenoid tubercle, the anterior face of which articulates with the posterior surface of the condyle. The articular disc is thin, of uniform thickness (it may be perforated in its central region) and conforms closely to the curvatures of the condylar and squamosal articular surfaces. The capsule is strong and tense both above and below the disc and is strengthened by medial and lateral ligaments. Because of its structure it is usually assumed that virtually the only movement that can take place at the joint is a simple vertical hinge (orthal) action about a transverse line passing through the long axes of the condyles, but observations of a slight lateral shift during the chewing cycle, as well as study of the wear facets on the teeth, indicate that a small degree of movement other than rotation must be possible. The nature of this movement is discussed further below.

The jaw symphysis remains unfused in the majority of carnivores but in some of the larger members of the order, including the big cats and bears, the two halves of the mandible are united at the symphysis by dense fibrous tissue and fibrocartilage or even by bony union so that the lower jaw becomes effectively a single, rigid, functional unit (Scapino, 1976).

The temporalis is much the heaviest of the jaw-closing muscles, in most carnivores averaging between 60 and 70 per cent of the total weight of this muscle group (Table 2). It is a bipennate muscle. The fibres of the deep head arise from the bony floor of the temporal fossa and are inserted into the medial aspect of the central tendon, while the fibres of the superficial head arise from the temporal fascia and are inserted into the lateral aspect of the central tendon. The central tendon is itself attached to the mandibular coronoid process. The masseter is usually the next largest muscle of the jaw-closing group. It too is divided into superficial and deep parts. The superficial portion takes origin from the anterior part of the zygomatic arch and its fibres run posteroinferiorly to be inserted in the region of the angular process, while the deep portion arises from the whole length of the zygomatic arch and its fibres run more vertically to be inserted into the lateral surface of the mandible at and near to the masseteric crest. The zygomaticomandibularis takes origin from the inner surface of the zygomatic arch. Its fibres run medially and

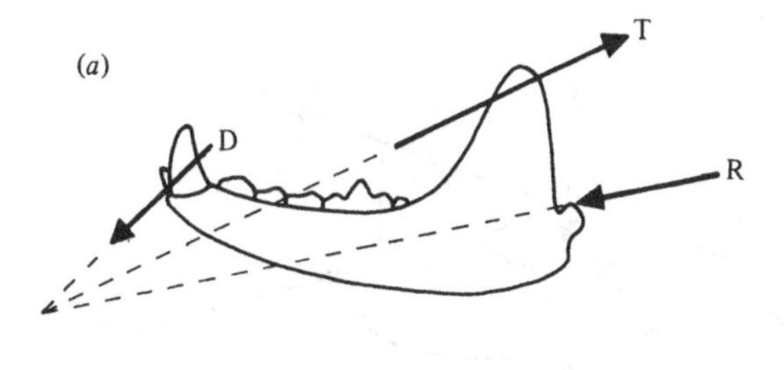

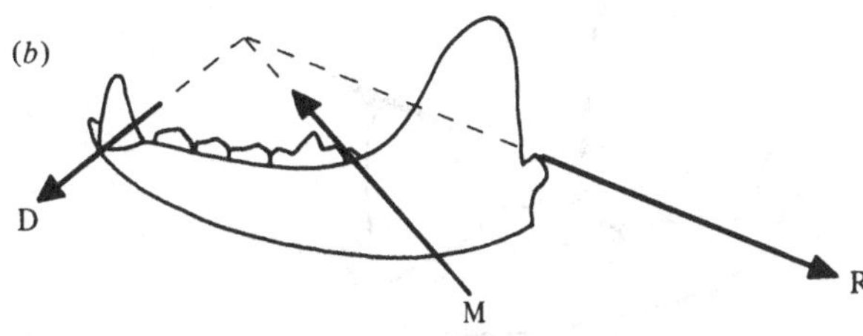

Fig. 62. Forces acting on lower jaw of *Martes* when holding struggling prey between canines. The triangles of forces are shown when the displacing force (D) at the canines is resisted: (*a*) by temporalis (T) alone; (*b*) by masseter (M) alone. R = reaction force at jaw joint. (After Smith & Savage, 1959.)

somewhat inferiorly to be inserted into the lateral surface of the coronoid process and adjacent regions of the mandibular ramus. In bears and pandas it is enlarged at the expense of the masseter muscle (Table 2) and its fibres have an almost horizontal course. The medial pterygoid is a rather small muscle but with a characteristic mammalian disposition. The virtual absence of anteroposterior translatory movements of the condyle in the carnivore jaw joint would appear to deprive the lateral pterygoid of any major function. This expectation appears to be borne out by the very small size of the muscle (Table 2). However, as discussed below, the muscle may not be completely without functional significance.

The arrangement of the jaw musculature of carnivores is related to the need to protect the jaw joint as well as to the need to produce large biting forces. Smith & Savage (1959) have provided an analysis of the force acting on the jaws during the capture of prey, and the implications regarding jaw muscle function, in the mustelid carnivore *Martes*. This force will act mainly in the region of the canine teeth and will have a forward component due to the struggles of the prey and a downward component due to the action of the muscles in closing the jaws. The resultant, therefore, will be in a downward and forward direction. In Fig. 62*a* is shown the triangle of forces for the jaw when it is assumed that the displacing force is resisted by the temporalis muscle alone. As can be seen, the balancing force at the jaw joint is a relatively small compression. In Fig. 62*b* the triangle of forces is given when it is assumed

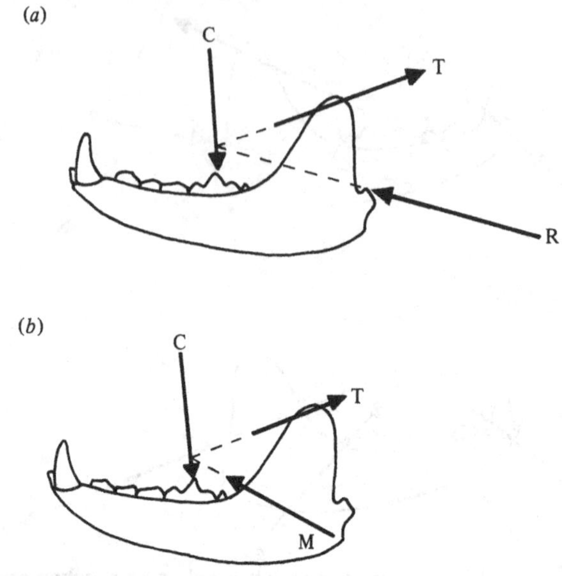

Fig. 63. Forces acting on lower jaw of *Martes* when biting on cheek teeth. The triangles of forces are shown when (*a*) temporalis (T) is alone used; (*b*) both temporalis and masseter (M) are used. C = bite force, R = reaction force at jaw joint. (After Smith & Savage, 1959.)

that only the masseter muscle is acting (the medial pterygoid can be included with the masseter since it acts in virtually the same plane). The balancing force at the jaw joint is now much greater and is in a potentially disarticulating direction. As might be expected, therefore, the temporalis is especially massive in carnivores. Moreover, its power is further enhanced by the possession of a bipennate structure and by prolongation of the coronoid process above the plane of the jaw joint, thus giving the muscle a long moment arm.

When the jaws are being used to slice food the force acting will be located principally at the carnassial teeth and will be directed, so far as the lower jaw is concerned, in a downward direction. In Fig. 63*a* the triangle of forces is shown when only the temporalis muscle is acting. The resultant force at the jaw joint is large and potentially disarticulating. However, if both the masseter (plus medial pterygoid) and temporalis are acting to elevate the mandible, there is little or no force at the joint (Fig. 63*b*). A principal function of the masseter is thus to protect the jaw joint when the cheek teeth are being used. For this purpose, the closer its line of action passes to the jaw joint, the better its effect will be, so accounting for the relatively short moment arm of this muscle.

An essentially similar conclusion is reached by Turnbull (1970) but on the basis of the earlier suggestion by Davis (1955) that the carnivore jaw cannot

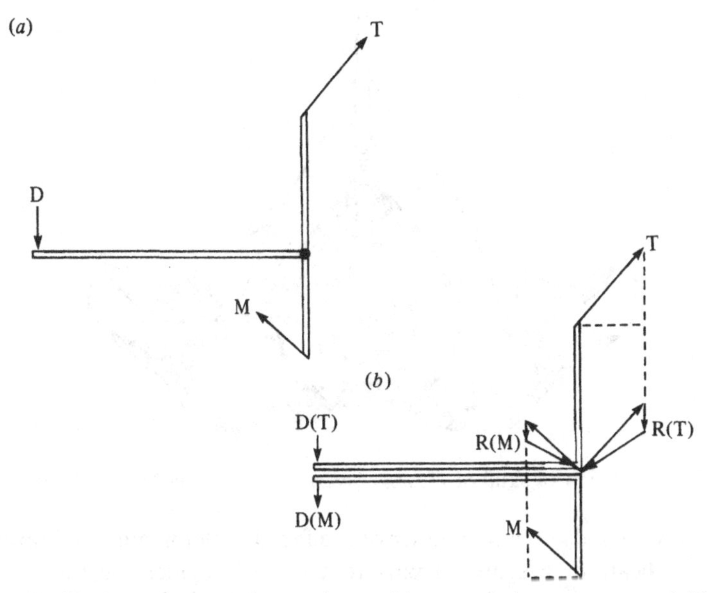

Fig. 64. To illustrate the lower jaw acting as (*a*) a couple-lever system and (*b*) as two bent levers (see text for description). Forces indicated as follows: M = masseter + medial pterygoid; T = temporalis; D = resistance force at dentition; R = resistance force at jaw joint. In (*b*) (T) and (M) indicate the proportion of the resistance forces at the dentition or joint resulting from the action of the masseter/pterygoid and temporalis, respectively. (After Turnbull, 1970.)

be regarded as a class III lever. Because the coronoid process projects well above the level of the jaw joint (relative to the occlusal plane) the effort force developed by the temporalis muscle can be considered as acting on the far side of the fulcrum from the resistance force at the teeth, the system functioning, therefore, as a bent class I lever (Fig. 64). The same argument applies to the masseter and medial pterygoid muscles on account of the well-developed nature of the angular process. The temporalis and masseter–medial pterygoid blocks might be considered, therefore, to produce an approximation to a couple (but see objections to viewing jaw muscles as producing couples discussed on pp. 160–161). Any couple action so produced would, as can be seen from Fig. 64*b*, result in a reduced reaction force at the joint, much as indicated by the Smith & Savage (1959) analysis. The interpretation of the mandible as a class I rather than as a class III lever could well apply to other mammalian groups where the coronoid and angular processes are well developed and orientated relative to the jaw joint and biting teeth such that the effort force is acting on the remote side of the fulcrum from the resistance force.

The presence of a large and powerful adductor jaw musculature is reflected in profound structural modifications of the carnivore skull which involve not

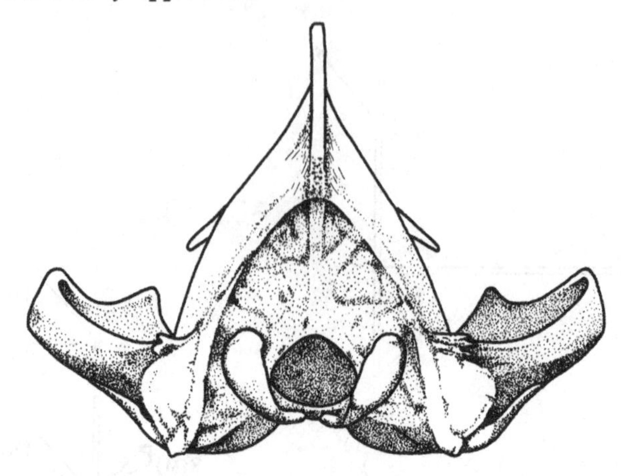

Fig. 65. Posterior view of skull of *Hyaena hyaena* to show sagittal and nuchal crests.

only the jaws but also the remainder of the facial skeleton and the braincase. These modifications are due in part to the need to increase the areas of attachment for the jaw and neck musculature and in part to strengthen the skull so that the very large stresses developed during jaw usage can be dissipated without endangering its structure.

In general, the enlargement of the attachment areas is achieved in such a way that other mechanical attributes of the muscles, such as their lines of action and their moment arms, are preserved or rendered more efficient for the particular functions of each muscle. It has already been noted how this has been achieved for the insertions of the temporalis, masseter and medial pterygoid by the development of prominent coronoid and angular processes. A similar, mechanically efficient arrangement is seen in the origins of these muscles. Thus, the cranial attachments of the masseter and medial pterygoid are accommodated by the presence of a massive zygomatic arch and lateral pterygoid plate while the area of origin of the temporalis muscle may be increased by the development of prominent sagittal and nuchal crests (Fig. 65). In some groups (e.g. *Hyaena* and *Crocuta*) the frontal paranasal air sinuses extend from the postorbital region beneath the sagittal crest as far posteriorly as the supraoccipital bone (Buckland-Wright, 1969; see also Chapter 7), acting, presumably, to increase the surface area of the cranium and also to offset the weight of the cranial superstructures. As well as providing extra area for the origin of the temporalis and nuchal muscles, the crests contribute to the strength of the skull and help resist the enormous deforming forces generated during jaw usage. Cranial crests are present also in some anthropoid primates where their manner of development and growth have been more intensively studied than in carnivores (see above) but it seems likely that similar morphogenetic influences underlie their appearance in both groups.

Unfortunately, detailed cinephotographic and cineradiographic analyses of jaw movements in carnivores have not yet been reported (although a few are in progress apparently – see, for example, de Vree & Gans, 1975; Hiiemae, 1978). Simple observation of feeding carnivores indicates that, generally, powerful use of the carnassial teeth takes place unilaterally, usually with frequent changes of side. The fact that the carnivore jaws are anisognathous must mean that in order to make maximum use of the carnassials some lateral shift of the mandible towards the side in use takes place, otherwise the lower teeth would pass too far medially from their opponents to produce an efficient slicing action. Examination of the wear facets on the teeth indicates that such a lateral shift occurs before the teeth come into intercuspal range.

Brodie (1934) suggested that this lateral shift takes place by the mandible being moved bodily to the working side, the condyles sliding along the line of their long (i.e. transverse) axes. The zygomaticomandibularis and lateral pterygoid muscles have the majority of their fibres passing horizontally in the frontal plane and could act, therefore, to produce this type of movement – movement of the mandible to, say, the left side being brought about by contraction of the left zygomaticomandicularis and right lateral pterygoid. Some of the deeper fibres of the deep masseter have a similar orientation to those of zygomaticomandibularis and may also be involved in producing lateral shift. This type of transverse mandibular movement is quite distinct from the rotation about a vertical axis through or close to one or other condyle seen in the more extensive transverse mandibular movements of the generalised and ungulate groups.

Sicher (1944) described the lateral shift as taking place during jaw opening, but from Hiiemae's (1978) preliminary report of her cine study of the domestic cat it appears, for this species at least, that the opening stroke is undeviatingly vertical and that it is during the closing stroke that the mandible makes its slight lateral deviation to the working side. During the subsequent power stroke the mandible moves vertically and slightly medially back towards the midline. As might be expected from the scissor-like action of the carnassial teeth, the amount of vertical movement during the power stroke is considerably greater in the cat than in mammals with a grinding type of chewing action. The total cycle time is of the order of 300 ms of which the power stroke occupies about 23 per cent. The relative duration of the power stroke is thus of aproximately the same order as that in *Didelphis*, *Tupaia* and *Galago*, each of which lacks a phase II in this part of the chewing cycle, as does the cat to judge from its slicing type of action.

Of the atypical omnivorous and herbivorous members of the Carnivora, the bears and pandas have received the greatest attention so far as the functional anatomy of the craniomandibular apparatus is concerned (e.g. Sicher, 1944; Davis, 1955).

In spite of their feeding habits the bears and pandas possess a dentition, jaws and jaw musculature which structurally are of an unmistakably carni-

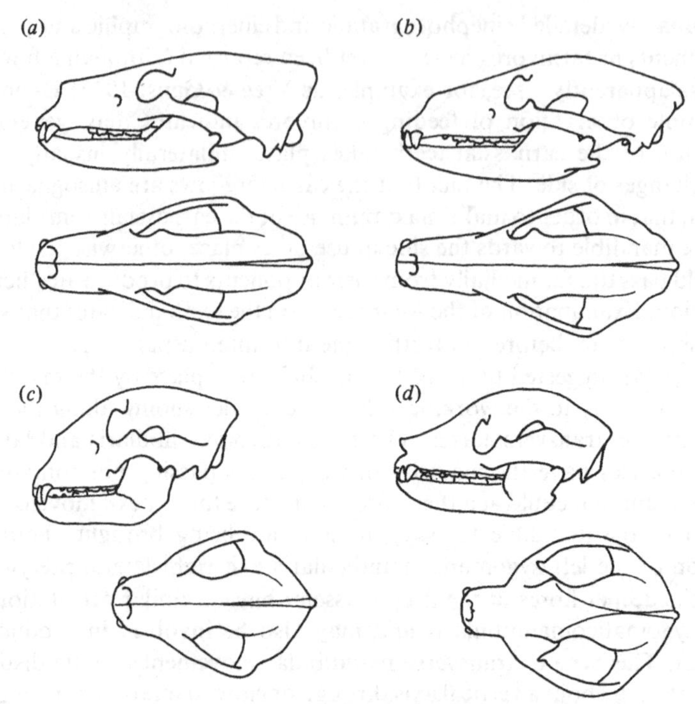

Fig. 66. Left lateral and dorsal views of skulls of: (*a*) *Thalarctos maritimus* (carnivorous bear); (*b*) *Ursus horribilis* (omnivorous bear); (*c*) *Tremarctos ornatus* (herbivorous bear); (*d*) *Ailuropoda melanoleuca* (panda – specialised herbivore). Not to scale. (After Sicher, 1944.)

vorous type. The canines are prominent and, to some extent, interlocking; the jaw joint has, in most respects, a typical carnivore construction and position; and the jaw musculature is modelled on the usual carnivore plan with the temporalis constituting the major part of its bulk (Table 2).

However, several smaller-scale modifications are present which, as Sicher (1944) has pointed out, can be arranged in an adaptational series leading from the more carnivorous members of the Ursidae, through the omnivorous members to the predominantly herbivorous members and thence to the highly specialised herbivorous pandas (this series not, of course, implying taxonomic relationships). Progressing through this series, the teeth become at first secobunodont and then fully bunodont. In addition, there is a widening of the zygomatic arch (Fig. 66) and an associated increase in the mass of the zygomaticomandibularis muscle (in *Ursus americanus* (omnivore) the muscle makes up 12 per cent of the jaw muscle mass compared to below 10 per cent in typical carnivores, while in *Tremarctos ornatus* (predominantly herbivorous) its relative mass is increased to 15 per cent, Table 2; the muscle is probably enlarged still further in the pandas but quantitative data appear to be unavailable). There is probably an accompanying increase in the proportion of the fibres of this muscle running in a horizontal direction in the frontal plane and this may also be true for the similarly orientated fibres of the deep masseter. Both bears and pandas have a lateral pterygoid muscle lying horizontally in the frontal plane in the characteristic carnivore fashion.

The increase in the occlusal area of the dentition in this series suggests that lateral movements of the mandible (to produce a grinding action) become correspondingly more important in passing from the omnivorous to the fully herbivorous types, which is, of course, functionally in keeping with the texture of the foodstuffs that have to be reduced. Yet examination of the jaw joint reveals that, in progressing through this series, the depth of the mandibular fossa increases, and hence the possible range of translatory movement of the usual mammalian type (i.e. produced by rotation about a posteriorly located vertical axis) becomes more limited. The solution of this apparent paradox may be found in Brodie's (1934) description in carnivores of a bodily shift of the mandible brought about by a simultaneous sliding of both condyles along the line of their transverse axes. This type of lateral movement, if present in bears and pandas, would not be restricted, indeed might be facilitated, by a deep mandibular fossa and well-fitting condyle. The capsule of the jaw joint is much slacker, especially between condyle and disc, in bears and pandas than in more typical carnivores and could well represent a structural adaptation to allow a wider range of side to side sliding.

Cinephotographic and cineradiographic studies of masticatory movements in the bears appear to be lacking, perhaps not surprisingly. Examination of the wear facets on the teeth, however, indicates that a lateral bodily movement of the mandible does take place. This type of movement has also been observed in the feeding giant panda (*Ailuropoda melanoleuca*) (Sicher, 1944). The structural modifications in the jaw musculature, especially in the zygomaticomandibularis, appear to be appropriate for producing this type of movement.

Davis (1955) has described the condyles in *Tremarctos ornatus* as making an angle of about 5° to 10° with the transverse plane of the skull (the medial condylar pole lying somewhat posterior to the lateral pole). He takes this finding, together with the observation that the canines in this species interlock to an extent that effectively prevents lateral shift anteriorly, to indicate that transverse chewing movements take place by the mandible rotating about a vertical axis located just anterior to the incisors. There would thus be little lateral movement in the incisor and canine region but quite considerable movement at the molars. This movement would be facilitated by the orientation of the axes of the condyles which lie not in the transverse plane but on the circumference of the circle of horizontal rotation.

The type of transverse jaw movement suggested for the bears and pandas is quite unlike that encountered in other mammals, principally the generalised and ungulate groups, which employ wide excursive movements of the mandible in chewing. This adoption of different mechanisms to meet essentially similar functional needs undoubtedly reflects the fact that the bears and pandas have become readapted to omnivorous or herbivorous diets from a more fully carnivorous state in which a slight degree of lateral sliding of the condyles may already have been present as part of the scissor-like action of the jaws. If the latter assumption is true the extension and exaggeration of this movement, to provide the necessary degree of transverse excursion for reducing vegetable foodstuffs, would seem a more likely evolutionary pathway than a return to chewing movements based on those of the generalised group of mammals.

6

THE AUDITORY REGION

TYMPANIC CAVITY AND AUDITORY BULLA

STRUCTURE OF THE AUDITORY BULLA IN LIVING EUTHERIANS

With the inclusion of the articular and quadrate, as the malleus and incus, in the mammalian middle ear, the tympanic cavity has become enlarged by the addition of a new chamber, the epitympanic recess, which houses the head of the malleus and the body of the incus. The essential elements in the walls of the tympanic cavity proper are the tympanic bone and pars tensa of the tympanic membrane laterally and the periotic bone medially, while the epitympanic recess lies more superiorly between the pars flaccida of the tympanic membrane and the tegmen tympani, a neomorphic addition to the periotic. The tympanic cavity may be further enlarged by distension of its floor to form a hypotympanic sinus, or by extension of its cavity (pneumatisation) into neighbouring bones, especially the squamous and the mastoid part of the periotic. When pneumatisation occurs the resulting accessory (epitympanic) sinuses usually open from the epitympanic recess. The middle ear cavity may be continuous or may be partly subdivided by septa or trabeculae. The ossicular chain and periotic fenestrae are contained entirely within the tympanic cavity proper and epitympanic recess.

The middle ear structures of mammals have become more completely enclosed by bone than is the case in other vertebrates. This has been achieved by the formation of the auditory bulla, a more or less complete bony capsule surrounding the tympanic cavity (the term bulla is used with differing meanings by different authors: to some it is just the ventral wall of the capsule, to others it is all the walls enclosing the tympanic cavity – the latter usage is adopted here). The advantage of having the tympanic cavity enclosed in this way may be that it prevents the pharynx and adjacent soft tissues from pressing against the lining of the enlarged cavity during mastication and swallowing, so altering the pressure of the contained air and hence interfering with auditory sensitivity.

The number of elements, additional to the periotic and tympanic, which may enter into the composition of the walls of the bulla varies widely between the different groups of mammals. As a result of the comprehensive studies by van Kampen (1905) and van der Klaauw (1929, 1931) it has become

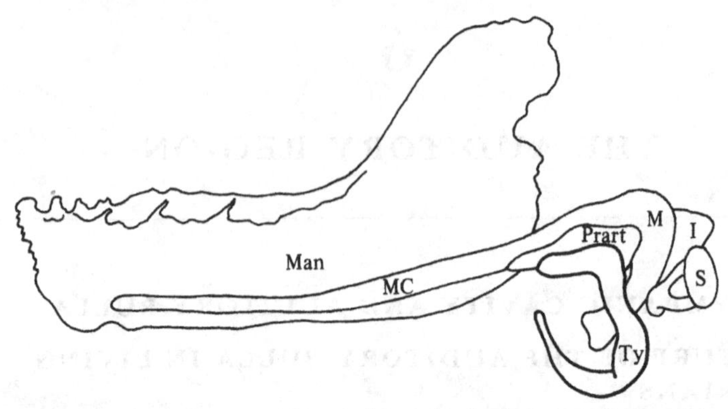

Fig. 67. Lower jaw of mammalian embryo to show, in semi-diagrammatic form, Meckel's cartilage and the developing tympanic and ear ossicles. For labelling see list of abbreviations.

recognised that one or more of the following elements become involved in the structure of the eutherian bulla.

1. Tympanic (ectotympanic). This dermal element, probably homologous with the angular of the reptilian lower jaw is usually, if not always, the earliest bone of the region to ossify. It begins to develop in the angle between Meckel's cartilage and the malleus and soon becomes U-shaped, with the gap between its two limbs located dorsally (Fig. 67). When it first develops the tympanic tends to lie nearly horizontal, below the periotic, but during ontogeny, in all but the most primitive mammals, it comes to occupy a more vertical plane. In its primitive form the tympanic was probably narrow in its mediolateral dimension but in most modern mammals it has expanded laterally to form a bony external acoustic meatus or medially into the ventral wall of the bulla or in both of these directions. Despite these expansions, the tympanic usually remains U-shaped, when viewed from the side, although in a few groups (e.g. microchiropterans, some rodents and carnivores) the dorsal gap may be completely or almost completely closed.

2. Periotic. The bone formed by ossification of the otic capsule consists, in mammals, of more or less distinct petrous and mastoid parts. The petrous part lies anterodorsally and is composed of very dense bone enclosing the structures of the inner ear. Its lateral surface, which forms the medial wall of the tympanic cavity, is pierced by the fenestrae vestibuli and cochleae, and possesses an eminence, the promontory, produced by the underlying cochlea. In insectivores and primates, and possibly edentates and whales, the internal carotid artery lies on or near the promontory. The artery is also present in this position in prenatal bats but subsequently disappears. In other cases, including rodents and ungulates, and also monotremes and marsupials, the internal carotid artery runs medially in the basicapsular groove between the petrous laterally and the basioccipital and basisphenoid medially (see

Appendix 1, p. 238). Descriptions of the position of the internal carotid artery in carnivores are not altogether consistent but the artery is generally considered to be medial in this group too. The petrous part of the periotic is usually hidden from view ventrally by the auditory bulla, but its superior surface, including the tegmen tympani, which forms the roof of the epitympanic recess, can be seen in the floor of the cranial cavity. More ventrally, a petrosal tympanic wing may be present extending into the floor of the bulla. The mastoid part, developed from the posterior region of the facial crest, lies posteroventrally, is formed of less dense bone than the petrous part, and can be seen on the external surface of the skull just lateral to the exoccipital. It may form the rear wall and hind part of the floor of the bulla. In the Hominoidea the mastoid is produced into a process for attachment of the sternomastoid muscle.

3. Entotympanic. Van der Klaauw defined as entotympanics all the skeletal elements, bony or cartilaginous, that lie in the ventral wall of the tympanic cavity and are ontogenetically primarily independent of the other elements in this region, except perhaps the tympanohyal and the cartilage of the Eustachian tube. Entotympanics are generally regarded as neomorphs in therian mammals. They frequently fuse with each other and with other bullar elements early in ontogeny so that their presence must be sought in the very young or fetal skull. Van der Klaauw's investigations showed that two entotympanics can frequently be distinguished – a caudal entotympanic, which grows in a rostral direction and is often connected in the initial stage of its development with the tympanohyal, and a rostral entotympanic, which develops in connection with the cartilage of the pharyngotympanic tube and grows in a caudal direction. Recent studies (see below) suggest that, while frequently not recognisable in the adult skull, entotympanics or associated elements have a wide distribution in eutherian (and possibly also metatherian) mammals. The ontogenetic development of the entotympanics is known for relatively few species (and even here descriptions tend to be vague and terminology confused, especially in the older studies), but it appears that generally these elements ossify in close association to cartilaginous anlages.

4. Tympanohyal. The tympanohyal, the most dorsal element apart from the stapes to ossify in Reichert's (hyoid arch) cartilage, is connected during early development with the crista facialis (crista parotica) on the lateral surface of the periotic.

5. Squamous. The entoglenoid part of the squamous is a frequent major contributor to the bony walls of the tympanic cavity, usually lying anterosuperiorly.

6. Basisphenoid and basioccipital. Tympanic wings may be present on these bones.

7. Exoccipital. The tympanic cavity may be bounded posteriorly by the exoccipital bone and its paroccipital process. In some forms (e.g. *Elephas*) the tympanic cavity may extend into the exoccipital.

8. Anterior process of malleus. This part of the malleus forms in the cartilaginous connection with the distal part of Meckel's cartilage by the spread of ossification from a dermal element, thought to be homologous with the prearticular of the reptilian lower jaw. In primitive mammals the anterior process lies in a sulcus, the sulcus malleolaris, on the anterior limb of the tympanic with which it may co-ossify. This connection is lost in more avanced mammals but the anterior process remains connected to the region of the petrotympanic suture by the anterior malleolar ligament formed from the perichondrium of Meckel's cartilage.

9. Spence's cartilage. As would be expected from its presumed homology with a segment of the extrastapes, Spence's cartilage accompanies the chorda tympani along the first part of its course in the tympanic cavity (see Figs. 20 and 45). It may remain cartilaginous or become ossified (as Bondy's '*chordafortsatz*'), when it is usually incorporated into the tympanic.

10. Alisphenoid. The alisphenoid is a major component of the marsupial bulla but, in eutherians, it is a large element in the bulla of only macroscelidids and some other insectivores.

These bullar elements can be subdivided into two groups according to whether they lie anterior or posterior to the opening of the pharyngotympanic tube. On the rostral side of this opening may be found the alisphenoid and squamous, while on the caudal side occur the caudal entotympanic, periotic, basisphenoid, basioccipital, exoccipital and tympanohyal. The tympanic, periotic, entotympanics, squamous and, less commonly, the basisphenoid and alisphenoid are the only elements to form major contributions to the bullar walls and usually only one or, at most, two do so in any particular species. The tympanic, periotic, entotympanics, tympanohyal and squamous may undergo variable degrees of fusion to form the composite temporal bone. Such fusion reaches its maximum extent in the higher primates.

Novacek (1977) has categorised the bullae of eutherians, on the basis of the elements making major contributions to its ventral wall, into the following six groups (Fig. 68).

Group 1 (entotympanic)

Caudal and rostral entotympanics have been found to occur in a large number of mammals, including members of the Insectivora, Tupaiidae, Edentata, Rodentia, Microchiroptera, Hyracoidea and Carnivora (van der Klaauw, 1931; Hunt, 1974; Novacek, 1977). A recent study of the development of the auditory bulla by Presley (1978, and personal communications) in a variety of non-therian, metatherian and eutherian mammals has added considerably to knowledge of the ontogeny of these elements.

In developing monotremes (*Tachyglossus*, *Ornithorhynchus*) and marsupials (*Didelphis*, *Trichosurus*) he found that Reichert's cartilage bridges across from the lateral wall of the sulcus facialis (see section on otic capsule, Chapter 2)

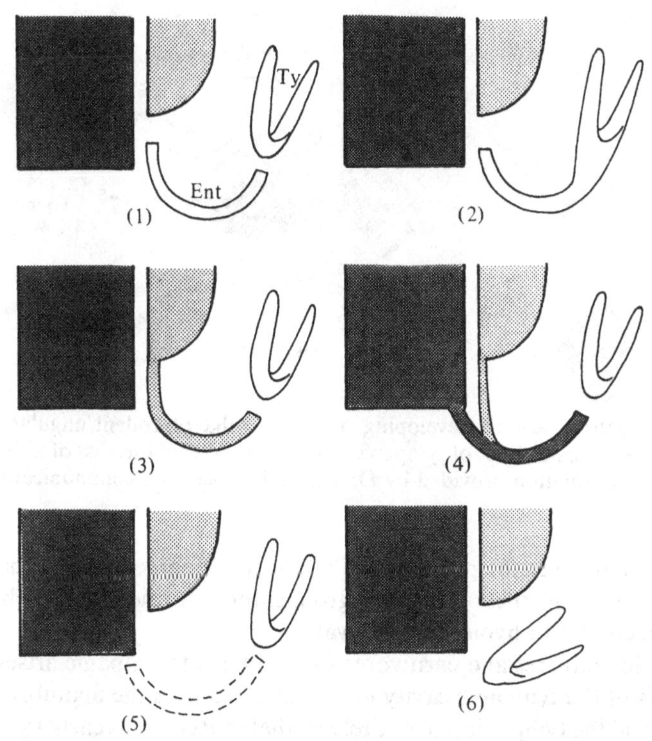

Fig. 68. Schematic representation of composition of auditory bulla according to Novacek: (1) ossified entotympanic bulla; (2) tympanic bulla; (3) petrosal bulla; (4) basisphenoid or alisphenoid bulla; (5) cartilaginous bulla; (6) bulla absent (membranous bulla). Dark shading = basisphenoid or alisphenoid; light shading = periotic; broken outline in (5) = cartilage. For labelling see list of abbreviations. (After Novacek, 1977.)

to abut medially on to a shallow outgrowth of the periotic. In *Didelphis* and *Trichosurus* the tympanic cavity extends posterior to this bridge to reach the fenestra cochleae but in the monotremes Reichert's cartilage delineates the posterior wall of the tympanic cavity. That part of Reichert's cartilage which contacts the periotic becomes fused with it to form its tympanic wing.

In the eutherian mammals studied by Presley (insectivores – *Erinaceus*, *Talpa*, *Potamogale*, *Petrodromus* and *Elephantulus*; rodents – *Rattus*, *Meriones*, *Apodemus*, *Microtus*; carnivores – *Felis*, *Mustela*; ungulates – *Ovis*, *Sus*; bats – *Tadarida*, *Myotis*, *Hipposideros*, *Rousettus*; primates – *Homo*) no medial contact was found between Reichert's cartilage and the periotic. In the bulla of rodents, fibrous connective tissue acts as a bridge between Reichert's cartilage and the periotic to form the floor of the tympanic cavity (Fig. 69). The mastoid component of the bulla later spreads into this part of the floor. In the ungulates an entotympanic may develop as an inconstant element in this connective tissue as it does more constantly in the edentate

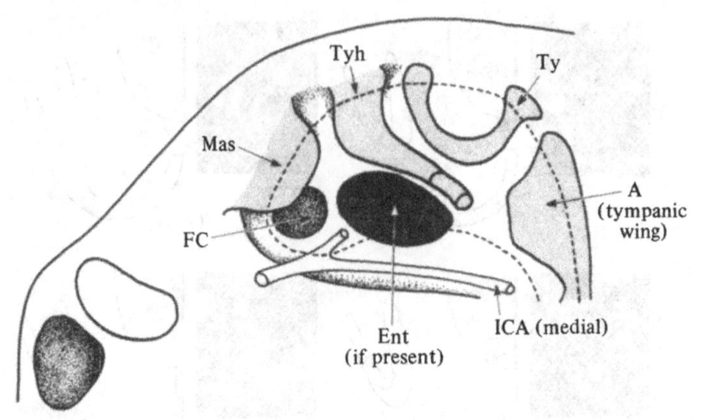

Fig. 69. Ventral view of developing auditory bulla of rodent/ungulate/edentate. Dashed line indicates limit of tympanic cavity. For labelling see list of abbreviations. (Based on information provided by Dr R. Presley, personal communication.)

Dasypus. Whatever the precise details of its development the caudal part of the bullar floor in these eutherian groups appears, therefore, to be related developmentally to hyoid arch derivatives.

In the insectivores and carnivores the caudal entotympanic arises close to the lining of the tympanic cavity in the same plane as the manubrium of the malleus and the tympanic and entirely medial to Reichert's cartilage (Fig. 70). In *Homo* no separate entotympanic develops but the floor of the tympanic cavity is formed by the tympanic wing of the petrosal in the same plane as that in which the entotympanic appears in insectivores and carnivores (see also below for development of the tympanic wing in prosimians). In both cases, therefore, this part of the bullar floor is more closely associated with derivatives of the mandibular arch than of the hyoid arch.

The rostral entotympanic always develops by chondrification and ossification of connective tissue in close relationship to the cartilage of the pharyngotympanic tube.

The development of the entotympanics is probably better known for the terrestrial carnivores than for any other major mammalian order as a result of the detailed and comprehensive study of the auditory bulla in this group by Hunt (1974). There are, in addition, several excellent accounts of the adult bulla of the terrestrial carnivore, notable amongst which are the early studies by Flower (1869) and Pocock (1928). The description which follows is based largely on the findings of these three authors.

The carnivore bulla is composed almost exclusively of the tympanic and caudal and rostral entotympanics. As in other mammals the tympanic is the first bullar element to ossify, appearing shortly before birth as a single centre. A short time after, the remaining bones of the bulla begin to ossify and there is usually only a brief period in which all the elements are both

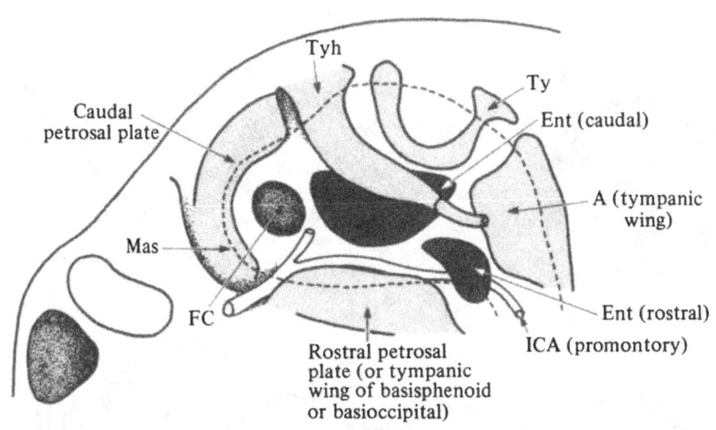

Fig. 70. Ventral view of developing auditory bulla of insectivore/bat/carnivore. Dashed line indicates limit of tympanic cavity. For labelling see list of abbreviations. (Based on information provided by Dr R. Presley, personal communication.)

present and unfused. The tympanic develops as an incomplete ring with the tympanic membrane stretched between its two limbs. The surface of the tympanic that faces towards the base of the skull is concave and helps floor the auditory bulla. The anterior horn of the tympanic is attached to the squamous posteromedial to the postglenoid foramen, while the posterior horn is attached to the mastoid process (formed by squamous and mastoid bones). At first the tympanic lies in the frontal plane but during development it rotates towards the sagittal plane (the maximum rotation in any species being about 45°). The petrotympanic fissure receives the anterior process of the malleus.

The rostral entotympanic is an independent element in the majority of, if not all, carnivores and is very constant in its shape and relationships. It is triangular in lateral view with its dorsal apex contacting the alisphenoid–petrous suture at which point is found a foramen (which Hunt identifies as transmitting the promontory branch of the internal carotid artery). The rostral entotympanic extends no further anteriorly than the middle lacerate foramen, while posteriorly it tapers to an edge which contacts but never extends beyond the promontory of the petrous. Thus, the locations of the posterior carotid and middle lacerate foramina give an indication of the posterior and anterior limits, respectively, of the rostral entotympanic.

The caudal entotympanic is usually single (although in *Ursus* two caudal entotympanics develop – two elements may be present initially in other carnivores, but fuse too early to be separately identified). The relationships of the caudal entotympanic, unlike those of the rostral element, vary between the different carnivore groups. Hunt (1974) classified the variations into five types. Briefly, these can be described as follows (Fig. 71).

(1) Type A. In this type the bulla is formed mainly by the tympanic (strictly

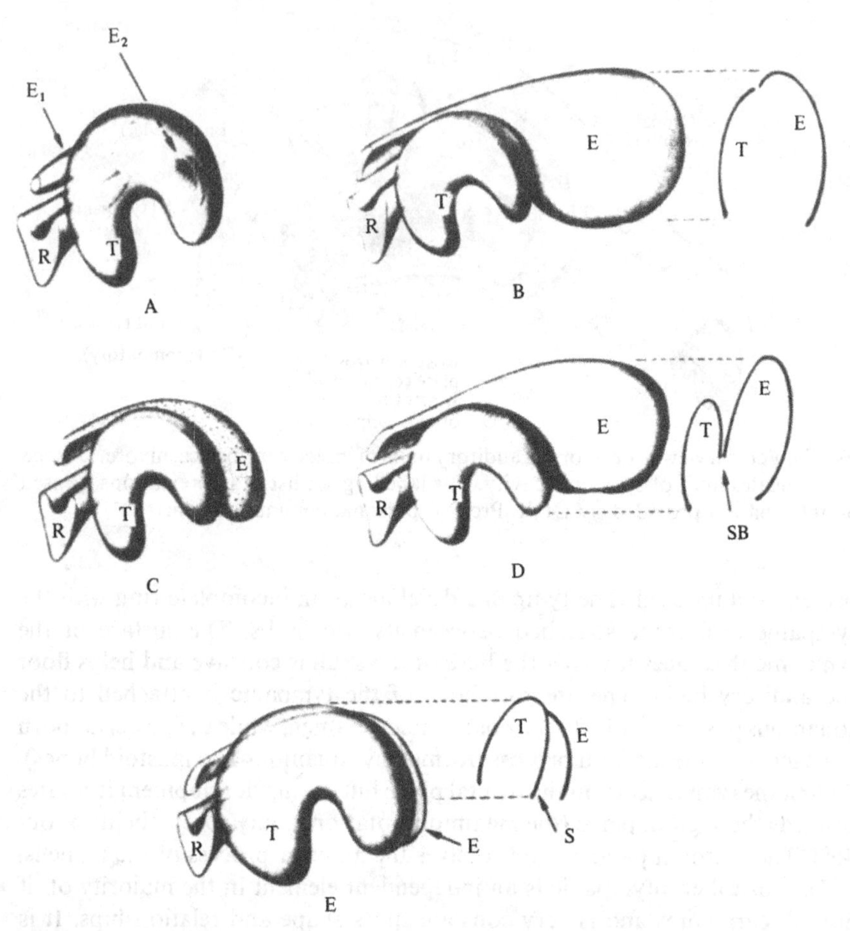

Fig. 71. The five types of auditory bullae in Carnivora. Bullae shown in lateral view with ventral surface uppermost and anterior to left. Each is drawn as though contributing elements are slightly displaced. E = caudal entotympanic; R = rostral entotympanic; T = tympanic; S = septum; SB = bilaminar septum. Transverse sections through centre of types B, D and E are shown at right (sections through types A and C would be similar to type B except that the caudal entotympanic would be smaller). In Type A two caudal entotympanics (E_1, E_2) are present. For description of types see text. (Reproduced with permission from Hunt, 1974.)

speaking, therefore, this type belongs to Novacek's Group 2, but is considered here for convenience in comparison with the remaining four types of carnivorous bullae). The caudal entotympanic is an elongated plate attached to the medial ege of the tympanic, intervening between the latter and the rostral element. The bulla does not extend posterior to the mastoid process due to lack of caudal entotympanic inflation and there is generally no invasion of contiguous skull bones by the middle ear cavity. This is probably the most

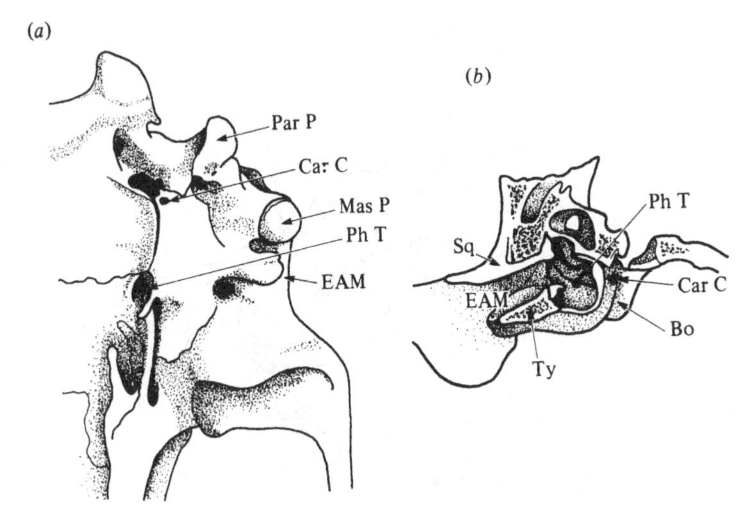

Fig. 72. The right auditory bulla of *Ursus horribilis*. (*a*) Ventral view; (*b*) coronal section, viewed from in front. Not to scale. For labelling see list of abbreviations. (After Flower, 1869.)

primitive type of bulla amongst living carnivores and is found in Ursidae (Fig. 72), pandas, Otariidae, Odobenidae and lutrine and mephitine Mustelidae (as well as in several extinct carnivores).

(2) Type B. This variety of bulla appears to have been derived from Type A by enlargement of the caudal entotympanic. The chamber of the bulla is inflated and partly subdivided by supporting struts (not to be confused with a true septum, see below). A bulla of this type is found in procyonids, musteline, guline and meline Mustelidae, Canidae and Phocidae. In those groups which inhabit arid and open country (*Nasua*, canids, some badgers), the bulla may be markedly inflated as a result of enlargement of the caudal entotympanic and invasion of the mastoid. As discussed below, this is probably an adaptation to low-frequency sensitivity.

(3) Type C. The African Palm Civet (*Nandinia binotata*) possesses a unique bulla structure but one of considerable interpretative significance in that the caudal entotympanic remains cartilaginous in the adult and thus demonstrates clearly the three elements of the bullar wall (strictly speaking, this type of bulla belongs to Novacek's Group 5, but again is considered here for ease of comparison). The middle ear cavity is an undivided chamber.

(4) Type D. In the remaining Viverridae and in the Felidae the cavity of the bulla is subdivided into anterolateral and posteromedial parts by a septum (Fig. 73). The anterolateral chamber is the tympanic cavity proper and is formed principally by the rostral entotympanic and tympanic, while the posteromedial chamber is formed entirely by the inflated caudal entotympanic. The dividing septum is bilaminar, being formed by inflections of the tympanic

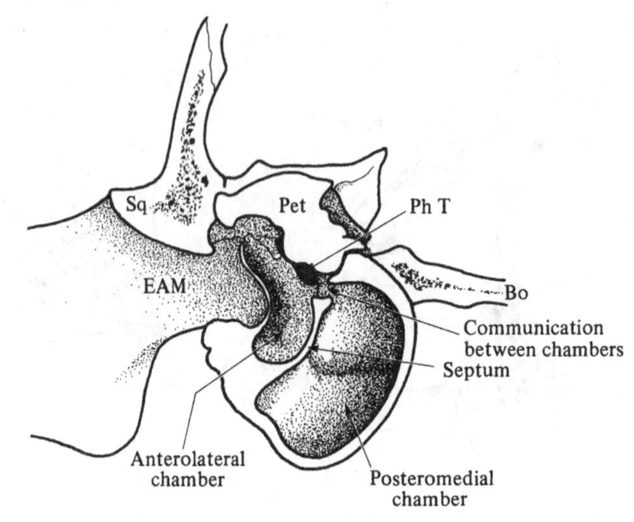

Fig. 73. Coronal section of left auditory bulla of the tiger viewed from behind. For labelling see list of abbreviations. (After Flower, 1869.)

and caudal entotympanic. The middle ear cavity does not invade other skull bones.

(5) Type E. The hyaenid bulla is also double chambered. In this case the septum is formed solely by an inflection of the posterior rim of the tympanic and is not, therefore, bilaminar as in other aeluroids. The tympanic is enlarged relatively more than the caudal entotympanic, and, as a result, the anterior chamber extends ventral to the posterior chamber. The latter communicates with a cavity in the mastoid.

The bulla of *Nandinia* may represent the primitive aeluroid condition from which the double-chambered structures of the viverrids and felids, on the one hand, and the hyaenids, on the other, were derived (but see section on primitive structure of bulla). The tendency towards hypertrophy of the bullar cavity seen in many of the groups, although achieved by somewhat different structural modifications, can be interpreted as a general adaptation to maximal auditory acuity at low sound frequencies together with a high level of sensitivity over a wide range of frequencies (see section on functions of the middle ear).

Group 2 (tympanic)

Although the tympanic often expands into the floor of the bulla it is the principal element in this region in a relatively restricted group including, as well as those carnivores listed in Group 1, Type A, some members of the Insectivora, Artiodactyla, Perissodactyla, Cetacea, Rodentia, Lagomorpha and Proboscidea (van der Klaauw, 1931; Kellogg, 1928; Osborn, 1936; Wood, 1962; Hunt, 1974; Webster, 1975; Novacek, 1977).

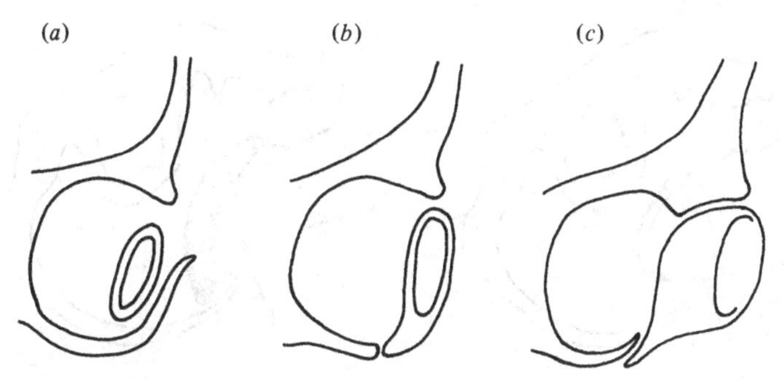

Fig. 74. Schematic representation showing various relationships of tympanic ring and auditory bulla in primates. (*a*) Lemuriform type in which ring enclosed within osseous bulla; (*b*) lorisiform and platyrrhine type in which ring remains exposed and contributes to formation of outer wall of bulla; (*c*) tarsioid and catarrhine type in which ring is produced laterally to form tubular acoustic meatus. (After Le Gros Clark, 1934.)

The development of the tympanic bulla of the Ursidae (described above) (Hunt, 1974) and of the heteromyid rodent *Dipodomys merriami* (Webster, 1975) has been the subject of particularly detailed description. In *Dipodomys merriami* the tympanic has at first a simple annular form but at about three days postnatally it expands medially to floor in the future tympanic cavity, and laterally to partly enclose the external acoustic meatus. It appears to be the sole component of the floor of the bulla, no evidence of a cartilaginous entotympanic being observed. In the first week of postnatal life it grows rapidly to complete the lateral wall and floor of the bulla (the roof being formed by the petrous and stylohyal). In the second week the sutures between the bullar elements fuse and resorption of the inner surface of the bulla, to enlarge the middle ear cavity, begins. The entire process of bullar formation and the marked hypertrophy of the middle ear chamber characteristic of the heteromyid rodents are virtually complete within two weeks of birth.

Group 3 (periotic)

A tympanic process of the petrous is well developed in some insectivores but only in primates does it form the major element in the floor of the bulla. In Lemuriformes the bulla is much inflated and the tympanic plate of the petrous is expanded lateral to the tympanic (so that the latter is sometimes described as being intrabullar) to contribute to the external acoustic meatus (Fig. 74*a*). The tympanic in Lorisiformes, by contrast, lies in the lateral wall of the bulla (i.e. is extrabullar), being somewhat expanded ventrally to meet the lateral edge of the petrosal tympanic plate (Fig. 74*b*). In some genera (e.g. *Loris* and *Nycticebus*) the tympanic is also produced laterally to form a short tubular external acoustic meatus. In *Tarsius* the bulla, which is considerably inflated, is formed from the petrous with the tympanic applied superficially to its outer

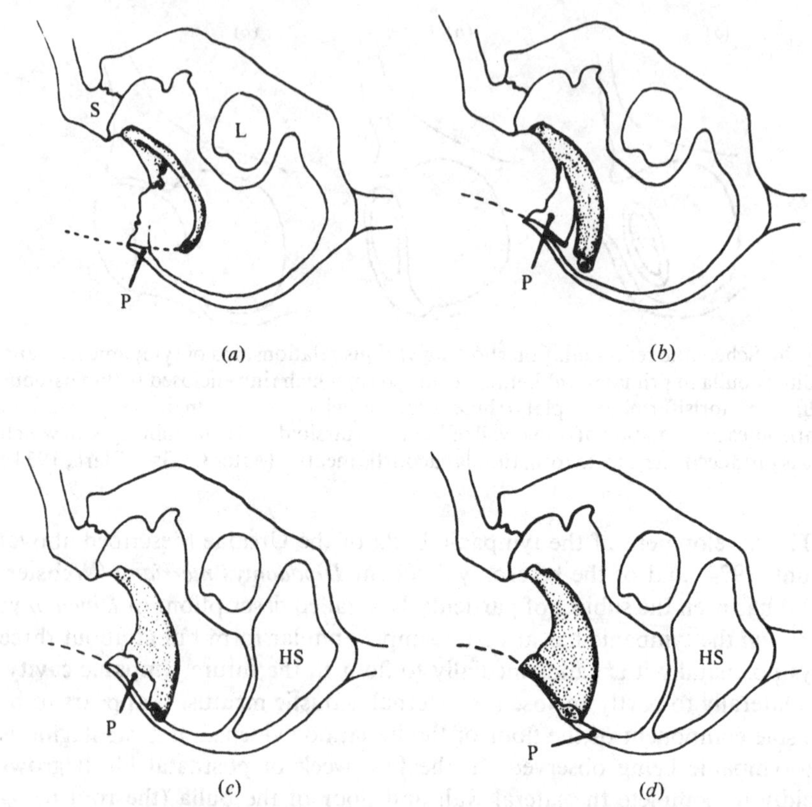

Fig. 75. Diagrammatic representations of coronal sections through right ear region of (*a*) *Lemur*; (*b*) *Microcebus murinus*; (*c*) *Loris*; (*d*) typical lorisiform (e.g. *Nycticebus*). The tympanic is stippled; the ectodermal surface of the external acoustic meatus and annulus membrane is indicated by a broken line. Abbreviations: HS = hypotympanic sinus; L = labyrinth of inner ear; P = petrous; S = squamous. (Reproduced with permission from Cartmill, 1974.)

surface and produced to form a tubular meatus (Fig. 74*c*). Amongst the Anthropoidea the bulla of the New World monkeys is expanded and the tympanic, which forms the lateral wall, is not produced into a bony meatus, while in the Old World monkeys, together with the apes and man, the bulla is but little inflated and the tympanic is produced laterally into a definite bony tubular meatus (Fig. 74*c*).

Considerable taxonomic weight has been given to the question of whether the tympanic is intra- or extrabullar, although it is still uncertain which of these conditions is the more primitive primate condition (see, for example, Le Gros Clark, 1934, 1959; van Valen, 1965; McKenna, 1966). However, the findings in recent studies by Cartmill (1974), Szalay (1972, 1974) and MacPhee (1977) indicate, on both morphological and developmental grounds, that the significance of the differences in the arrangement of the tympanic

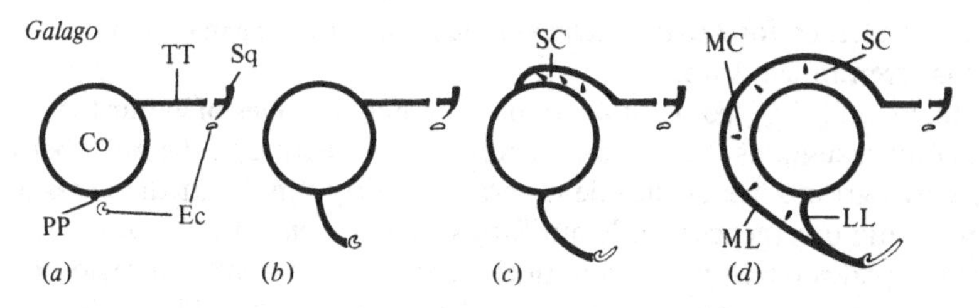

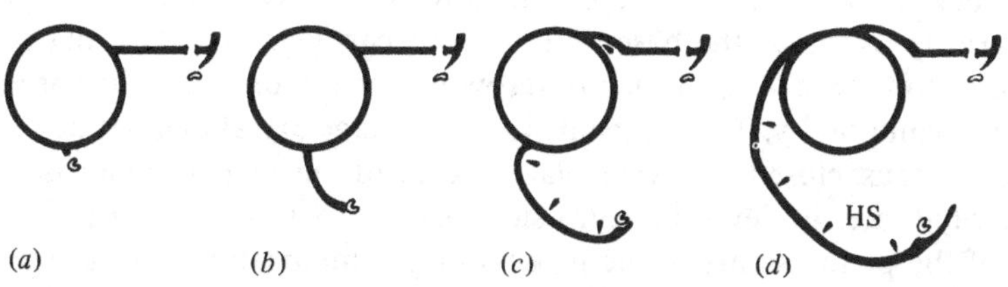

Fig. 76. Sequence of bullar ontogeny in *Galago* and *Microcebus* illustrated by schematic cross-sections through centre of promontory: (*a*) = late prenatal; (*b*) = neonatal; (*c*) = young postnatal; (*d*) = juvenile. Abbreviations: Co = inner ear; Ec = ectotympanic; HS = hypotympanic sinus; LL = lateral lamella of petrosal plate; MC = mastoid cavity; ML = medial lamella of petrosal plate; PP = petrosal plate; SC = supracochlear cavity; Sq = squamous; TT = tegmen tympani. Arrows indicate major sites of pneumatic activity. (Reproduced with permission from MacPhee, 1977.)

between the lemuriforms and the remaining groups is probably less than has often been assumed. The lemuriform tympanic is not really intrabullar, as is usually stated; this misleading appearance is produced by the bulla bulging laterally, ventral to the external acoustic meatus. In the typical modern lemur the tympanic contacts the squamous and petrous bones only at its anterior and posterior extremities; its central arc is not, however, free but is attached to the bullar part of the meatus by an annulus membrane (Fig. 75*a* and *b*). Furthermore, in some lemuriforms (e.g. *Microcebus murinus*) the annulus membrane is replaced to some extent by bony plates derived from the tympanic medially and the petrous laterally, to produce a condition not so different from that found in *Loris* (Fig. 75*c*), where the petrous participates, with the tympanic, in the formation of the inferior margin of the meatus. In more typical lorisiforms the petrous no longer participates in this margin (Fig. 75*d*).

Similarly, in tarsiiforms there is a range of tympanic relationships. In *Tarsius*, as described above, the tympanic is superficially placed, but in the extinct *Necrolemur* this element is located deep within the tympanic cavity. In the latter form the tympanic is continued into a shelf which appears to be an ossified annulus membrane. The bulla bulges laterally, ventral to this shelf, and ventral, therefore, to the external acoustic meatus, in a manner reminiscent of that seen in lemuriforms.

MacPhee (1977), from a study of developmental stages of lemuriforms and lorisiforms, suggests that the differences in the structure of the bulla between these two groups are explicable in terms of fairly simple (at the gross level) ontogenetic divergences. In lemuriforms (*Microcebus* and *Propithecus*) two petrosal plates develop, one anteriorly from the promontorial region of the petrous and the other posteriorly from the mastoid region. The anterior plate is a dermal outgrowth into the floor of the tympanic cavity. The posterior plate is first formed in cartilage and is ossified from a centre in the pars canalicularis. It develops in the hind wall of the tympanic cavity, passing posteriorly as far as the base of Reichert's cartilage. The two plates soon coalesce to form a single broad outgrowth in the floor and posterior wall of the tympanic cavity. The tympanic is at this stage extrabullar. It lies ventral to the petrous, close to the ventrolateral edge of the coalesced petrosal plate. Somewhat later in development, osteoclastic activity begins on the internal face of the plate as part of the process of pneumatisation and, at the same time, the edge of the plate begins to enlarge ventrally and laterally to the tympanic, which thus takes up its so-called intrabullar position (Fig. 76). The annulus membrane appears to be no more than a continuation of the lining of the external acoustic meatus whose 'intrabullar' position is produced, like that of the tympanic, to which it is attached, by the lateral growth of the petrosal plate. A plate of bone, the linea semicircularis, projects upwards to a variable degree from the petrosal plate to form an arc of bone immediately ventral to the tympanic. It is presumably the same structure as that described as projecting from the petrous into the annulus membrane (see above).

The early development of the petrosal plate in lorisiforms (*Galago*, *Loris*) is essentially similar but at a later stage it fuses with, instead of overgrowing, the tympanic. Before this fusion takes place the petrosal plate and the tympanic cavity proper have attained virtually adult size. Pneumatisation takes place from the epitympanic recess into the petrous and mastoid parts of the periotic. From here it proceeds anteriorly and posteriorly into the petrosal plate which thus becomes divided into medial and lateral lamellae. The lateral lamella (longitudinal septum) occupies the original position of the petrosal plate before it became pneumatised (Fig. 76).

A further interesting observation made by MacPhee (1977) is that the fibrous articulation between the petrosal plate and tympanic, which possesses all the layers of a typical suture (see Chapter 8) in the lorisiforms, is lacking the cellular middle layer in fetal lemuriforms (i.e. the articulation is really just a close approximation of two periosteal membranes). As a result the interface between the tympanic and petrosal plate may lack the stability of a normal suture and so allow the overgrowth of the latter element.

MacPhee suggests that his developmental findings provide a ready explanation of the relationship between the intra- and extrabullar conditions, whichever is deemed to be the more primitive for primates. If an extrabullar structure was the original condition, the intrabullar state could have been achieved by a reduction in sutural tissue formation between petrosal plate and tympanic and increased pneumatisation of the tympanic cavity proper in the fetal stages of the ancestral lemurs. If, on the other hand, the intrabullar condition was the original one, the extrabullar arrangement would be the result of acquisition of a full sutural union of tympanic and petrosal plate and decreased pneumatisation in the fetal stages of ancestral lorises.

Group 4 (basisphenoid)

Tenrec and erinac insectivores alone, among the major living mammalian groups, possess a bullar floor in which a tympanic process of the basisphenoid forms the dominant element (van der Klaauw, 1931; Butler, 1948; Novacek, 1977). A tympanic wing of the periotic may also be present and considerable variation is encountered in the relative expansion of these two components of the bullar floor between the different genera of these two families.

Whether or not the production of the tympanic wing of the basisphenoid is morphogenetically very different from the production of a discrete entotympanic element is not known; Presley's (1978) findings in *Potamogale*, *Petrodromus*, *Erinaceus* and *Elephantulus* (described above) suggest that the two processes are similar in their development.

Group 5 (cartilage)

A ventral bullar wall composed of cartilage is rare amongst adult extant eutherians, being known to occur in only the megachiropterans *Pteropus*, *Acerodon* and *Boneia*, the edentate *Dasypus* and the viverrid *Nandinia* (van der Klaauw, 1931; Hunt, 1974). Its occurrence appears to be due, in each case, to a failure of ossification of the entotympanic elements.

Group 6 (membrane)

In a few mammals, including monotremes, soricids and some talpids (van der Klaauw, 1931; Novacek, 1977), the floor of the bulla is composed of a fibrous connective tissue membrane, containing neither cartilaginous nor ossified elements, and the tympanic lies in a more or less horizontal plane beneath the periotic. In the trichechid Sirenia the bullar floor is also membranous but the tympanic lies in a more vertical plane than in the monotreme or insectivore forms.

Numerous groups of fossil mammals have been described as lacking an ossified bullar floor but, as van der Klaauw (1931) has pointed out, in many of these cases this may be a misleading appearance caused by damage to the bulla or by loss of bony elements during fossilisation.

PRIMITIVE STRUCTURE OF THE AUDITORY BULLA

The two essential elements in the wall of the tympanic cavity are, as already noted, the tympanic and periotic bones. It seems probable, from the structure of the auditory bulla seen in such primitive living mammals as insectivores, edentates and prosimians, that the tympanic cavity in the earliest eutherian mammals had a medial wall and roof hollowed out from the periotic and a lateral wall formed by a simple ring-like tympanic, which was inclined at less than 45° to the horizontal (Fig. 68(6)). It has been widely assumed that the floor of the tympanic cavity, in the region between the tympanic and periotic, was unossified and consisted of either a fibrous connective tissue membrane or, alternatively, cartilage. Novacek (1977) has provided cogent reasons for rejecting the view that the floor was cartilaginous. These include observation that: (1) a cartilaginous bulla occurs in few adult living mammals and, where it does occur, appears to be derived rather than a primitive character; (2) several eutherians, which are judged to be primitive for other morphological reasons, have ossified entotympanics; and (3) cartilage seems to have evolved as an embryonic adaptation and its occurrence in the adult is more usually the ontogenetic retention of the embryonic condition than the phylogenetic retention of an ancestral stage.

It seems more likely, therefore, that the auditory bulla in the ancestral mammals had a floor composed of membrane similar to that still found in living monotremes and in some primitive eutherians. Novacek (1977) suggests that the concomitance between the enlargement of the entotympanic and the increasing inclination of the tympanic during ontogenetic development might be an indication that these two tendencies arose phylogenetically as correlated modifications of the ancestral condition.

The widespread distribution of the entotympanic elements amongst placental mammals and their presence in forms considered primitive on other morphological grounds are strongly suggestive that they were an early eutherian acquisition. The presence in some marsupials of an element (the tympanic wing of the periotic), having much in common with the caudal entotympanic, might be due to the common ancestor of placentals and marsupials already possessing this bullar component, although it seems more probable, in view of the relative rarity of the entotympanic in marsupials, that the two elements have been independently derived.

On the basis of such considerations Novacek (1977) has postulated the following possible pathways for the evolution of the varieties of bullae. He starts from the assumption that the common ancestor (*a* in Fig. 77) of marsupials and placentals and the most primitive members of these two groups (*b* and *c*) had a membranous bullar floor and an only slightly inclined tympanic. Early in eutherian evolution a partial bony floor to the bulla was formed by the development of the rostral and caudal entotympanics (*e*). From this condition evolved the fully formed entotympanic bullar floor (*f*) (the

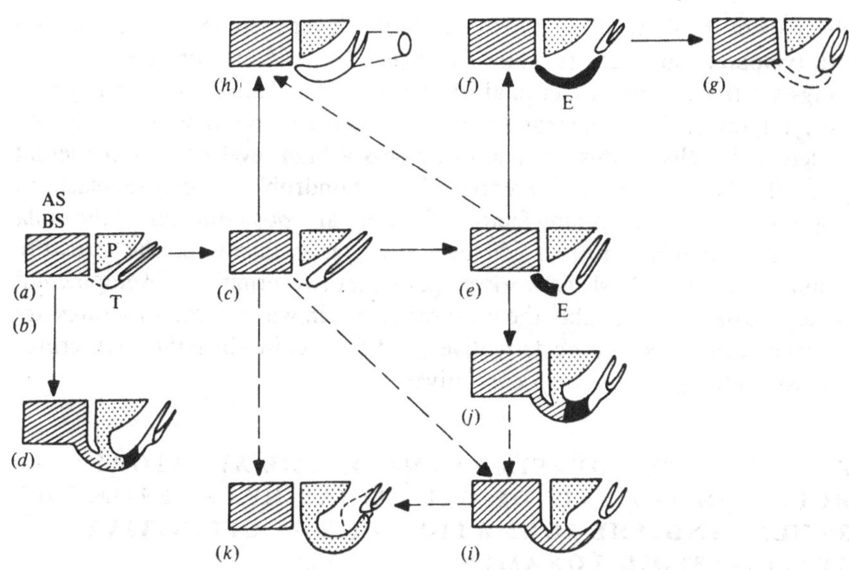

Fig. 77. Hypothetical scheme for the evolution of the auditory bulla in eutherians and metatherians. See text for explanation of stages. Broken arrows indicate alternative pathways of derivation. All are posterior views of right side. Abbreviations: AS, BS = alisphenoid, basisphenoid; E = entotympanic; P = periotic; T = tympanic. (Reproduced with permission from Novacek, 1977.)

cartilaginous bulla found in *Nandinia*, *Dasypus* and three megachiropteran genera (*g*) may be secondary derivations from this condition – but see above regarding *Nandinia*). In macroscelidids a specialised condition (*j*) may have been derived from (*e*) by contributions to the bulla formed by tympanic wings developed from the basisphenoid, alisphenoid, petrous and tympanic. The tympanic type of bulla (*h*) may have evolved directly from the early eutherian condition (*c*) or by loss of the entotympanic from the early entotympanic stage (*e*). The basisphenoid bulla (*i*) may have been derived directly from the early eutherian stage (*c*) or possibly from the macroscelidid condition (*j*) by loss of the entotympanic. The primate petrosal bulla (*k*), in Novacek's opinion, was most probably evolved directly from the primitive eutherian bulla (*c*), but the developmental evidence presented by Presley (see above) appears more consistent with a derivation from the entotympanic type of bulla (*e* or *f*) by coalescence of the entotympanic elements with the petrous so that they appear to ossify as a tympanic wing of the latter element.

The great variation encountered in the structure of the walls of the bulla, even in closely related genera, indicates considerable plasticity in the ontogenetic development of this region of the skull and emphasises the need for caution when attaching phylogenetic significance to bullar morphology. R. Presley (personal communication) describes a layer of mesodermal tissue between the tympanic and periotic (or basicranium) in which develop the elements of the floor of the bulla. Within this layer there may arise separate entotympanic cartilages (usually subsequently ossifying) or the tympanic wings of the contiguous cranial elements, either initially as cartilage or ossifying directly as membrane bones. It could well be, therefore, that the skeletogenic cells of this mesoderm possess a high level of developmental plasticity, being able to differentiate into chondroblasts or osteoblasts in response to local determining factors. The floor and posterior wall of the bulla are mammalian neomorphs whatever the precise details of the structure contained therein. Presley's work suggests that two principal developmental patterns exist: one, possibly the most primitive, in which these structures are associated with second arch derivatives, and another in which these structures are associated with first arch derivatives.

EFFECT OF THE DEVELOPMENT OF THE AUDITORY BULLA ON THE EXIT OF THE FACIAL NERVE FROM THE SKULL, AND THE FORMATION OF THE DEFINITIVE STYLOMASTOID FORAMEN

As described in Chapter 2, the facial nerve of mammals leaves the cranial cavity by perforating the tegmen tympani which represents the secondary wall of the otic region of the cranial cavity and is a newly acquired mammalian feature. After passing through the tegmen the nerve reaches the medial wall of the tympanic cavity where it turns to run posteriorly first above the stapes and then below the crista facialis. In this part of its course it is enclosed within the bony facial canal which develops by means of a supplementary ossification of the petrous without being preformed in cartilage. The exit from the facial canal is at the stylomastoid foramen. Primitively, this foramen was probably formed by the mastoid laterally, the petrous medially and posteriorly, and the floor of the bulla anteriorly.

With the changes associated with the development of the auditory bulla, the stylomastoid foramen has been displaced peripherally by the enclosure of the part of the facial nerve which is located immediately below the primitive stylomastoid foramen in a short sleeve of bone formed by the attachment of the tympanohyal to the periotic bone. The structure of this region has attracted much attention from comparative anatomists, notably Flower (1871), Howes (1896), van Kampen (1905), van der Klaauw (1929) and Sprague (1943, 1944). It appears that the early generalised mammalian condition was one in which the tympanohyal passed distally, from its attachment to the crista facialis, through the tympanic cavity to leave by the primitive stylomastoid foramen. Here it lay anterior to the facial nerve (a relationship termed protrematic by Howes). This arrangement is found, with but few modifications, in bats where the tympanohyal itself or a cartilaginous extension may pass through the stylomastoid foramen to connect with the stylohyal (Sprague, 1943).

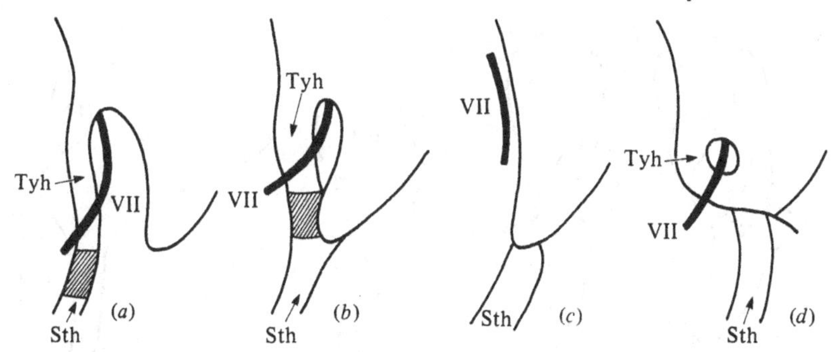

Fig. 78. To show stages in the formation of definitive stylomastoid foramen (see text for explanation). For labelling see list of abbreviations. (Based on van Kampen, 1905.)

Probably an early modification of this primitive arrangement was for the tympanohyal to become the actual anterior border of the stylomastoid foramen, instead of merely one of its occupants. Such an arrangement is still found in tenrec and erinac insectivores (Sprague, 1944). But in many of the more advanced mammals there has been a tendency for the attachment of the hyoid skeleton to the skull to develop a more posterior relationship to the sylomastoid foramen (the opisthotrematic position). As Sprague has demonstrated, the tympanohyal remains essentially protrematic in position and it is only the stylohyal which becomes opisthotrematic. The modifications involved are shown in Fig. 78, which is based on van Kampen's (1905) well-known illustration of the stages in the development of the opisthotrematic condition. In stage *b* the distal end of the tympanohyal is in contact with the mastoid (sometimes the petrous, or the paroccipital process in forms, such as the rabbit, where it is well developed) and, as a result, the stylohyal appears to articulate with structures behind the stylomastoid foramen. In stage *c* the tympanohyal has become indistinguishably fused with the periotic or paroccipital elements while in *d* only the distal end of the tympanohyal is so fused. The effect of these modifications is to lengthen slightly the facial canal by the addition of an extra collar of bone outside its original aperture and thus to create a new (definitive) stylomastoid foramen. Each of these stages is found in one or more species of insectivore (Sprague, 1944). In forms with an enlarged bulla the tympanohyal may become surrounded by this structure to lie within a sheath (vagina processus hyoidei). The distal end of the tympanohyal then projects from the sheath to articulate with the stylohyal, which may thus appear to be attached to the undersurface of the bulla. This condition is found in many eutherians, including the Tupaiidae and Macroscelididae amongst insectivores and in members of the Carnivora, Artiodactyla and Primates.

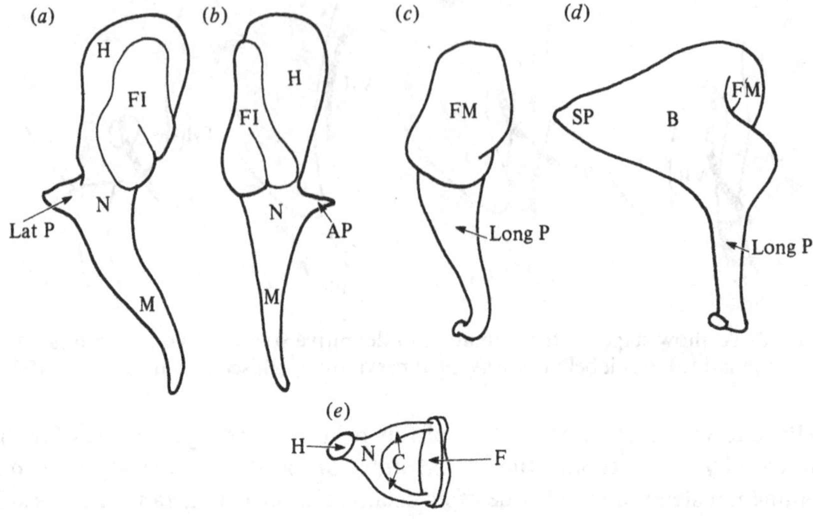

Fig. 79. The human ear ossicles. (*a*) Posterior view of left malleus; (*b*) medial view of left malleus; (*c*) anterior view of left incus; (*d*) medial view of left incus; (*e*) superior view of left stapes. Abbreviations: AP = anterior process; B = body; C = crura; F = footplate; FI = facet for incus; FM = facet for malleus; H = head; Lat P = lateral process; Long P = long process; M = manubrium; N = neck; SP = short process.

THE OSSICLES

Although the detailed proportions of the middle ear ossicles vary considerably from one eutherian group to another, their fundamental structure is remarkably consistent and allows a general plan to be described as a basis for the consideration of the functional significance of the ossicular chain and its variations, which is given in the next section.

The malleus (Fig. 79) possesses a head, manubrium, and anterior and lateral processes. The manubrium is embedded in the pars tensa of the tympanic membrane. In primitive mammals the anterior process connects with a sulcus (the sulcus malleolaris) on the upper surface of the anterior limb of the tympanic bone close to the petrotympanic (Glaserian) fissure through which the chorda tympani emerges. In the Insectivora this articulation is usually a fibrous ankylosis but in groups with well-developed sensitivity to high-frequency sound, especially Microchiroptera and also in Cetacea, it is more frequently an osseous ankylosis. The connection is lacking in many advanced mammalian groups but here the anterior process may remain attached by the anterior ligament of the malleus to the region of the petrotympanic fissure. It appears that a direct connection is often present in immature stages but is lost as age advances. This is the situation encountered in man where a firm connection is present in the child but, with increasing age, the bone of the anterior process becomes reduced and replaced by ligament. The lateral process of the malleus projects towards and may connect with the pars flaccida of the tympanum. The head of the malleus bears the articular facet for the incus.

The incus (Fig. 79) consists of a body, bearing the facet for articulation with the malleus, a long process which articulates with the head of the stapes and a short process which is attached by the posterior incudal ligament to a fossa (the fossa incudis) on the posterior wall of the epitympanic recess.

The stapes (Fig. 79) comprises two crura, a head for articulation with the long process of the incus, a neck and a footplate which occupies the fenestra vestibuli. The stapedial artery passes through the intercrural foramen. In most mammals this artery degenerates and is absent in the adult (see Appendix 1, p. 238), possibly because its pulsations would interfere with the conducting efficiency of the ossicles, but in bats the artery persists throughout life and may be very large (presumably its relatively slow pulsations do not interfere with the ultrasonic sensitivity characteristic of this group).

The incudomalleolar and incudostapedial joints are typically of the synovial variety, but in forms where a stiff ossicular chain is advantageous in conferring high-frequency sensitivity these joints may be fibrous. The stapedial footplate is usually attached to the margins of the fenestra vestibuli by an annular ligament. The anterior fibres of the annular ligament are generally longer and thinner than those situated posteriorly, indicating that most of the footplate movement takes place at its anterior margin.

Although the precise manner of functioning of the ossicles varies from one mammalian group to another, in association with adaptations of the ear to different styles of life (some examples of which are discussed in the next section), the basic features of the mechanism retain much in common and can be summarised as follows. The manubrium of the malleus is activated by movements of the tympanic membrane and constitutes the force lever arm of the ossicular chain. Consequent upon the movements of the manubrium, the malleus and incus rotate together about an axis which passes through the short process of the incus and the anterior process and anterior ligament of the malleus. The long process of the incus, the resistance lever arm of the system, and the stapes are therefore moved in the same direction as the manubrium. Since the area of the tympanic membrane is greater than that of the stapedial footplate and the force lever arm is usually about double the length of the resistance lever arm, the footplate moves with a greater pressure but over a smaller distance than does the membrane.

A variety of functions has been assigned to the two middle ear muscles, the stapedius and tensor tympani. In placental mammals the former muscle arises from the posteromedial wall of the tympanic cavity close to the fenestra vestibuli and is inserted into the neck of the stapes, while the tensor tympani takes its origin from the wall of the pharyngotympanic tube, or from this plus the floor of the nearby fossa in the petrous bone, or, in some species, from the floor of the fossa alone, and is inserted into the malleus close to the root of the manubrium. They are both involved in reflexes activated by auditory stimuli. Probably the most favoured theory has been that the two muscles act principally to protect the inner ear from sound of excessive intensity. This is consistent with the morphology of the muscles since it is apparent from their origins and insertions that their contraction will damp down movements of the ossicles. It is not clear, however, how the muscles could protect the inner ear from sudden loud sounds since it is doubtful if their reflex contraction would be speedy enough to achieve this. Furthermore, it seems likely that in natural (as opposed to man-made) environments the need for protection from loud sounds would arise but rarely. In view of these difficulties a number of authors (e.g. Haan, 1957) has returned to the old theory that the principal function of the stapedius and tensor tympani muscles is to act as a frequency selection device, by regulating the resonance frequency of the middle ear structures. Since contraction of the muscles will increase the rigidity of the ossicular mechanism, it will result (for reasons discussed below) in better transmission of high-frequency sound and suppression of low-frequency sound. By preventing slippage at the joints between the ossicles it may also decrease amplitude distortion of low-frequency sound (further discussion of the function of the middle ear muscles is beyond the scope of the present work; the interested reader is referred to Dallos, 1973, and Anderson, 1976).

FUNCTIONS OF THE MIDDLE EAR AND THE SIGNIFICANCE OF VARIATIONS IN ITS STRUCTURE AND IN THE PROPORTIONS OF THE AUDITORY BULLA

The principal function of the middle ear in non-aquatic mammals is to transmit the small vibratory changes in air pressure, which constitute airborne sound, to the detector mechanism of the inner ear with the minimum of energy loss. It involves, therefore, the transference of vibratory energy from an air to a fluid medium. If the acoustical impedances (see Appendix 2, p. 239) of two media are widely different, the transference of this type of energy between them will be very incomplete; most of the energy will be reflected at the interface. Air, being both less dense and less stiff, possesses a much lower acoustical impedance than the cochlear fluid. Although the middle ear structures have been traditionally described as functioning as a pressure transformer, serving to increase the pressure at the stapedial footplate relative to that at the eardrum, it is, in fact, more accurate to regard them as an impedance transformer which matches the acoustical impedance of the cochlear fluid with that of air (see, for example, Wever & Lawrence, 1954; Dallos, 1973; Webster & Webster, 1975; Tonndorf & Khanna, 1976).

The middle ear transformer mechanism acts by modifying both of the factors in the acoustical impedance ratio (see Appendix 2). As already noted, pressure at the stapedial footplate is increased because its area is less than that of the tympanic membrane and because the manubrium lever arm is longer than the incudal lever arm, while conversely velocity at the footplate is decreased by the configuration of the lever arms. The acoustical impedance of air is 41.5 dynes s/cm^3 while the impedance of the mammalian cochlea is probably of the order of 5600 dynes s/cm^3 (Zwislocki, 1965). For complete transmission of all the sound energy arriving at the tympanum the middle ear mechanism would need to transform the cochlear impedance to 41.5 dynes s/cm^3 as measured at the tympanic membrane.

It cannot, however, be assumed that the middle ear structures are themselves without impedance (i.e. that they match the impedances of cochlea and air with no loss of energy). In fact, they have an intrinsic impedance which is due to their mass, stiffness and friction. Of these three factors, friction is very small so that the impedance of the middle ear is due almost entirely to mass and stiffness. The mass of the middle ear structures is mainly that of the drum and ossicular chain, while their stiffness is related to the rigidity of the drum, the degree of fixation of the ossicles and the volume of air in the tympanic cavity (this last affecting stiffness by modifying the effective rigidity of the drum). Since the energy loss due to mass and stiffness is dependent on the frequency of the sound vibrations, with the loss from mass increasing and that from stiffness decreasing with increasing frequency, there will be one particular frequency (the resonance frequency) at which these two factors cancel each other out. The impedance of the middle ear is then that arising solely from friction and is, therefore, very low. At the resonance frequency the transfer of energy across the middle ear is maximal. Since the impedance of the cochlea is due, in large measure, to friction, it is independent of frequency. Thus, the middle ear plays the dominant role in determining the frequency of maximum auditory sensitivity.

In addition to acting as an impedance matching device, the relatively robust middle ear serves to protect the very delicate inner ear from excessively high energy input. As already described, the principal protective action is probably achieved by the reflex contraction of the middle ear muscles on exposure to sound, clinical evidence suggesting (in man at least) that contraction of stapedius is the more important in this respect. Further protection is provided by slippage at the incudomalleolar and incudostapedial joints, thus reducing the energy flow along the ossicular chain. Both of these protective actions are most effective at low frequencies.

The mammalian inner ear is capable of being stimulated by sound waves conducted through the bones of the skull as well as through the tympanic membrane and ossicular chain. In non-aquatic mammals, transmission through the ossicular mechanism is far more efficient than that through the bone pathway. So far as the great majority of such mammals is concerned, bone conduction is, therefore, of largely theoretical interest but in man this type of conduction has considerable clinical significance, which has led to it being extensively investigated (see, for example, Tonndorf, 1966). This topic is discussed more fully in the section dealing with aquatic mammals where bone conduction may well be of much greater physiological importance.

As already emphasised, the morphological characteristics of the middle ear structures are the principal limiting factors in determining the range of auditory acuity. The inner ear does show structural modifications in some mammalian groups which seem to be adaptations for increasing sensitivity in specific parts of the range, but in other cases the cochlea appears to be relatively unspecialised without restricting auditory acuity. For further details of cochlear structure in various mammals, the reader is referred to the extensive studies by Pye (1966*a*, *b*, 1967, 1970, 1977, 1979).

Webster & Webster (1975, 1978) have provided a detailed analysis of the function of the middle and inner ear in the Heteromyidae, which illustrates the principles just discussed. In these small rodents the mass of the middle ear structures is small, while their stiffness is very low as a result of thinning of the tympanic membrane, reduction in the number of ligaments and size of the muscles attached to the osicles, thinning of the annular ligament and enlargement of the tympanic cavity. This great reduction in stiffness is an adaptation to low-frequency sensitivity which is important in avoidance of predators. For example, in *Dipodomys merriami* maximum sensitivity (as judged by cochlear microphonic potentials) occurs at 500 Hz whereas in cats and guinea-pigs the corresponding figure is about 1000 Hz. Previous studies (Webster & Strother, 1972 – again using cochlear microphonics) have shown a strong correlation between the volume of the heteromyid middle ear cavity and low-frequency sensitivity, as would be expected because of the effect of volume on stiffness. It seems likely, therefore, that the greatly increased size of the middle ear cavity in these rodents is part of the adaptation to low-frequency sensitivity. The cochlea also exhibits structural features enhancing low-frequency sensitivity. Because of the very small mass of the ear ossicles, sensitivity does not decline drastically at higher frequencies, so that heteromyid rodents possess, in addition to low-frequency acuity, a good range of frequency sensitivity. Similar considerations seem to apply, at least so far as the middle ear is concerned, in the gerbilline rodents (Lay, 1972), in small primates (lemuriforms, lorisiforms and tarsiiforms – Cartmill, 1974) and possibly also in carnivores (Hunt, 1974). In all of these groups the auditory bulla is found to be markedly inflated in those species where low-frequency acuity appears to have a high survival value.

In those carnivores and other mammals where the bullar cavity is divided into two partly separated chambers by the presence of an incomplete septum, it seems likely that the secondary chamber acts as a Helmholtz resonator. At frequencies above the resonance frequency of the secondary chamber the acoustic mass of the air in the opening between the two cavities has a very high reactance and so eliminates the acoustic effect of the air in the secondary chamber. At frequencies below the resonance frequency, however, the reactance is low and the effective acoustic volume of the bulla then becomes that of the tympanic cavity proper and secondary chamber combined. The middle ear in these animals possesses, therefore, at one and the same time, the bullar volumes appropriate to either high- or low-frequency hearing.

There are two (at least) mammalian groups in which the auditory apparatus has assumed entirely new (amongst mammals) functions. These are the bats and the cetaceans. In the former group the ear has become part of an echolocation system which has largely taken over from vision as the principal locating sense as an adaptation to flying in the dark; in the latter group the ear has undergone profound structural modifications in its adaptation to underwater hearing and navigation.

The echolocative mechanisms of bats have been described in detail by Griffin (1958), Pye (1968) and Henson (1970). Although not falling within the main context of the present work, it is necessary, in order to understand the structural modifications seen in the chiropteran middle ear, to summarise briefly the other mechanisms involved in echolocation. The orientation sounds of bats are produced within the larynx and emitted through the mouth and nose (Novick & Griffin, 1961). Their frequency is in the ultrasonic range, between 12 kHz and 150 kHz (the upper frequency limit of human hearing is about 20 kHz; in many other mammals, however, this limit is considerably higher, e.g. in other anthropoid primates it is about 30 kHz, in the cat about 50 kHz, in the rat about 90 kHz; while in the whale it is of the same order as in bats, see below). Ultrasonic sound is particularly suitable for use in echolocation because its short wavelength gives great resolving power and directionality (since it tends to radiate as a beam in one direction). It is, however, more rapidly attenuated than lower-frequency sound. A further navigating advantage may accrue from the ease with which ultrasonic sound can be distinguished against a general background of lower-frequency noise. The echolocative sounds of bats are emitted in pulses which have a repetition rate of generally less than 10 per second but which may rise to 200 per second during the capture of prey. The pulse durations are usually of the order of one-fiftieth of a second. In some bats (Vespertilionidae and Molossidae) there is an orderly frequency sweep of about one octave during the pulse, which may be an important character of the emitted sound in allowing the bat's auditory apparatus to distinguish it from interfering noise of similar ultrasonic frequency. It is probable that all microchiropteran species and at least a proportion of megachiropteran species use ultrasonic echolocation. Ultrasonic sound sensitivity has also been demonstrated in rodents (Ralls, 1967) and insectivores (Gould, Negus & Novick, 1964) where it may be important in echolocation as well as communication.

In order to use the echoes produced by the reflection of the emitted sound from surrounding objects as locating stimuli, the bat's sense of hearing must be highly directional in all planes of space and, during the capture of prey, must presumably be able to focus on a particular echo source. The external ears of the Microchiroptera are extremely variable in their size and shape but in many cases have a highly complex form with elaboration of the tragus and antitragus. The funnel-shaped arrangement so produced may help to increase sound pressure at the drum but a more important function is probably to reduce sensitivity to sounds emanating from directions other than that in which the external ears are pointed and so increase directional discrimination.

However, as in other mammals, directionality of hearing is doubtless largely dependent (so far as the peripheral mechanisms are concerned) upon the fact that the two ears function independently and so can detect intensity and phase differences in the sounds received at the right and left ears.

The cochlea of bats is generally enlarged, especially in the insectivorous Microchiroptera, and has a structure suggestive of sensitivity to high-frequency sounds (Pye, 1966*a*, *b*, 1967, 1970). Studies of cochlear microphonic potentials (see Henson (1970) for review and bibliography) have shown that the maximum amplitude of the potentials recorded at the fenestra cochleae varies with frequency, being less at very high frequencies (70–100 kHz) than in the lower ranges, but this diminution with increasing frequency appears to be less than for other mammals. The ossified otic capsule in the Microchiroptera is generally lacking a firm union with the rest of the skull. In many species the basicapsular fissure is wide so that there is no contact between the periotic and the basioccipital bone and the gaps between the adjacent bone surfaces may be occupied by blood sinuses or loose connective or adipose tissue. The resulting isolation of the cochlea probably serves to reduce bone conduction of the sounds created at the larynx during pulse emission.

As already noted, the middle ear structures are the crucial factors determining frequency sensitivity. Numerous investigators have studied the chiropteran middle ear in an attempt to correlate its structure with its role as a high-frequency conductor system. Hinchliffe & Pye (1969) and Henson (1970), in particular, have provided detailed surveys of the morphology of this structure in a wide variety of bat species. In general, the chiropteran middle ear resembles that found in many insectivores and rodents belonging to Group 1 (entotympanic) of Novacek's (1977) classification. In most microchiropteran species, ossification of the bullar floor is incomplete, the entotympanic remaining cartilaginous and failing to reach the periotic so that the intervening part of the floor remains membranous. Usually, the bulla is only moderately inflated, with virtually no trabeculation, and there is no pneumatisation of neighbouring bones. The tympanic cavity is of simple shape, possessing just the two basic subdivisions of tympanic cavity proper and epitympanic recess.

The ear ossicles are small and lightly built, often with deep grooves to reduce their mass further. The incudomalleolar joint is usually of the synovial type but the incudostapedial joint is more frequently of a fibrous nature. As in other mammals with a well-developed sensitivity to high-frequency sound (e.g. cetaceans, some rodents and carnivores), the ossicles in the Microchiroptera are attached quite tightly to the bones surrounding the tympanic cavity. The malleus is fixed by its anterior process to the sulcus malleolaris of the tympanic bone by a bony ankylosis; the incus is attached to the incudal fossa on the posterior wall of the epitympanic recess by a strong posterior incudal ligament; the stapes is tightly bound down to the margins of the fenestra vestibuli by a well-developed annular ligament. The ossicular mechanism possesses, therefore, a considerable degree of stiffness which is further enhanced by the large mass of the middle ear muscles and the relatively uninflated nature of the bulla.

The microchiropteran tympanic membrane is considerably thinner than that of other mammals possessing a similar membrane area. In general, the surface area of the drum, the size of the ossicles and the area ratio of drum to stapedial footplate are all smaller, while the mechanical advantage arising from the relative lengths of manubrium and incudal lever arms is somewhat greater in Microchiroptera which emit very high frequency pulses (50–125 kHz) than in those utilising frequencies below 50 kHz. The effect of the reduced area ratio is to diminish pressure at the footplate and hence the impedance-transforming ability of the middle ear. The effects of an ossicular lever arm ratio of increased mechanical advantage are to increase pressure but to reduce the rate of volume of displacement at the cochlea, both of which act to increase impedance-transforming ability. The ossicular lever arm ratio is thus a more potent factor than the area ratio in its effect on impedance matching. The arrangement seen in the high-frequency Microchiroptera allows, therefore, a small area (and hence mass) of the tympanic membrane to be achieved with probably little or no loss in impedance matching.

The combination of an ear drum and ossicles with small masses, and a high degree of stiffness of the ossicular mechanism found in the Microchiroptera emitting very high frequency pulses will, because of the nature of the relationship between the energy losses due to these factors and sound frequency, increase the resonance frequency and so enhance acuity to very high frequency sound.

The size of the tensor tympani and stapedius muscles is relatively very large in all bats but especially in the Microchiroptera. Relative to the mass of the drum and auditory ossicles the mass of the tensor tympani is twice as great and that of stapedius 7 times as great in *Myotis lucifugus* as in the cat (Wever & Vernon, 1961), while relative to body weight the volume of the two middle ear muscles combined is some 1.5 to 6 times greater in a variety of Megachiroptera than in the shrew *Cryptotis parva* (Henson, 1970). Studies of the physiological characteristics of the middle ear muscle reflexes indicate that these differ from one bat species to another but that, in general, the middle ear muscles perform both protective and analytical functions as well as adding incidentally to the stiffness of the ossicular chain (further consideration of this topic is beyond the scope of the present work, but the interested reader is referred to the excellent reviews by Griffin, 1958, and Henson, 1970).

Although the functional adaptations in the cetacean ear appear, at first sight, to be quite different from those just recounted for the bats, there is now much evidence that the dolphins and porpoises, and probably also the larger toothed whales, use high-frequency echolocation (commonly termed echonavigation or sonar in the underwater situation) in a manner essentially similar to that of bats. Such a method of navigation has manifest advantages, as compared to one based on sight, for animals that live by catching prey in water frequently obsured by sediment and other particulate matter and which often dive to depths where the amount of light penetrating is very low even in the clearest water conditions. Although precise navigation is a less imperative need in the whalebone whales so far as feeding is concerned, it is still required in order that these very large creatures can avoid underwater obstacles. Detailed investigations of voice production and hearing in these huge creatures are lacking, but it seems likely that the whalebone whales (except possibly the grey whale *Eschrichtius glaucus*) do not emit high-frequency sounds equivalent to those produced by odontocetes. It remains possible, however, that they use sound in the sonic range as an echolocative device.

Echonavigation has been most intensively investigated in the bottlenose dolphin (*Tursiops truncatus*) (see, for example, Kellog, Kohler & Morris, 1953; Kellogg, 1961; Tavolga, 1964). This animal, like many if not all odontocetes, produces a series of characteristic noises including whistles and clicks (the latter, each lasting less than 10 ms, are emitted characteristically in a rapid sequence of variable pulse repetition rate, giving an impression of a sound like a squeal or bark). The whistle appears to have a frequency range of 7 to 15 kHz, while the click, although being predominantly in the sonic range, contains vibrations above 20 kHz and possibly extends as high as 120 kHz. It seems likely, therefore, that the dolphin uses a wide range of sound, probably in a complementary fashion – the sonic frequencies, since they travel over great distances but have poor resolution, being used to detect distant objects, and the ultrasonic frequencies, since they have high resolution but are attenuated rapidly with distance, being used to examine nearby objects in detail. The sonic frequencies are doubtless also used in communication. The clicks appear to be produced within the diverticula of the nasal passage below the blowhole (see Chapter 7), while the whistle and other lower-frequency sounds may be produced within the larynx, although the precise mechanism of their production is unknown.

In order to make use of echolocative sounds it is essential that the ear be capable of determining the direction of the sound it is perceiving with great accuracy. In terrestrial mammals, directionality, as we have seen, is partly a function of the external ear and partly (and probably more importantly) related to the ability of the two ears to work independently and so detect differences in timing, phase and intensity of the sound arriving at the right and left sides of the head. Underwater, an external ear can have no sound-gathering function since physically its constitution (and, therefore, its acoustic impedance) differs little from the medium in which it is immersed. A further difficulty in determining the directionality of noises in a water medium, so far as the typical air-adapted mammalian ear is concerned, arises from the fact that because of the close impedance matching between the medium and the structures of the body, underwater sound (in the sonic range) will cause the whole head to vibrate and thus effectively prevent the right and left ears from acting independently. The inability of the human ears to determine sound direction underwater has been attested by several observers (e.g. Haan, 1960). Because of the poor impedance matching between the typical air-adapted mammalian ear and the water medium, there is also a reduction of hearing sensitivity underwater. In submerged men, for example, the sensitivity threshold is raised by about 30 dB at the most sensitive frequency. The air-adapted middle ear is, in fact, not merely superfluous but actually a hindrance to underwater hearing.

These problems of hearing underwater are rendered more acute by the nature of sound propagation in water, which differs in several important respects from that in air. Sound waves travel about four times faster in sea water than they do in air, and their transmission is affected by pressure (depth), temperature, salinity and the presence of sediment. Because water is a much heavier and more rigid medium than air, sound pressure must be about 60 times greater under water to produce the same sound intensity as in air. The amplitude of the waves in water is smaller by a corresponding amount than in air. As well as being reflected from the bottom or from any suspended object, sound waves are also reflected from the surface because of the poor impedance matching of air and water. Thus sounds may travel in complex paths, involving several reflections from surface or bottom, especially in shallow water. In deep water, however, such factors produce little distortion and low-frequency sound waves will travel over very great distances, owing to their very low attenuation in the fluid medium.

Clearly, in order to be used as an acoustic analyser in an echolocative system, the ear must have undergone quite drastic structural modification during the evolution of the Cetacea to readapt it from its function as a receptor for airborne sound to that of a water-borne sound receptor. As might be expected, the external pinna, now rendered functionless, has been lost, a process doubtless accelerated by the need for streamlining. The external acoustic meatus is a very narrow sinuous canal in the toothed whales. In the whalebone whales the meatus begins as a narrow opening, usually becomes reduced to just a strand of connective tissue in the blubber layer and then opens out again into a patent and expanded canal (filled with a plug of keratin and cholesterol) just before the tympanic membrane is reached. The part of the meatus in the blubber layer is surrounded by muscular tissue representing the auricular muscles of other mammals while the deeper parts of the canal are partly surrounded by cartilage.

The cetacean middle ear shows equally marked structural modifications. Detailed descriptions of this region have been provided by Haan (1957) and Fraser & Purves (e.g. 1960*a*, *b*), who also provide summaries and extensive bibliographies of the previous literature. The auditory bulla appears to be composed entirely of the tympanic bone (i.e. is of the Group 2 type) and is located ventral to the petrous, which is an exceedingly dense and massive element. The two are connected, more or less firmly, to form the petrotympanic unit. In the whalebone whales the connections are by means of two bony pedicles, one situated anteriorly and the other posteriorly, while in the toothed whales the connections tend to be fibrous rather than bony and the anterior connection is small and may be absent. Part of the mastoid component of the periotic is fused with the petrotympanic while the remainder is incorporated into the squamous.

The most striking modification of the cetacean ear, however, is undoubtedly the trend towards isolation of the petrotympanic unit from the remainder of the skull by the presence of large air-containing extensions from the middle ear cavity and nasopharynx. These pneumatic sinuses may include: (1) the pterygoid sinus, a large sac opening into the nasopharynx and extending forwards ventral to the bones of the cranial base to occupy the hollow between the palatine and pterygoid bones; (2) the anterior sinus, that part of the pterygoid sinus extending forwards of the pterygoid bone; (3) the peribullary sinus, occupying the space which separates the petrotympanic unit from the neighbouring bones of the cranial base; (4) the posterior sinus, occupying the concavity of the paroccipital process; (5) the middle sinus, situated under the zygomatic process, posteromedial to the mandibular fossa. The tympanic cavity communicates with the system of pneumatic sinuses through a hiatus epitympanicus and the tympanoperiotic fissure. The sinuses in the Odontoceti are, in most instances, greatly enlarged and the pterygoids are deeply excavated and expanded. The isolation of the petrotympanic from the rest of the skull is, therefore, virtually complete, any remaining connections being ligamentous. In the Mysticeti the degree of pneumatisation is much less and a certain amount of sutural connection with the remainder of the braincase is maintained, especially posteriorly (see Fraser & Purves, 1960*b*, for detailed comparative descriptions).

During the course of development the sinuses enlarge progressively to occupy spaces left between the petrotympanic and the basioccipital and squamous bones by a downward displacement of the petrotympanic relative to the skull base. Continued expansion of the pneumatic sinuses is achieved by removal of bone tissue from the neighbouring elements of the skull. The vascular elements of the removed bone, however, remain intact to form large plexuses in the walls of the pneumatic sinuses. The plexus in the wall of the peribullary sinus adjacent to the basioccipital bulges through the tympano-periotic fissure into the tympanic cavity where it is termed the corpus cavernosum tympanicum. The plexuses are supplied by the internal maxillary artery.

The walls of the pneumatic sinuses are composed of tough fibrous tissue that is especially well developed on the externally orientated surfaces to which the jaw muscles gain attachment. They are lined with a mucous membrane containing numerous mucous glands. The vascular plexuses lie within the fibrous layer. The interior of the sinuses is subdivided by trabeculae and is occupied by an albuminous foam consisting of bubbles of air trapped within

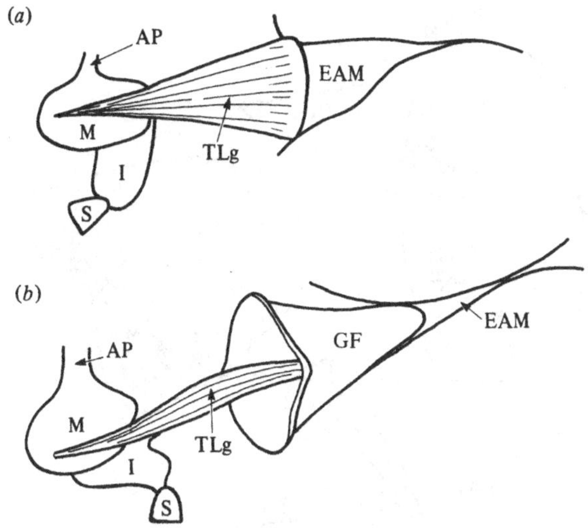

Fig. 80. To show the tympanic ligament of the cetacean middle ear. (*a*) Toothed whale; (*b*) baleen whale. Abbreviations: AP = anterior process; GF = glove finger; TLg = tympanic ligament; for other labelling see list of abbreviations. Not to scale. (After Fraser & Purves, 1960*b*.)

a continuous phase formed by an emulsion of oil and mucus in which the oil droplets are of uniform size, having a diameter of about 1 μm.

The cetacean tympanic membrane has a very characteristic structure. In the toothed whales the drum has a concave external surface with an internal extension, the triangular tympanic ligament, projecting from the centre of the drum to attach to the much reduced manubrium of the malleus (Fig. 80*a*). In the whalebone whales the membrane is convex externally, being generally likened to a glove finger pointing into the proximal part of the external acoustic meatus (Fig. 80*b*). From the upper surface of the internal surface of this glove finger a fibrous tympanic ligament extends to the manubrium of the malleus. This ligament is closely comparable with the triangular ligament of odontocetes and, in both cases, the ligament approaches the manubrium at a very acute angle.

The ossicles of the cetacean ear display a number of morphological differences from the arrangement typical of eutherians. The malleus (Fig. 81*a*, *b*) has a rather massive appearance. It is attached by a stout anterior process to the margin of the tympanic (at the petrotympanic fissure, as in other mammals where this connection is found). This attachment is buttressed by a strong projection, the sigmoid process, of the tympanic. The manubrium is greatly shortened, being little more than a tubercle on the head, and is of approximately the same length as the long process of the incus. It receives the attachment of the acutely orientated tympanic ligament into its tip. Apart from its attachment to the tympanic and articulation with the incus, the malleus lacks connections, either bony or ligamentous, with other skeletal

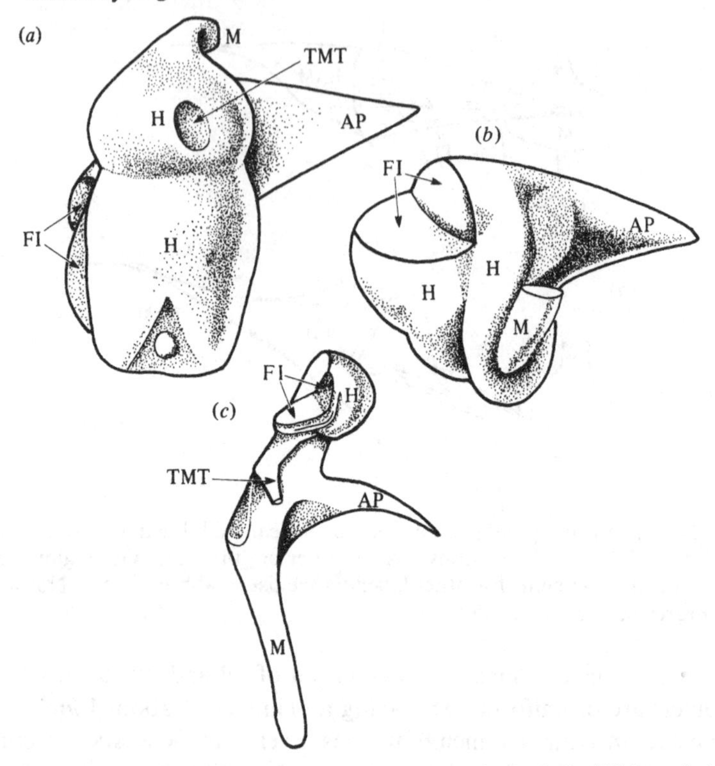

Fig. 81. (*a*) Medial, (*b*) superior view of malleus of toothed whale (*Globicephala*); (*c*) medial view of malleus of cat. TMT = process for attachment of tensor tympani; other abbreviations as in Fig. 79. Magnification of (*c*) approximately 1.5 × that of (*a*). (After Haan, 1957.)

elements of the bulla. The incus is likewise relatively massive and articulates at a very precisely fitting joint with the malleus. The attachment of its short process to the wall of the bulla is usually well developed. The stapes is small but heavily built and its footplate is very firmly attached to the margins of the fenestra vestibuli. The tensor tympani muscle is present but often mainly tendinous.

The manner in which the cetacean ear functions has been the subject of investigation and speculation by comparative anatomists since the middle years of the eighteenth century. The theories that have been advanced (succinctly summarised by Haan, 1957) are numerous and, in many instances, imaginative but most have been rendered obsolete by the more recent studies.

One of the most convincing attempts to relate the various morphological features peculiar to the cetacean ear to its function as an underwater acoustic analyser has been made by Fraser & Purves (1960*a*, *b*). They pointed out that the presence of the air-filled sinuses has the effect of acoustically isolating the petrous, bulla and external meatus from the other tissues of the head since virtually all the sound energy will be reflected at the air–water (flesh) interface. This acoustic isolation is preserved at all pressures encountered, including those during deep diving, because of the properties of the foam occupying the pneumatic sinuses and the presence of the vascular plexuses. Fraser & Purves have shown that while foam of the consistency found in the sinuses will become reduced in volume under increased pressure (they used pressures up to 100 kPa), an equilibrium is eventually reached in which no further diminution occurs in bubble size because of the rigidity of the continuous phase. As hydrostatic pressure increases, the vascular plexuses will become engorged with blood, thus providing a mechanism for reciprocal filling of the sinuses as the volume of the foam is decreased. Similarly, distension of the corpus cavernosum tympanicum will compensate for the reduction of the air volume in the tympanic cavity as pressure increases and so maintain the pressure equilibrium between the cavity and the exterior. The large surface area presented by the dispersed oil droplets in the foam and the high absorption coefficient of nitrogen in oil are together sufficient to ensure that during diving the nitrogen within the sinuses will be absorbed into the foam rather than entering the blood of the vascular plexuses. The efficiency of foam in damping vibrations increases as the diameter and separation of the contained bubbles diminish; thus acoustic isolation is maintained during diving, despite the reduction in volume of the foam.

Frazer & Purves have provided evidence, by means of acoustic probes in material preserved by deep freezing, that the cetacean external acoustic meatus, although narrow or even completely closed, is capable of acting as a muscularly controlled, acoustically efficient, vibration conduit through which sound travels with less attenuation than it does through the surrounding soft tissue. Vibrations transmitted through the bones of the skull at points remote from the meatus will be largely reflected at the interface between the walls of the tympanic cavity (and surrounding sinuses) and the contained air. Hence, the sound transmitted through the meatus will be dominant over sound transmitted along any other path.

The characteristic morphology of the cetacean ear ossicles can be related to the nature of sound propagation underwater and, in particular, to the greater pressure amplitude of the sound waves. Fraser & Purves have provided an elegant demonstration, by means of a model of the middle ear mechanism of the fin whale, of the way in which the ossicular mechanism is accommodated to this increase in pressure amplitude. In the model the tympanic ligament is simulated by a transducer, activated by a variable frequency oscillator, attached to the tip of the manubrium of the malleus in such a way that the angle made between the transducer and the manubrium can be varied. The malleus and incus occupy their usual relationships to each other while the stapes is represented by the stylus of a crystal pick-up connected to an oscilloscope. The amount of deflection of the stapes is greatly influenced by the angle between the transducer and the manubrium,

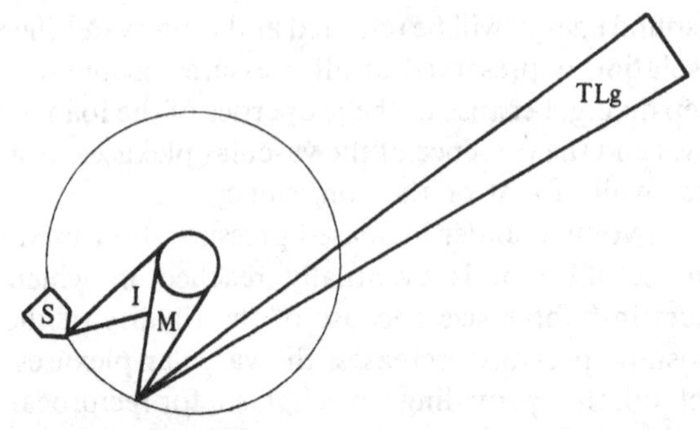

Fig. 82. To illustrate the crank mechanism described by Fraser & Purves (1960*a*, *b*). The more acute the angle at which the tympanic ligament approaches the malleus, the greater the magnification of movement of the stapes. TLg = tympanic ligament; for other labelling see list of abbreviations.

being much greater when the angle approaches that found in the relationship between the tympanic ligament and the manubrium in the intact ear (less than 5°) than when it is 90°. The authors suggest that the manubrium is behaving like a crank mechanism in which the amount of rotation increases as the angle of approach of the piston to the crank decreases (Fig. 82). The displacement amplification achieved by this mechanism is of the order of 30:1 and the pressure reduction conversely 1:30. If the middle ear of the cetacean is compared with that of man, taken as an example of a terrestrial mammal, it is found that the area ratio between drum and footplate is of the order of 30:1 in both because, although the tympanic area in the cetacean is reduced relative to that in man, the footplate area is reduced commensurately. However, the ratio of the lever arms in the cetacean ear is 1:1, in comparison with 2:1 in the human ear, a modification which by itself would result in the pressure at the fenestra vestibuli of the whale being only half that in man. This is compensated for by the fact that the pressure amplitude of a sound wave in water is some 60 times greater than that in air while the tympanic ligament – manubrium crank mechanism reduces pressure by about 30 times. The net effect of these two factors is equivalent to the 2:1 pressure magnification produced by the human lever arm ratio so that the pressure amplitude at the inner ear of the fin whale is approximately the same as that in man.

Although Fraser & Purves's analysis of auditory function in the whales has received wide acceptance, there is by no means universal agreement over some of the details of their theory. In particular, doubts have been expressed about the role of the external acoustic meatus as a sound conduit and the effect of the surrounding muscle on its conducting ability. These questions have considerable significance in interpreting the way in which the direction of sound is perceived. As described above, determining sound direction underwater is virtually impossible with the typical ear of terrestrial mammals because the external pinna would be rendered functionless and the close

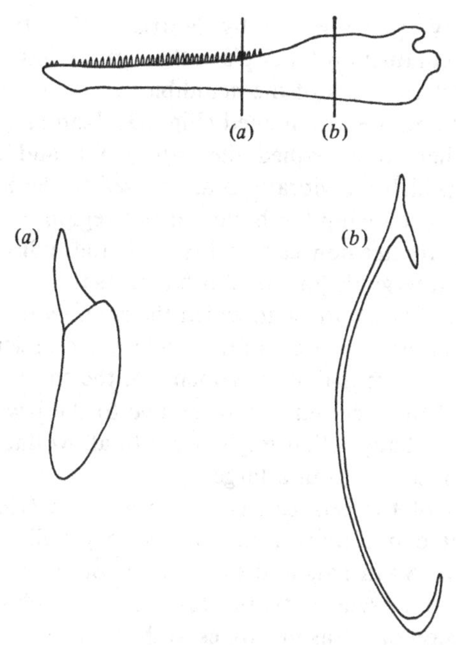

Fig. 83. To show pan bone of dolphin mandible. (*a*) Coronal section through region of dentition; (*b*) coronal section through region of pan bone. Approximate position of sections shown above.

impedance matching between water and the tissues of the head would make it impossible for the two ears to be able to work independently. The acoustic isolation of the petrotympanic in the cetacean, by contrast, enables the two ears to function independently and thus allows directionality of underwater sound perception. Fraser & Purves have proposed that the presence of external acoustic meatus capable of acting as sound conduits would enhance time, phase and, particularly, intensity discrimination between right and left sides by effectively increasing the distance between the two areas at which the sound is received (the distance between the openings of the right and left meatus being considerably greater than that between right and left petrotympanics).

However, Haan (1957) has suggested that the external acoustic meatus in whales has little or no acoustic function. His principal reason for this view is that blubber and water differ little in their ability to transmit sound. Since a considerable length of the meatus is surrounded by blubber, Haan believes that it is unlikely that the meatus will provide a preferential path for sound conduction (although Fraser & Purves's practical demonstration seems to prove otherwise). For Haan, therefore, the crucial point at which sound waves are received into the ear system is the base of the tympanic ligament, not the opening of the meatus.

A third possibility, put forward by Norris (1964, 1969), is that in odontocetes, sound conduction takes place through the skin and superficial blubber over the posterior part of the mandible. The bone in this region of the jaw is expanded and greatly thinned (Fig. 83), being almost paper-thin in some species (where it is termed the pan bone), and could act as a membrane through which the vibrations are passed to the intramandibular fat body. The latter, occupying the hollowed-out region medial to the pan bone and possessing an extension leading directly to the auditory bulla, would then act as a passive wave guide (fat possibly having better sound transmission properties than other tissues) through which the sound vibrations reach the bulla and hence the anterior process of the malleus and ossicular chain. The advantage of such a system is that movements of the snout, by varying the angle of incidence of the received sound relative to the jaw surface, would produce transmission changes that might serve to allow fine angular discrimination of echoes received from a target.

In his description of the cetacean ear ossicles Haan (1957) emphasises the stiffness of their connections with the auditory bulla and margin of the fenestra vestibuli. As in bats and other terrestrial mammals with well-developed high-frequency sensitivity, this feature is undoubtedly an adaptation to the transmission of ultrasonic tones to the inner ear. However, in the cetaceans the need for stiffness in the ossicular chain is enhanced by the fact that, to obtain good transmission of sound waves arriving through water, where the sound pressure is much greater than in air, the mass of the ossicles must be proportionately greater than in an air-conducting system. Since mass and stiffness affect the resonance frequency in opposite directions, there is thus a further need for increased stiffness to offset the greater ossicular mass. Hence, the situation encountered in whales represents a compromise in which the ossicles are heavy enough for good sound transmission but not so heavy as to make the reception of very high frequencies impossible.

It is instructive to compare these characteristic features of the cetacean auditory apparatus with those of the pinniped carnivores. In many ways the hearing problems in these amphibious mammals are even more acute than those in the fully aquatic whales because of their need to hear well both in water and in air. Information about the structure and function of the pinniped ear is rather sparse, but the studies by Møhl (1968 – this reference also contains a review of the previous literature), Ramprashad, Corey & Ronald (1972) and Repenning (1972) allow certain broad conclusions to be drawn.

Møhl (1968) has described the air and water audiograms for the harbour seal (*Phoca vitulina*). Its hearing sensitivity in air is some 15 dB inferior to that in water, while maximum sensitivity occurs at about 12 kHz in air compared to 32 kHz in water. The upper frequency limit of hearing (with pitch discrimination) in water is of the order of 60 kHz while the upper limit in air is probably much lower. It is clear, therefore, that the phocid ear is much better equipped for the reception of water-borne than of airborne sound.

Nonetheless, hearing ability on land is still quite good. The fall of 15 dB in hearing sensitivity between water and air can be compared, for example, with the fall of 30 dB in hearing sensitivity of the fully air-adapted ear of man when submerged. Møhl has also demonstrated that the phocid seals have good directional hearing, both in air and under water. Such information as is available regarding the hearing capabilities of the otariid seals and the odobenids indicates that these too hear well, and with good directional sense, in air and water.

The mechanism of hearing in air by the pinnipeds appears to follow the general mammalian pattern. The high-frequency cut-off at a relatively low level noted in the harbour seal (and presumably present in other phocids) is probably attributable to the large mass and rather loose suspension of the ossicles. The inner ear itself, like that of whales and bats, appears to be well adapted to high-frequency hearing, but presumably this adaptation is prevented from becoming operative in air because of the characteristics of the middle ear.

The mechanism of pinniped hearing under water is still uncertain. There appear to be three possibilities. First, water-borne sound could be transmitted through the tympanic membrane and ossicular chain in the same way that airborne sound is conducted. It has already been pointed out that this is an inefficient mechanism underwater because of the poor impedance matching between the water environment and an air-adapted middle ear mechanism; however, there are reasons, to be discussed below, for believing that this mechanism may operate in pinnipeds below certain depths. The two other possibilities are both varieties of bone conduction. Repenning (1972) terms these resonant reaction and conductive reaction.

In resonant reaction the whole skull moves together with the surrounding water under the influence of the water-borne sound. The ossicles and cochlear fluids participate in this motion but, because of inertial effects, at amplitudes and phases different from those of the skull. The inertial effect on the ossicles will result in slight relative motion between the footplate of the stapes and the fenestra vestibuli. In conductive reaction, sound waves which cross the water–flesh and the flesh–bone interfaces pass through the bones of the skull to the petrous which is thus slightly distorted. This distortion of the bony capsule about the cochlea causes adjustments in the cochlear fluids. Since the fenestra vestibuli is probably less compliant than the large fenestra cochleae and since the distorting walls on the vestibular side have a larger surface area than those on the tympanic side, the tendency will be for greater fluid displacement in the scala vestibuli than in the scala tympani, thus tending to displace the cochlear partition towards the latter. Repenning (1972) suggests that the Phocidae display structural modifications of the ear indicating that they are adapted for hearing underwater principally, although not exclusively, by resonant reaction, while the Otariidae and Odobenidae have structural specialisations for hearing principally by conductive reaction.

The pinna is completely missing in the Phocidae, doubtless to reduce hydrodynamic resistance. The external acoustic meatus is long, narrow and tortuous and filled with wax-covered hairs. An area of vascular cavernous tissue is present beneath the lining of the meatus. The auricular cartilage is modified to surround the meatus and it seems likely that the auricular musculature, also much modified, can control the opening and possibly also the closing of the canal. The ossicles are massive, as in the Cetacea, but not so rigidly suspended as in that fully aquatic group. They are of characteristic form – the malleus lacks an anterior process, the incus possesses a greatly enlarged upper part of the body (the 'head'), and the stapes has an enlarged footplate. In the harp seal (*Pagophilus groenlandicus*) the ossicles are attached by a single fold of mucous membrane from the roof of the epitympanic recess (Ramprashad, Corey & Ronald, 1972). The tympanic bulla, composed of entotympanic and tympanic ossifications (Group 1, Type B), is moderately inflated although not more so than that of many fissiped carnivores. The fenestra vestibuli is large and, since the tympanic membrane is of normal mammalian size, the area ratio between membrane and fenestra is low. The petrous tends to be loosely attached to the neighbouring cranial bones, apart from the squamomastoid, and has a large mass of bone forming its apex on the side of the cochlea opposite the squamomastoid. The cochlea capsule is itself thin-walled. The fenestra cochleae is greatly enlarged and in some phocids opens outside the tympanic cavity through a foramen located, posterior to the stylomastoid foramen, at the bulla–mastoid junction.

Repenning (1972) suggests that the great mass and relative slackness of the phocid ossicles represent an adaptation for an inertial effect during resonant reaction to water-borne sound and that the presence of the large 'head' of the incus, which presumably lies above the rotational axis, augments this effect. The large size of the fenestra cochleae may well serve to augment further ossicular inertia at the fenestra vestibuli by reducing containment of the cochlear fluids. In addition, the basal whorl of the phocid cochlea is greatly enlarged which would serve to enhance inertial effects resulting both from ossicular movement and from direct action on the cochlear fluids.

Although the most obtrusive characters of the phocid ear thus appear to reflect adaptation to resonant reaction, those features believed to augment conductive reaction are also well developed. In particular, the squamous part of the temporal bone is large and may act to accentuate the selective reflection effect and so enhance directional hearing (see below) while the detachment of the petrous from bones other than the squamomastoid would presumably restrict the source of stimulation of the cochlea by conductive reaction to that received through the squamous route. Furthermore, the thin walls of the cochlear capsule and the mass of bone at the apex of the petrous appear to provide an ideal construction for maximum distortion by conducted sound waves.

All pinnipeds possess distensible cavernous tissue within the mucous membrane lining the tympanic cavity. This tissue reaches its maximum development in the phocids. During diving (which may be to depths of some hundreds of metres) it is presumed that the cavernous tissue becomes distended with blood and so helps equalise the pressure within the tympanic cavity. Repenning (1972) has suggested that at depths of about 100 m this cavernous tissue and also that in the external acoustic meatus would be distended to a point where it would contact virtually all of both surfaces of the tympanic membrane and that in this condition the membrane would act as though immersed in water. No reflective barrier would then exist to prevent the membrane–ossicle mechanism from reacting to water-borne sound in much the same way that it reacts to airborne sound. The reduced area ratio between tympanic membrane and fenestra vestibuli may then act as a protective device to avoid overstimulation at the cochlea from the greater pressure of water-borne as compared to airborne sound.

The major contrasts between the ears of otariids and odobenids, on the one hand, and those of the phocids, on the other, are related by Repenning (1972) to the respective specialisations for hearing predominantly by conductive and by resonant reaction. In the otariid seals the fenestra vestibuli and cochleae are both enlarged (the latter opening into the tympanic cavity). The tympanic membrane is extremely small and the area ratio consequently much reduced. The ossicles are not enlarged relative to those of fissiped carnivores. Cavernous tissue is present in the lining of the tympanic cavity but is not so well developed as in phocids. The auditory bulla is composed largely of the tympanic ossification (see Group 1, Type A). As in phocids, the petrous is detached from all bones save the mastoid and possesses a relatively massive apical portion.

The most distinctive feature of the otariid acoustic apparatus is the enlarged mastoid process which has flattened external surfaces facing laterally and ventrally. The amount of sound reflected at an interface between two media is related to the difference in the speed with which sound is transmitted in the media and the angle of incidence of the sound. The critical angle of incidence at a water (flesh)/bone interface is about 35° from the normal to the interface; sound incident at a greater angle than this will be reflected. The structure of the otariid mastoid, with its large external surface, may represent an adaptation to accentuate selective reflection. This, together with the mastoid–petrous fusion, the thin-walled cochlear capsule and petrous apical mass, would presumably enhance the efficiency of sound reception by conductive reaction as well as adding directionality to the sense of hearing.

In the Odobenidae the auditory structures display many similar specialisations to those encountered in the otariid seals, apart from a rather bigger area ratio due to the large size of the tympanic membrane without a corresponding enlargement in the area of the fenestra vestibuli, and a more restricted development of the cavernous tissue in the lining of the tympanic cavity. Like the otariids, the odobenids have greatly enlarged mastoid processes. It seems likely, therefore, that underwater hearing adaptation in odobenids, as in otariids, consists of improvements in the reception of sound by conductive reaction. The small amount of cavernous tissue and the large tympanic membrane may represent retention of primitive features and would both appear to act as limitations to the depth of diving.

APPENDIX 1: THE INTERNAL CAROTID ARTERY

It will be recalled that primitively the internal carotid artery enters the cranial cavity by passing through the hypophysial foramen in close proximity to the pituitary gland (Chapter 2). In land vertebrates the opening for the artery is located either in the lateral part of the basisphenoid (in amphibians, reptiles, monotremes, marsupials and some insectivores and chiropterans) or close to the middle lacerate foramen (in remaining eutherians). The carotid circulation in mammals has undergone a series of changes in its pattern of distribution which, because of the proximity of the vessels to bone, can sometimes be traced in fossil forms.

In typical reptiles the common carotid gives rise after a short course to a small, external carotid artery (to tongue and pharynx) and then continues, as the internal carotid, towards the head to enter the cranioquadrate passage. Near the base of the skull the internal carotid gives off the large stapedial artery (so called because it perforates the stapes – a relationship explained by its development from an anastomosis between first and second arterial arches) which supplies, through supraorbital, infraorbital and mandibular branches, the extrabulbar contents of the orbit (the eyeball itself being supplied by the ophthalmic branch of the intracranial part of the internal carotid) and the jaws. Before entering the cranial cavity through the foramen in the basisphenoid, the internal carotid gives off the palatine artery to the roof of the mouth. The carotid circulation in the early mammals is unknown but, so far as the ontogeny of the cranial arterial system of living mammals is a guide, it appears likely that it followed the typical reptilian pattern.

Bugge (1974) has proposed a possible scheme of phylogenetic changes by which the carotid arterial patterns encountered in modern mammals might have been derived. According to this scheme, the early mammalian condition was for the internal carotid to bifurcate into medial and lateral divisions, the point of division lying near the posteromedial corner of the promontory from where the medial division continued along the basicapsular groove. The lateral division underwent further furcation into stapedial and promontory branches. The stapedial branch was still a large artery supplying extrabulbar orbit and jaws while the promontory branch continued across the promontory to enter the cranial cavity through an opening in the sphenoid. Bugge suggests that in some living mammalian groups the promontory branch has been lost while in others it is the medial division that is missing – in the former cases the internal carotid artery represents the persisting medial division and runs medial to the periotic while in the latter the internal carotid is the promontory.

This widely accepted view of the origin of the two types of internal carotid artery has recently been challenged by Presley (1979) who has studied the embryonic development of both the medial and promontory types of internal carotid artery in a wide variety of monotremes, marsupials and eutherians. He found that the artery is in all cases a single vessel derived from the dorsal aorta, being located initially in the mesoderm directly inferior to the cochlea. There is no evidence of separate vessels which could give rise to medial or promontory arteries. It appears rather that the adult position of the artery is due to relative growth causing the vessel to move either medially to the region of the basicapsular groove or laterally to the region of the promontory.

The circle of Willis, into which the internal carotid arteries feed, frequently establishes anastamotic channels with the branches of the external carotid and vertebral arteries. In insectivores, rodents and primates, for example, the supply to the circle of Willis is typically shared about equally between the internal carotid and vertebral systems. In other cases, however, the supply to the circle is largely through the external carotid or vertebral channels and the internal carotid is much diminished or even absent. Thus, in the Artiodactyla and the Aeluroidea there has been a marked tendency for the principal arterial supply to the brain to be taken over by branches of the external carotid system with a consequent diminution in the importance and size of the internal carotids.

A further modification to the blood supply to the head has been brought about in many eutherians by the branches of the stapedial artery being 'captured' by the external carotid as a result of enlargement of anastamotic channels between the two arteries. The jaws have thus come to be supplied through the external carotid artery, the anastamotic channel (now the first part of the maxillary artery) and the infraorbital and mandibular branches of the stapedial artery (now branches of the maxillary artery). The stem of the stapedial artery is either lost altogether or greatly reduced (the principal exceptions being in certain species of monotremes, insectivores, bats, prosimians and rodents). The atrophy of the stapedial artery appears to be related to the development of the auditory bulla, possibly because the closed bony canals in which it became enclosed were incapable, for structural reasons, of being enlarged to a size adequate to meet the arterial demands of the enlarged masticatory apparatus. It is also possible that the pulsations of a large artery so close to the middle ear structures were deterimental to the development of the highly acute hearing typical of many mammalian groups.

APPENDIX 2: ACOUSTICAL IMPEDANCE

An alternating force applied to a medium will cause it to oscillate along the line of application of the force. The amplitude of the oscillations will be related to the magnitude of the applied force, and inversely to the mass and stiffness (together called the reactance) and the friction (the resistance) of the medium. Mechanical impedance is defined by the ratio between the applied force and the resultant velocity of movement in the medium. In acoustics the applied force is the alternating pressure of the air and the resulting velocity is best defined as the rate of volume displacement. Therefore, acoustical impedance (Z) is given by the ratio of pressure to the rate of volume displacement (volume velocity) and is expressed as dynes/cm^3/s (= dynes s/cm^3). Since the rate of volume displacement is inversely related to reactance and resistance, an increase in these factors will increase impedance and vice versa. The effect of resistance on acoustical impedance is independent of the frequency of the sound waves but the effects of the two components of reactance are modified, although in opposite directions, by the frequency – at high frequencies more of the applied force is taken up in moving mass than in overcoming stiffness; at low frequencies more of the applied force is used up in overcoming stiffness than in moving mass.

The better the impedance matching of two sound conducting media, the more complete will be the transfer of energy between them and the less will be the amount of energy reflected. The difference between the impedances of air and water is such that only 0.1 per cent of the energy of a sound wave of 1 kHz reaching the interface will be transferred to the fluid medium, the remaining 99.9 per cent being reflected. For a more detailed discussion of acoustical impedance, see Brooks (1976).

7

THE NASAL REGION

The nasal fossae in land vertebrates make up the first part of the respiratory pathway and provide also the distribution area for the peripheral olfactory apparatus. During its passage through the fossae the inspired air is moistened, cleansed of coarse particulate matter, subjected to olfactory analysis and, in homoiothermic animals, warmed. The fundamental configuration of the fossae is remarkably constant throughout the great majority of mammalian groups, the structure of this part of the skull being largely uninfluenced, apparently, by factors such as total body size, type of diet and mode of life which have had such profound modifying effects upon the morphology of other cranial regions. The main exceptions to this generalisation are provided by those mammalian groups where highly specialised respiratory requirements prevail (principally the Cetacea) or where olfaction has become of reduced importance (notably the Anthropoidea).

The floor of the mammalian nasal cavity (comprising right and left fossae) is formed by the hard palate and the roof by the anterior extension of the cranial base (including the cribriform plate of the ethmoid and part of the frontal) and the nasal bones. The principal skeletal elements in the lateral wall of the nasal cavity are the premaxilla, maxilla, palatine, frontal, ethmoidal labyrinth (lateral mass), nasoturbinal, maxilloturbinal and lacrimal. The cavity is divided into two by a median septum which is bony in its posterior part but cartilaginous anteriorly. The septum represents the medial walls of the original capsules together with an incorporated part of the interorbital septum. Posteriorly, the cavity ends at the paired posterior nasal apertures lying one either side of the posterior margin of the median septum, while anteriorly it begins at the anterior nares in the fresh state but in the macerated skull, where the anterior cartilaginous part of the walls and septum is missing, the opening into the nasal cavity is formed by the single anterior nasal aperture.

The nasal cavity is lined by three types of epithelium – respiratory ciliated, stratified squamous and specialised olfactory. The areas of distribution of these three types vary considerably between species but, in general, the anterior part of the cavity is lined by stratified squamous and the postero-superior part by olfactory epithelium. The remainder bears respiratory epithelium. The area of mucous membrane is increased in reptiles and birds but more especially in mammals by the formation of scroll-like bony plates – the turbinals – which project into the lumen of the nasal cavity from its lateral walls. In some reptiles, notably the crocodiles, a prominent, complex scroll, the concha, is present which is believed to correspond to the maxilloturbinal of mammals. The upper surface of the reptilian concha is covered with olfactory epithelium but in mammals the maxilloturbinal has been transposed forwards, as part of the great development of the nasal cavity, and now lies entirely within the respiratory part of the nasal cavity. The area of olfactory mucosa in mammals is increased, in some species enormously, by the presence in the posterior part of the nasal cavity of the ethmoturbinals which form part of the labyrinths of the ethmoid. The turbinals of mammals, therefore, serve both to increase the efficiency of the warming, cleasing and moistening actions of the nasal mucosa and to enhance the sense of smell.

Opening from the nasal cavity of placental mammals are the paranasal air sinuses. These are essentially extensions of the nasal cavity, being lined predominantly by the respiratory type of mucous membrane. Their number and extent vary widely. Generally they are located within the bones immediately adjacent to the nasal cavity but in some groups they are much more extensive. Numerous functions have been attributed to the sinuses but, as will be discussed below, none is totally convincing.

It will be recalled that in the early mammal-like reptiles the nasal fossae still possessed a reptilian structure, consisting of relatively small cartilaginous capsules opening externally on to the surface of the snout through the external nostrils, and internally into the mouth through the internal nostrils. The major modifications by which the nasal cavity of the later mammal-like reptiles became greatly enlarged, especially in its anteroposterior dimensions, have been traced in Chapter 4. This enlargement was undoubtedly related to the increased respiratory requirements associated with the emergence of a mammalian type of metabolism. It must also have been of advantage in allowing an increase in the size of the peripheral olfactory apparatus, a factor which may well have been crucial in the emergence of the earliest mammals which, it is believed, were nocturnal creatures and greatly dependent, therefore, upon the possession of an acute sense of smell. The smaller-scale changes involved in the evolution of the characteristic mammalian nasal cavity, including the origin of the ethmoid and its turbinal elements, are poorly known because of the sparsity and imperfect preservation of fossilised remains of the nasal region of the skull of the later mammal-like reptiles and early mammals.

TYPICAL MAMMALIAN NASAL CAVITY

A nasal cavity constructed along essentially similar lines occurs in insectivores, rodents, carnivores, bats, ungulates, tree shrews and prosimians – in fact, in the majority of terrestrial mammals – and is only slightly modified in the Sirenia (see, for example, the detailed descriptions by Paulli, 1900*a*, *b*, *c*;

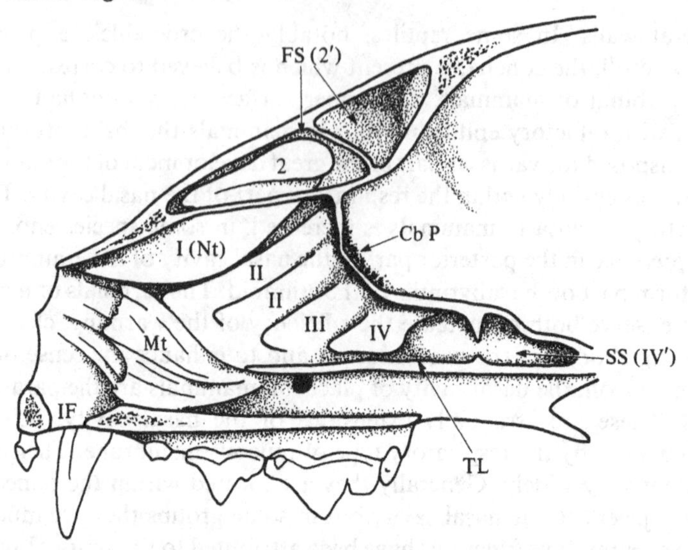

Fig. 84. Sagittal section of cat skull to show, semi-diagrammatically, the right lateral wall of the nasal cavity. Endoturbinals, ectoturbinals and sinuses numbered according to scheme described in text. For other labelling see list of abbreviations.

Dieulafe, 1906; Anthony & Iliesco, 1926; Negus, 1958; Cave, 1973). In these groups the cavity is a long, almost tubular chamber (Fig. 84). The anterior part of its roof, formed by the nasal bones, tends to slope anteroinferiorly while the posterior part, formed principally by the cribriform plate of the ethmoid, frontal and body of the sphenoid, slopes posteroinferiorly. The size of the cribriform plate and its number of perforations vary widely between mammalian groups in relation to the area of olfactory mucosa. The floor of the nasal cavity, composed of the palatine processes of the premaxillae and maxillae and horizontal processes of the palatine bones (with additional elements in some species, e.g. the pterygoid processes in the great ant-eater), occupies the horizontal plane. It is pierced by the incisive foramina, the margins of which are formed by the palatine processes of the premaxillae anteriorly, laterally and medially and the maxillary palatine processes posteriorly. Each transmits a nasopalatine duct by which the nasal cavity is in communication with the mouth. The nasal cavity usually has its maximum vertical dimension opposite the posterior extremity of the nasal bones.

The major structural elements in the lateral wall of the cavity are the premaxilla, maxilla and frontal and the perpendicular plate of the palatine but these are overlaid on their nasal surfaces by the lacrimal, ethmoidal labyrinth, nasoturbinal and maxilloturbinal and are, consequently, to some extent obscured from view in the sagittally sectioned skull. There is much species variation in the size of the bones surrounding the nasal cavity and, therefore, in the detailed composition of its boundaries (e.g. see Dieulafe, 1906).

As compared with the typical reptilian condition, the new elements in the mammalian nasal region are the ethmoid and the repositioned vomer. The former element, in its fullest development, consists of perpendicular plate (with its superior projection, the crista galli), cribriform plate and right and left labyrinths each consisting of lateral plate and the ethmoturbinals (some authors apply the term ethmoturbinal to the complete labyrinth; here it is limited to the curved plates of bone projecting into the nasal cavity). The ethmoid thus contributes to the roof and posterior parts of the lateral walls of the nasal cavity and to the nasal septum, as well as to the floor of the anterior part of the cranial cavity and medial walls of the orbits. Over part of its extent the lateral surface of the lateral plate is opposed to the posterosuperior part of the medial surface of the maxilla and the adjacent surface of the palatine bone, while from its medial surface are dependent the ethmoturbinals (the most comprehensive description of the variation in the ethmoid from one mammalian group to another is still that of Allen, 1882).

The area of olfactory mucosa is generally co-extensive with that of the ethmoturbinals. As might be expected, therefore, the number, size and shape of these elements vary widely between species in association with the degree of olfactory acuity. According to Dieulafe (1906) the olfactory area is augmented by the presence of the ethmoturbinals by a factor of approximately 5 in the sheep, 2.5 in the dog and 1.34 in man. A very detailed description of the ethmoturbinals in each of the major mammalian groups has been provided by Paulli (1900*a*, *b*, *c*), based on examination of whole-head sections. His findings form the principal basis of the following account.

Each ethmoturbinal consists of a lamella (the olfactory plate) projecting medially into the nasal cavity from the lateral ethmoidal and cribriform plates (Fig. 85). The lamellae may undergo repeated branching, especially in keen-scented species, and towards their free extremities undergo some degree of inrolling to form olfactory folds. The ethmoturbinals are usually arranged, depending on how far they project medially into the nasal cavity, into two or more rows, the elements forming the more lateral rows being termed ectoturbinals and those in the most medial row endoturbinals. In a sagittally sectioned skull the endoturbinals form prominent, somewhat obliquely lying, rows in the posterior part of the nasal cavity (Fig. 84).

For identification purposes Paulli denoted the endoturbinals by roman numerals in an anterosuperior–posteroinferior sequence while the ectoturbinals he identified in a similar sequence but by arabic numerals. Confusion has arisen over the designation of the endoturbinals because (1) some authors include the nasoturbinal, which usually ossifies at least in part from the ethmoid, while others do not and (2) there is divergence between authors as to whether the numbering should be applied to all olfactory plates or only to those taking independent origin from the lateral plate (as will become apparent, the two do not correspond because of the tendency of the lamellae of certain endoturbinals to split close to their origin to give more than one

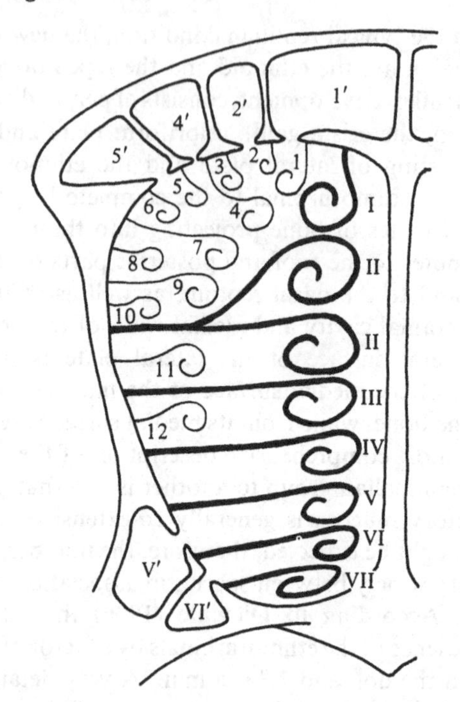

Fig. 85. Schematic representation of coronal section (parallel with cribriform plate) through nasal cavity of *Tapirus* to show ectoturbinals (thin lines, arabic numerals), endoturbinals (thick lines, Roman numerals) and relationship between paranasal air sinuses and ethmoturbinals. Turbinals and sinuses numbered according to scheme described in text. (After Paulli, 1900*b*.)

olfactory plate). In the following account the procedure of many of the early comparative anatomists, including Paulli, will be adopted in which the nasoturbinal is included in the endoturbinal series and the numbering is applied to fully independent lamellae only. There is also discrepancy between authors in the designation of the ectoturbinals. Allen (1882), for example, appears to restrict this term to the turbinals lying in the interspace between the nasoturbinal (endoturbinal I) and the succeeding endoturbinal (II) while Paulli (1900*a*) calls any ethmoturbinal not reaching close to the median plane an ectoturbinal. Again, the terminology of Paulli will be used here.

In many macrosmatic species the more posterior ethmoturbinals are located within a recess formed by a horizontal shelf of bone, the transverse lamina. This is made up of two plates, ossifying in the posterior lamina transversalis, which project medially from the lateral ethmoidal plates of each side to articulate with the wings projecting laterally from the vomer. It thus divides the nasal cavity posteriorly into upper and lower compartments (Fig. 84). The upper compartment, housing the ethmoturbinals, is a capacious but blind olfactory recess lying below the cribriform plate and usually prolonged posteriorly by a hollowing out of the anterior surface of the body of the presphenoid. In species with widely separated orbits, including the cat and the lemur, Cave (1973) has described a series of complex cavities within the upper compartment associated individually with pairs of ectoturbinals and serving, presumably, to increase the olfactory area. The lower compartment, or nasopharyngeal meatus, is respiratory in function, being continuous anteriorly with the inferior meatus (see below) and terminating posteriorly at the posterior nasal aperture. The transverse lamina is especially well developed in carnivores, particularly in members of the Canidae.

The presence of the transverse lamina excludes the greater part of the olfactory area from respiratory air currents. Olfaction must then depend upon diffusion from the scent-laden inspired air or upon the projection of puffs of air into the recess by sniffing. The advantage of housing the olfactory area within a recess may be that the air within it will not be washed out during exhalation and will, therefore, be retained longer for olfactory analysis.

As illustrated by Paulli's extensive studies, species variation in the number and form of the ethmoturbinals can be readily correlated with the importance of the olfactory sense in the life of the animal. Much of the variation described by Paulli has no significance beyond this and the following brief summary of his findings will, therefore, be limited to an account of the general features of the ethmoturbinals in the major mammalian orders and will exclude details of differences between individual species.

1. *Insectivores, chiropterans and hyracoids*

In insectivores the ethmoid is extensive, occupying a considerable part of the nasal cavity, and is characterised by possessing four endoturbinals. The lamella of endoturbinal II (endoturbinal I being the nasoturbinal) is split to form two olfactory plates (the total number of which is, therefore, five). There is also a row of small ectoturbinals, the number of which varies but does not exceed three (ectoturbinals 1 and 2 lie between endoturbinals I and II and ectoturbinal 3 lies between endoturbinals II and III). The olfactory region is prolonged posteriorly into the body of the presphenoid, the recess so formed being completely filled with ethmoturbinals. The arrangement of the ethmoturbinals in the Hyracoidea and most Chiroptera follows that of insectivores closely.

2. *Carnivores*

The size of the olfactory region of the nasal cavity is relatively very large in all fissiped carnivores, the ethmoid occupying the greater part of the nasal cavity and extending in a posterior direction into a deep recess in the body of the presphenoid. Typically in the canids, viverrids, hyaenas and felids, there are, as in insectivores, four endoturbinals with the lamella of endoturbinal II split to give a total of five olfactory plates (Fig. 86*a*). In the ursids and mustelids the number of endoturbinals remains typically four but the number

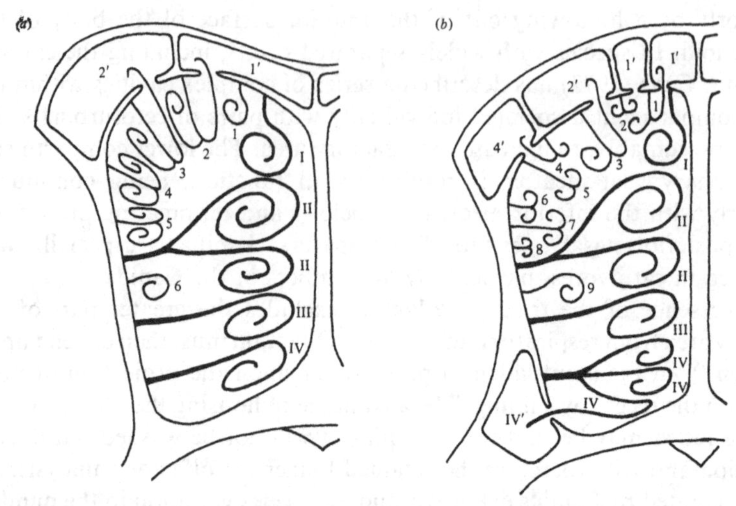

Fig. 86. Schematic representation of coronal section (parallel with cribriform plate) through nasal cavities of (*a*) big dog; (*b*) *Ursus arctos*. Not to scale. Turbinals and sinuses numbered according to scheme described in text. (After Paulli, 1900*c*.)

of olfactory plates is increased up to six or seven by splitting of the lamella of the fourth (which may be split into three) as well as that of the second endoturbinal (Fig. 86*b*). The surface area of the olfactory plates is much increased by their complicated rolling and folding. In procyonids the splitting of endoturbinal IV is more complete than in bears or mustelids so that six independent endoturbinals are formed in which endoturbinals IV, V and VI correspond to the original endoturbinal IV.

The number and arrangement of the ectoturbinals vary greatly between even closely related species and indeed within individual species. This is particularly true of the dog. In general, the number of ectoturbinals is low (usually between five and six) but is considerably increased (up to nine) in the bears, procyonids and some mustelids (e.g. the badger), where these elements may be arranged in two or more rows (Fig. 86*b*). The majority of ectoturbinals is located between endoturbinals I and II.

In some carnivores, parts of certain of the ectoturbinals may extend into the paranasal sinuses opening off the olfactory region (Figs. 84, 86). This extension appears to take place during postnatal growth, the sinuses being initially 'empty'. The maxillary sinus is never involved.

In their description of the carnivore nasal cavity Anthony & Iliesco (1926) agree with Paulli regarding the number of endoturbinal olfactory plates but the figures they give for the numbers of ectoturbinals are generally greater than those of the latter author. Such discrepancies between authors is not uncommon, arising, presumably, from differences of opinion as to what actually constitutes a distinct and separate ectoturbinal.

The ethmoid of the pinniped carnivores is much reduced in size compared to that of land carnivores but still possesses five endoturbinals with six olfactory plates and up to eight ectoturbinals.

3. *Rodents*

The typical ethmoid of rodents is similar to that of insectivores. There are four endoturbinals with the lamella of the second split to form two olfactory plates. The number of ectoturbinals is low and these elements are arranged in a single row. The olfactory region is prolonged posteriorly by a hollowing out of the body of the presphenoid. Amongst the rodents examined by Paulli, the porcupine provided the principal exception to this general description: in this rodent there are five endoturbinals (with the second again divided into two leaves) and six ectoturbinals.

4. *Edentates*

The number and arrangement of the ethmoturbinals differ greatly from one edentate species to another so that it is not possible to describe a typical condition. The number of endoturbinals ranges between five and eight (usually with the second and sometimes the fifth or sixth split to form, in each case, two olfactory plates). The ectoturbinals tend to be few in number but again there is much species variation. The nasal cavity is extended posteriorly deep into the body of the presphenoid, the recess so formed being filled, in most species, by the lower olfactory folds.

5. *Ungulates*

The ungulate ethmoid is extensive and possesses a complex structure with a very large number of much folded ethmoturbinals, especially ectoturbinals (Table 3). As in so many other morphological attributes, a distinction can be made between the artiodactyls and perissodactyls in the arrangement of this part of the nasal cavity although in this particular feature the pigs appear to be more closely comparable to the perissodactyls than to their own order.

The artiodactyl ethmoid is characterised by five endoturbinals with endoturbinal II split into two leaves (Fig. 87*a*). The increase in the number of endoturbinals has probably been produced by division of the lamella of the original endoturbinal IV. In the skulls of many cattle the fourth ectoturbinal is much enlarged and appears in the row of endoturbinals between the first and second olfactory folds. In perissodactyls the number of endoturbinals has been increased still further to six or more (probably by further splitting of the original endoturbinal IV or its derivatives) with a corresponding increase in the number of olfactory plates (Figs. 85, 87*b*). The pig possesses seven endoturbinals with eight olfactory plates. The number and arrangement of the ectoturbinals and the relationship of these elements to the endoturbinals vary greatly between and even within species (for further details of the number of these elements, see Table 3, and for examples of their

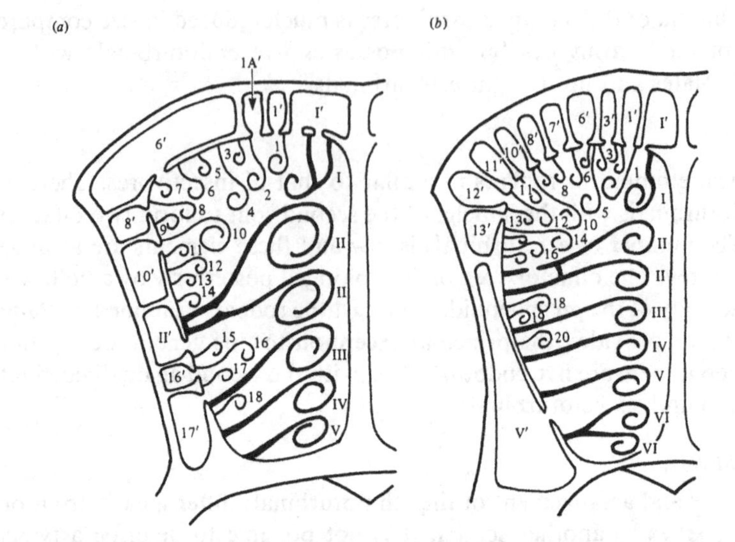

Fig. 87. Schematic representation of coronal section (parallel with cribriform plate) through nasal cavities of (*a*) *Bos*; (*b*) *Rhinoceros*. Not to scale. Turbinals and sinuses numbered according to scheme described in text. (After Paulli, 1900*b*.)

arrangement, see Figs. 85 and 87). In many ungulates the ethmoturbinals are pneumatised (see below).

6. *Elephants*

In the African elephant the ethmoid possesses seven endoturbinals but eight olfactory plates, endoturbinal II being split (Fig. 97*a*). There are 19 ectoturbinals arranged in two rows (the medial row comprising ectoturbinals 1, 5, 10, 13, 18 and 19). All the ethmoturbinals have much secondary folding.

7. *Prosimians*

The prosimian ethmoid is relatively unreduced compared to that of anthropoid primates, possessing a structure very similar to that seen in insectivores. There are four endoturbinals, again with the lamella of the second split into two leaves. The number of ectoturbinals is reduced to two or less. The ethmoturbinals tend to be of a rather simpler form than in many of the very keen-scented mammals, usually having just a single roll.

On the basis of these findings Paulli (1900*c*) suggested that the endoturbinals in the different mammalian orders can be regarded as homologous structures. So far as placental mammals are concerned, it seems that four endoturbinals with the lamella of the second divided to form two virtually independent olfactory plates represents an early, primitive condition. There is little departure from this structure in insectivores, hyracoids, bats, rodents,

TABLE 3. *The number of endoturbinals, olfactory folds and ectoturbinals in ungulates. Data from Paulli (1900b)*

	Number of endoturbinals[a]	Number of olfactory folds	Number of ectoturbinals	Medial row of ectoturbinals comprises:
Artiodactyls				
Cattle	5	6	18	1, 2, 4, 10 (or 12), 16, 18
Sheep	5	6	13	1, 3, 7, 11, 13
Cervus elaphus	5	6	20	1, 3, 7, 13, 19
Camel	5	6	13	1, 3, 7, 10, 12
Goat	4	6	13	1, 3, 7, 11, 13
Pig	7	8	20	1, 4, 8, 12, 18, 20
Perissodactyls				
Horse	6	6	31	1, 2, 5, 9, 17, 20, 22, 25, 27, 29, 31
Rhinoceros sondaicus	6	8	20	1, 5, 10, 14, 18, 20
Tapirus americanus	7	8	12	1, 4, 7, 9, 11, 12

[a] Including nasoturbinal.

prosimians and many carnivores. An increase in the number of olfactory plates or completely separate endoturbinals, such as is found in bears, mustelids, procyonids and ungulates, appears to have been achieved by the splitting of endoturbinal IV, in addition to that of endoturbinal II.

Paulli further suggested that the primitive placental arrangement is itself derived from a still earlier condition in which five independent endoturbinals were present. This arrangement is still found in marsupials. The number of endoturbinals was reduced, according to Paulli, to the four typical of placentals by fusion of the lamellae of the second and third of the original series. The splitting of endoturbinal II to form two olfactory plates, as occurs in the great majority of placentals, may well represent, therefore, a residual feature resulting from the development of this element from two originally quite separate endoturbinals.

The nasoturbinal (endoturbinal I) usually projects from high up on the side wall or from the roof of the nasal cavity, articulating with the turbinate crest of the nasal bone and the medial surface of the maxilla. It is typically a prominent element extending well forwards while posteriorly it may be interposed between the maxilloturbinal and the remaining ethmoturbinals or continue above the latter (Fig. 84). It ossifies, at least in part, from the ethmoid. Curving posteroinferiorly from the posterior part of the nasoturbinal is the uncinate process which forms the anterior border of the opening of the maxillary sinus.

The ectoturbinals are much more variable structures than are the endoturbinals, both in their number and arrangement, and there appears to be no possibility of homologising them between all the different placental orders (they are, in Paulli's words, 'analogous' rather than 'homologous' structures).

Ossification of the ethmoid is subject to considerable interspecific variation (see de Beer (1937) for details). There are generally several centres for the ethmoturbinals, lateral plate and cribriform plate. In primates, rodents and carnivores there is a separate mesethmoid ossification centre for the perpendicular plate but in many other mammalian orders the perpendicular plate is ossified in continuity with the presphenoid (in fact, as described in Chapter 2, the mesethmoid and presphenoid centres in these two groups may be homologous, differing only in their initial site and extent of spread). Each of the ethmoid ossification centres appears within the corresponding part of the nasal capsule, spreads by endochondral ossification and eventually fuses with its neighbours.

The maxilloturbinal, unlike the ethmoturbinals, is concerned in mammals mainly with air-conditioning, having an epithelial covering predominantly of the non-olfactory type. It occupies the inferior part of the nasal fossa lying below and anterior to the nasoturbinal and projecting from the medial surface of the maxilla. It is represented at the cartilaginous stage by an inrolling of the ventral edge of the side wall of the nasal capsule anterior to the crista

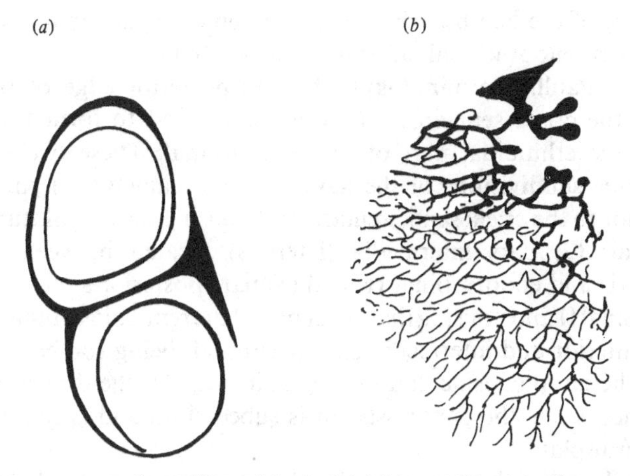

Fig. 88. Cross-section of maxilloturbinal of (*a*) bison; (*b*) seal. Not to scale.

semicircularis and is ossified from a single centre which may eventually fuse with the maxilla.

Although an element probably equivalent to the maxilloturbinal is present in some reptiles (notably the crocodiles and alligators) and birds, it is only in the mammals where it reaches its full development. Here its surface area is usually greatly increased by branching and rolling. In many rodents, ungulates and carnivores, for example, the maxilloturbinal, after projecting for a short distance into the nasal cavity, bifurcates into dorsal and ventral laminae which are rolled upon themselves in a scroll-like manner (Fig. 88*a*). In other mammals, especially the seal and walrus, an even greater surface area is obtained by repeated branching of the maxilloturbinal (Fig. 88*b*).

The maxilloturbinal and nasoturbinal subdivide the anterior part of the nasal fossa into three incompletely separated channels – the dorsal, middle and ventral meatus. The dorsal meatus lies above the nasoturbinal while the middle meatus is located between the two turbinals and is bounded laterally by the maxilla. Both lead back to the olfactory recess. The ventral meatus is usually the largest of the channels. Lying below the maxilloturbinal, it communicates with the nasopharyngeal meatus, and thus with the posterior nasal apertures, and is the principal respiratory pathway. The nasolacrimal canal is formed by the maxilla, lacrimal and maxilloturbinal and opens at the anterior end of the ventral meatus. A common meatus is sometimes described. This is merely the slit-like channel between the medial extremities of the turbinals, on the one hand, and the septum, on the other, into which the other three meatus open.

As was described in Chapter 2, the lateral wall of the amniote cartilaginous nasal capsule contains three elements. In anteroposterior sequence these are the cupola anterior, paranasal cartilage and lamina orbitonasalis. The free anterior edge of the lamina orbitonasalis becomes expanded and ossified to form the first endoturbinal after the nasoturbinal (i.e. endoturbinal II, according to Paulli's terminology); the free posterior edge of the cupola anterior is the crista semicircularis which is ossified to form the uncinate process of the ethmoidal part of the nasoturbinal. These two landmarks separate three subdivisions of the nasal capsule, namely: (1) pars anterior lying anterior to the crista semicircularis and housing the maxilloturbinal and nasoturbinal; (2) pars intermedia (lateralis) located between the crista semicircularis and endoturbinal II; and (3) pars posterior situated posterior to endoturbinal II and housing that element and the remaining ethmoturbinals. The pars anterior and intermedia are precerebral, being roofed over by the tectum of the nasal capsule and in the adult stage by the dermal nasal and frontal bones, while the pars posterior is subcerebral and is roofed over by the cribriform plate.

The lateral wall of the pars posterior lies in a sagittal plane slightly lateral to that of the pars anterior. In consequence, when the free anterior border of the former (i.e. the lateral plate close to the attachment of endoturbinal II) extends forwards during development to meet the lateral wall of the pars intermedia, it does so lateral to the crista semicircularis. In the interval between endoturbinal II and the lateral wall of the pars intermedia is developed dorsally a frontoturbinal recess and more ventrally a maxillary recess (lying posterolaterally) and a frontal recess (lying anterodorsally). The maxillary recess is the area where the excavation which eventually forms the maxillary sinus first begins. Similar relationships are preserved in the adult animal, the opening of the maxillary sinus into the nasal cavity being located between the lateral ethmoidal plate and the uncinate process.

The bony part of the nasal septum is formed by the perpendicular plate of the ethmoid superiorly and vomer inferiorly with minor contributions being made by crests on neighbouring bones (including the nasal, frontal, sphenoidal and maxillary) with which the perpendicular plate of the ethmoid and the vomer articulate.

The probable phylogenetic origin of the vomer has been fully discussed in Chapter 4. It is characteristically a flat, thin bone of trapezoidal shape. The superior border articulates with the ventral surface of the body of the presphenoid and the inferior border with the nasal crest on the upper surface of the hard palate. The posterior border is free and separates the posterior nasal apertures. The anterior border is grooved; the upper half of the groove articulates with the perpendicular ethmoidal plate while the lower half, in the fresh state, receives the inferior margin of the septal cartilage.

The vomeronasal organ in mammals consists typically of a pair of blind tubular diverticula lying beneath the mucous membrane in the floor of the nasal cavity, one each side of the vomerine part of the septum. The organ is lined by ciliated columnar epithelium over its lateral walls and by olfactory

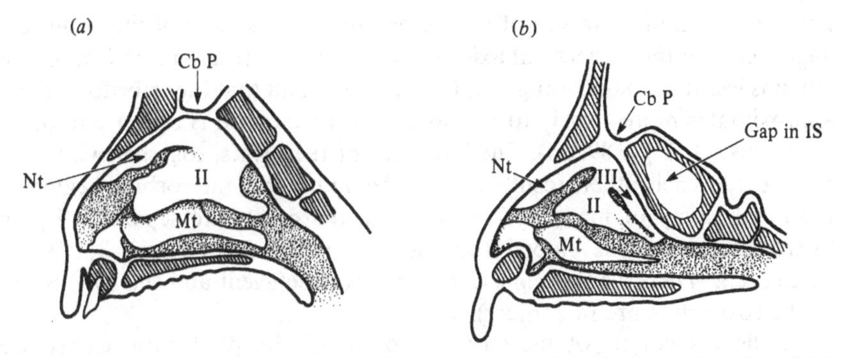

Fig. 89. Sagittal sections of skull to show right lateral wall of nasal cavity of (*a*) *Macaca*; (*b*) *Saimiri*. Not to scale. Endoturbinals numbered according to scheme described in text. For labelling see list of abbreviations. (After Cave, 1973.)

epithelium elsewhere. Opening into the lumen are numerous glands. Each diverticulum opens by a small orifice anteriorly.

There is considerable interspecific variation in the degree of development of the vomeronasal organ. It is prominent in rodents, lagomorphs, ungulates and carnivores but rudimentary in aquatic mammals and man. The orifices of the organ may open into the nasal cavity (rodents, lagomorphs and ungulates) or into the nasopalatine canal (carnivores). The organ is surrounded to a varying degree by a capsule which is usually composed of cartilage (the paraseptal cartilage, see Chapter 2).

DEPARTURES FROM THE TYPICAL MAMMALIAN CONDITION

ANTHROPOID NASAL CAVITY

In the anthropoid primates (including man), and in the tarsiers, the nasal cavity is much reduced in size and its internal architecture is radically modified compared to the characteristic mammalian condition just described (Fig. 89). The reduction affects principally the anteroposterior length of the cavity so that it is a short but relatively high roofed chamber. The turbinals are decreased in size and number and also in their degree of branching and rolling with the result that the nasal mucous membrane, especially that concerned with olfaction, is much diminished in surface area (Dieulafe (1906) for example, gives ratios for the olfactory to respiratory areas of the nasal mucosa of approximately 0.3 in man, 0.4 in monkeys, 0.7 in the rat and 1 in the cat). There is no transverse lamina and consequently no olfactory recess. These changes undoubtedly reflect the decreased importance of olfaction in the vision-dominated life of the tarsier and higher primates. Because of the changed spatial relationships of the nasal cavity and braincase, consequent upon the diminution in size of the former and enlargement of the latter, the angle between the basicranial axis and its anterior extension (see Chapter 8) remains less than 180° throughout life with the result that the cribriform plate approximates more closely to the horizontal plane than is the case in other mammals (cf. Figs. 89, 84). The large size of the orbits, together with their closeness to each other, results in there being no true interorbital region in the nasal cavity of many species. Instead, the orbits are separated by an interorbital septum developed from the presphenoid (see Chapter 2; in some species, e.g. *Tarsius* and *Saimiri*, the septum is deficient and the soft tissues of the two orbits are in contact).

Excellent accounts of the structure of the anthropoid nasal cavity are provided in the early studies by Seydel (1891) and Paulli (1900*c*) as well as in the more recent studies by Cave (1948, 1973). In the New World monkeys that have been studied there are usually three endoturbinals which are small and of simple form, lacking the rolling characteristic of these elements in non-primate mammals. In some forms (e.g. *Callithrix*) a rudimentary ecto-turbinal may be present but this is usually no more than a low ridge on the lateral plate. The number of endoturbinals in Old World monkeys is variable, being, for example, usually two in *Cercopithecus cynomolgus* and *Presbytis entellus*, three in *Mandrillus leucophaeus* and three or four in *Papio hamadryas*. As in the New World monkeys, these elements are all small and of simple form and may be incompletely separated from each other at their origin from the lateral plate. Ectoturbinals are usually completely lacking. The cribriform plate tends to be short and narrow in all monkeys and may even be lacking (resulting in the presence of a single olfactory foramen) in the skull of some Old World forms.

The nasoturbinal (endoturbinal I) is usually present, being best developed in the New World monkeys. In catarrhines, as in hominoids, this element is often reduced and may be represented by no more than a low ridge of mucous membrane (termed the agger nasi) and the uncinate process which nonetheless still forms the anterior margin of the opening of the maxillary sinus into the main part of the nasal cavity. The maxilloturbinal persists in anthropoids and although reduced in size and complexity compared with that of non-primates, the reduction is usually less pronounced than that of the ethmoturbinals.

The structure of the nasal cavity in the Hominoidea is better known than that of any other primate group, having been comprehensively studied by Jones (1938), Cave & Haines (1940) and Cave (1949, 1973) while very detailed accounts of the nasal fossae in man are available in all the standard anatomical texts (e.g. *Gray's Anatomy*, 1973).

The most striking features of the human nasal cavity are its greatly reduced size and the smallness and simplicity of the turbinals (Fig. 98*a*). The maxilloturbinal (the inferior nasal concha of human anatomy) is the least reduced in size but it lacks the complex, multirolled structure typical of non-primates. The nasoturbinal is represented by the agger nasi, a poorly defined curved ridge produced by the anterior part of the ethmoidal crest on the medial surface of the frontal process of the maxilla, and the uncinate process. The agger nasi tends to be more prominent in the child than in the adult. Despite some disagreement amongst early comparative anatomists (see Paulli, 1900*a*), it seems probable that endoturbinals II, III and IV are represented respectively by the unconvoluted middle, superior and highest nasal conchae. There are three meatus, termed inferior, middle and superior, lying below and lateral to the inferior, middle and superior concha, respectively. Above the superior concha is the sphenethmoidal recess, the lateral wall of which is distinguished in about 20 per cent of cases by the highest nasal concha and the related supreme meatus. Olfactory mucosa is confined to the area over the superior nasal concha, the opposed part of the septum and the intervening roof.

Despite the small size of the endoturbinals, the human ethmoid in its overall proportions is relatively larger than in other hominoids and of a more primitive form. Its rather large size is due to the presence of numerous air cells within the labyrinths. The medial surface of each labyrinth consists of a thin lamina (the medial plate) which begins above at the cribriform plate and ends below by becoming the middle nasal concha. The middle group of ethmoidal air cells occupies a rounded swelling, the bulla ethmoidalis, on the lateral wall of the middle meatus (Fig. 98*a*). The bulla probably represents the first ectoturbinal of the non-primate nasal fossa (although the relative positions of the ethmoturbinals have changed, the bulla, in lying on the far side of endoturbinal II (the middle nasal concha) from endoturbinal III (the superior concha), has the same relationship with these elements as has the first ectoturbinal).

The lateral surface of the labyrinth is composed of the orbital plate. Projecting posteroinferiorly from the anterior part of this plate is the uncinate process which ossifies in the crista semicircularis. This process crosses the maxillary hiatus (a large gap in the medial wall of the maxillary sinus which is much reduced in size in the articulated skull by the lacrimal bone, inferior concha and ethmoidal labyrinth as well as by the uncinate process) to articulate with the ethmoidal process of the inferior concha (Fig. 98*a*). The concave upper edge of the process is free. Between this edge and the inferior surface of the bulla is a curved gap, the hiatus semilunaris, which communicates laterally with the semilunar groove. The maxillary sinus opens into the floor of the groove (the size of the opening being still further reduced from that in the articulated skull by the presence of the mucous membrane). The structure of this region is, therefore, recognisably similar to the equivalent region of the non-primate nasal cavity.

The hiatus semilunaris extends forwards and upwards as a curved passage termed the infundibulum. In the majority of individuals the infundibulum is

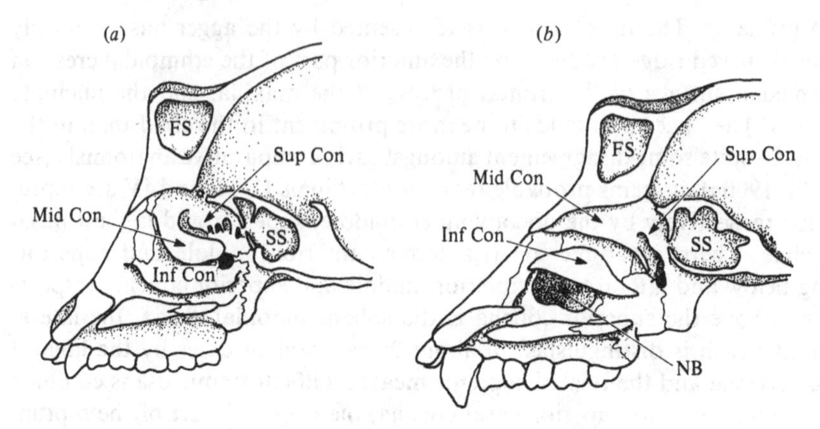

Fig. 90. Sagittal sections of skull to show right lateral wall of nasal cavity of (*a*) chimpanzee; (*b*) gorilla. Not to scale. For labelling see list of abbreviations. (After Cave & Haines, 1940.)

continued superiorly as the frontonasal canal which traverses the anterior part of the ethmoidal labyrinth. The frontal sinus communicates with the middle meatus either directly through the infundibulum or, in most instances, through the frontonasal canal and infundibulum.

According to the descriptions of Cave & Haines (1940) and Cave (1949) the structure of the nasal cavity in the chimpanzee and gorilla appears to be similar, in most essentials, to that of man (Figs. 90, 98). The principal differences are that in the two African apes (1) the nasoturbinal and endoturbinals are rather better developed, (2) the maxilloturbinal, as well as forming the inferior concha, has a vertical plate (equivalent to the tiny lacrimal and ethmoidal processes of the human inferior concha) which makes up a considerable part of the lateral wall of the nasal cavity, (3) the inferior concha is double rolled, (4) the number of ethmoidal air cells is less, being usually three only, (5) the ethmoidal bulla shows considerable individual variation in size, being indistinguishable in some specimens, and (6) the uncinate process may end freely instead of articulating with the inferior concha. In the gorilla the nasolacrimal canal is dilated, the lower portion often enormously so to produce a nasolacrimal bulla. The lateral wall of the bulla is formed by the maxilla and its medial wall by the vertical plate of the maxilloturbinal.

The nasal cavity of the orang-utan possesses only two well-developed conchae, namely the maxilloturbinal and endoturbinal II (Caves & Haines, 1940). A poorly developed nasoturbinal and endoturbinal III may also be present. The uncinate process is absent as are, therefore, the hiatus semilunaris and the semilunar groove.

The general arrangement of the nasal cavity in the gibbon is the most primitive of all the Hominoidea and exhibits considerable interspecific variability (Cave & Haines, 1940). In *Hylobates leuciscus* the nasoturbinal, maxilloturbinal and endoturbinals II, III and IV are well developed and the lateral wall of the middle meatus possesses an uncinate process and an ethmoidal bulla with a hiatus semilunaris and semilunar groove. However, in *Hylobates hoolock* and probably also in *Hylobates concolor* and *Symphalangus syndactylus* the uncinate process and bulla are weakly developed or absent and consequently there is no hiatus semilunaris or semilunar groove. The number of endoturbinals is also variable, for endoturbinal IV may be absent and endoturbinal III rudimentary. There is no doubt that the presence of a full complement of endoturbinals and of an uncinate process and ethmoidal bulla is to be regarded as the more primitive condition.

NASAL CAVITY IN CETACEA

The whole of the respiratory system in the Cetacea has undergone profound structural modification as part of the adaptation to marine life. Inhalation can take place only while surfacing and yet, during this period, which may be of short duration, sufficient air must be taken in to meet metabolic requirements throughout the subsequent dive which may be made to great depths and in some species apparently for time spans lasting up to one hour. Furthermore, mechanisms must exist to reduce or resist the effects on the respiratory system of the large hydrostatic pressures developed during the dive.

The structure of the nasal cavity shares in this modification (for descriptions of the structural modifications in the remainder of the respiratory system the reader is referred to Slijper, 1962). It has become adapted virtually exclusively to the respiratory function, olfaction being apparently of little or no value in such fully aquatic creatures.

In order to increase the speed of air displacement, and thus shorten the potentially hazardous period during which the whale must remain at the surface, the nasal cavity has become relatively short and wide and the turbinals have been lost or much reduced. As a result of these structural modifications, together with adaptations elsewhere in the respiratory system, up to 90 per cent of the air in the lungs can be replaced in one sequence of exhalation and inhalation, compared with an average of about 15 per cent in terrestrial mammals. Moreover, the nasal cavity has come to ascend at a steep angle through the skull so that the external nostril is located on the top rather than at the front of the head, thus enabling inhalation to take place without the head being lifted out of the water. In the sperm whale, where the presence of the mass of low-density spermaceti has altered the orientation of the head so that the snout has become the highest point, the nasal cavity still ascends almost vertically in the skull but is connected to the exterior by a passage that runs forward, through the spermaceti case, to open at the tip of the snout.

The epithelium lining the nasal cavity appears to be more or less completely devoid of glands and cilia. In odontocetes, olfactory receptors are absent or present in only rudimentary form while in mysticetes, olfactory epithelium is present over just a small area where a number of small conchae may also be found. Correspondingly, the olfactory nerves are absent or rudimentary in odontocetes and very small in mysticetes.

Before dealing with the bony part of the nasal passage, it is necessary to describe briefly the general structure of the cranium in whales (the auditory region has been described in Chapter 6). Although there is much species variation, as well as major differences between the toothed and baleen whales in the structure of the cranium, the following features are shared to some degree by all living cetaceans (for further details, see Miller, 1923; Kellogg, 1928). (1) Superficially, the most striking feature is the telescoping of the braincase produced by the overriding of adjacent bones. In both groups of whales this has resulted in a considerable reduction in the anteroposterior dimension of the braincase, especially in the temporal region. As already emphasised, the mutual relationships of the bones of the mammalian cranium remain remarkably constant despite the changes in their relative size and shapes such as are frequently encountered between even closely related mammalian groups. The cetaceans provide the principal exception to this general rule, for the telescoping of the braincase has resulted in the cranial bones developing extreme variations not only in their proportions but also in their relationships and articulations with each other. As might be expected in view of the telescoping phenomenon, the sutures of the braincase are of the squamous variety rather than being of the dentate type more commonly seen in the skulls of other mammalian orders. (2) Another impressive feature of the cetacean skull is the development of the facial bones, particularly the maxillae, premaxillae and mandible, to form an enormous protruding beak bearing the powerful dentition of the odontocetes and the baleen plates of the mysticetes. (3) The supraorbital processes of the frontal are extensive while the zygomatic bones, by contrast, are usually much reduced. The orbital cavity, lying below the supraorbital process and above the zygomatic bone, is small. (4) The vomer is a much larger bone than in other mammals and is extended far anteriorly and posteriorly. (5) The mesethmoid and presphenoid regions are occupied by a continuous ossification (see Chapter 2).

The nature of the telescoping differs between the two groups of whales. In the odontocetes there has been a tendency for the anterior cranial elements to override the more posterior ones (Fig. 91). As a result the slope of the cranial bones is convergent to an apex located above the anterior limit of the braincase. The proximal part of the maxilla spreads back over the supraorbital process of the frontal to approach or even meet the supraoccipital behind the orbit. Together with the underlying and expanded supraorbital process of the frontal, this part of the maxilla extends laterally to roof over the temporal

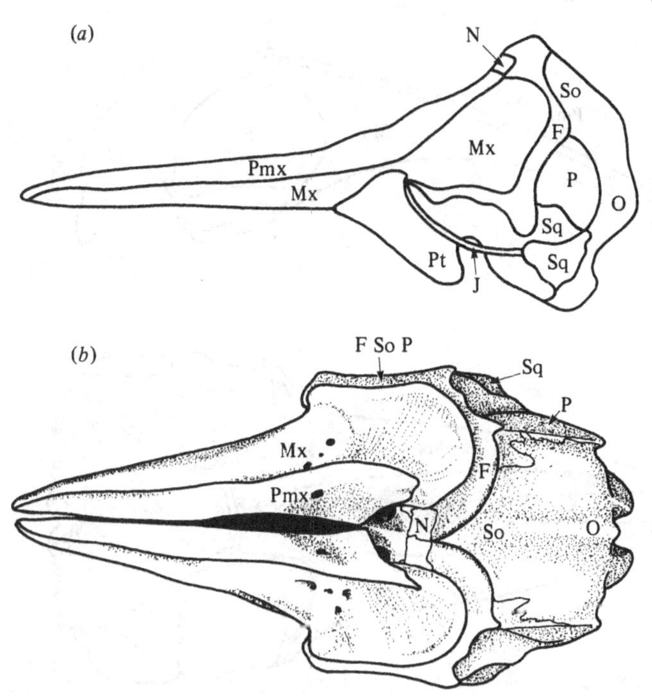

Fig. 91. The skull of the toothed whale. (*a*) Diagrammatic representation of telescoping, left lateral view; (*b*) dorsal view of skull of *Tursiops truncatus*. For labelling see list of abbreviations.

fossa. The anterior surface of the braincase above the external nasal aperture is thus formed into a smoothly concave area. The frontal bone is so compressed that in a median section of the skull its vertical height exceeds its anteroposterior length. The supraoccipital does not extend beyond the anterior limit of the braincase.

The telescoping of the mysticete skull has occurred in the opposite direction from that in the toothed whales, the tendency here being for the posterior elements to override the anterior ones (Fig. 92). In consequence, the bones of the braincase slope forwards to an apex located in front of the anterior limit of the cranial cavity. The maxilla ends posteriorly by abutting on to the supraorbital process rather than overriding it. Instead, it is the bones of the occipital and temporal regions which have grown forwards, resulting in the supraoccipital reaching forwards to the apex of the cranium and in the parietal overriding the supraorbital process of the frontal.

The upper jaw (or rostrum) consists of greatly elongated maxillae and premaxillae. As a result of this elongation the premaxillae retain their contribution to the margin of the external (i.e. anterior) nasal aperture, despite the posterior migration of the latter. The nasal bones, too, retain their

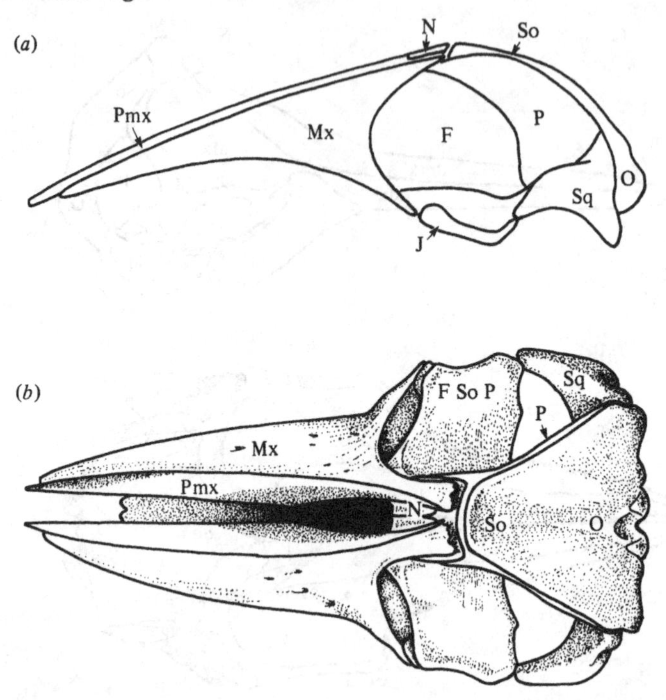

Fig. 92. The skull of the baleen whale. (*a*) Diagrammatic representation of telescoping, left lateral view; (*b*) dorsal view of skull of *Balaenoptera borealis*. For labelling see list of abbreviations. (*b* after Kellogg, 1928.)

place in or close to the margin of this aperture by having migrated far to the rear.

The details of the structure of the nasal cavity in the toothed whales vary somewhat from species to species but can be exemplified by the arrangement found in the bottlenose dolphin (*Tursiops truncatus*) (Fig. 91). The external nasal aperture lies directly above the interior aperture but between these two openings the nasal passage is bent into an arc, convex forwards, around the front of the braincase. In consequence, the roof and floor of the typical mammalian cavity have become reorientated to form the curved posterior and anterior walls, respectively, of the odontocete nasal passage. The boundaries of the nasal cavity are completed by anterolateral walls that curve round on either side from the wide posterior to the narrower anterior wall. The hard palate is greatly reduced in size.

The posterior wall is formed by the continuous mass of the mesethmoid–presphenoid complex which extends vertically over the anterior surface of the frontal. There is no cribriform plate and ethmoturbinals are lacking. The nasal bones are compressed to small nodules and lie on the sloping anterior surface of the braincase posterior to the external nasal aperture in articulation with the frontal. The anterior wall is formed by the maxillary and palatine bones. The anterior wall continues round on either side in a smooth curve to become continuous with the anterolateral walls. These comprise superiorly the premaxillae and inferiorly the maxillae, palatines and greatly enlarged pterygoid processes. The bony nasal septum is formed by the vomer which is greatly extended anteroposteriorly, articulating with the presphenoid behind and being interposed for a considerable distance between the maxillae in front. The septum is continued superiorly in cartilage. Traces of the maxilloturbinal persist.

The nasal cavity of the baleen whales departs less drastically from the typical mammalian condition than does that of the toothed whales (Fig. 92). The general direction of the nasal passage is upwards and forwards, rather than directly upwards as in odontocetes, and the frontal and nasal bones still form part of the roof (posterosuperior wall) of the cavity. The cribriform plate and rudimentary ethmoturbinals are present, indicating the persistence of the olfactory sense, even if much diminished. The palatine remains an extensive, plate-like bone that forms a small hard palate as well as entering into the lateral wall of the nasal cavity. The nasal septum is formed inferiorly by the vomer and superiorly by cartilage. As in odontocetes, the vomer is extended anteroposteriorly, articulating with the presphenoid behind and being interposed between the maxillae in front.

The nasal passage continues from the external bony aperture through the overlying soft tissues to open at the blowhole. In mysticetes the nasal septum is continued, in cartilage, through this distal part of the nasal passage right up to the blowhole so that two nostrils are apparent at or just below the surface of the head. The nasal passage can be closed, and thereby rendered watertight, by two large plugs attached to the septum. In the toothed whales the nasal septum is not continued right up to the blowhole which is, therefore, undivided. The distal part of the odontocete nasal passage consists of a system of air sacs and soft tissue plugs and folds which appears to be unnecessarily complex for just a sealing device.

A detailed account of the structure of this part of the nasal passage in the dolphin (*Tursiops truncatus*, *Delphinus delphis* and *Stenella plagiodon*) has been provided by Lawrence & Schevill (1956). They describe the cartilaginous nasal septum as extending upwards through most of the extent of the distal passage which is thus divided, except for its final uppermost part, into paired cavities. The posterior wall of the passage which lies close to the anterior surface of the cranium is relatively fixed while the anterior wall is mobile and closely adaptable to the posterior wall. Paired plugs are attached to the cartilaginous septum just above the external bony aperture. Under control of the nasal plug muscles these are capable of separating the distal nasal passage into upper and lower parts. Opening into the lower part are the large right and left premaxillary sacs which lie immediately adjacent to the upper surfaces of the premaxillae. The upper part of the passage receives the openings of the rather smaller but highly distensible paired vestibular sacs. These lie posterolateral to the blowhole and within its lips. Below the openings of the vestibular sacs the upper passage is narrowed to a transverse slit. Here open the small tubular and connecting sacs. The tubular sacs are U-shaped and lie horizontally with the connecting sacs leading from them as small appendices. Surrounding the distal nasal passage and its air sacs is the multilayered blowhole muscle.

There is now considerable evidence that the high-frequency sounds used by odontocetes in echolocating originate in the system of plugs and sacs of the distal nasal cavity but the exact mechanism of their production is uncertain despite numerous attempts at detailed analysis (e.g. see Tavolga, 1964). It seems that these sounds can be produced without an escape of air from the blowhole, implying that they result from the movement of air between the sacs. Presumably, the sacs are filled from the lungs and then local muscular control permits air to be passed from one sac to another, in the process producing the high-pitched clicks used in echolocating. There must also be a mechanism whereby the air within the nasal passage and sacs can be recycled so that sounds can continue to be produced even during prolonged dives.

A prominent and intriguing feature of the odontocete skull is the asymmetry of the bones in the region around the external nasal aperture, the bones on the left side of the aperture tending to be smaller than those on the right (Fig. 91). In some species the aperture itself has been displaced to the left. This lack of symmetry is not equally pronounced in all toothed whales, being especially marked in the narwhal, bottlenosed whale and sperm whale. Asymmetry is absent or insignificant in fossil odontocetes from the Oligocene and even in modern forms it does not appear until quite late in ontogeny (after the chondrocranial stage) and becomes progressively accentuated during post-natal growth. There seems little doubt, therefore, that it is a relatively recently acquired feature, brought about by variations in growth rates which are quite localised in their effect (neither the braincase nor lower jaw being affected). There have been numerous attempts to explain its appearance, some ingenious but none totally convincing.

Perhaps the most widely accepted explanation has been that put forward as long ago as 1910 by Lillie. He pointed out that, while an elongated epiglottis is a cetacean characteristic, this structure is especially long in odontocetes where, together with the arytenoid cartilages and associated muscles, it forms a cylindrical tube, a continuation of the larynx, projecting upwards into the nasopharynx above the soft palate. In the sperm whale and other odontocete species the tube does not lie centrally but over to the left-hand side. Lillie suggested that this arrangement, which he believed to be of advantage because it leaves a spacious pharyngeal passageway to the right of the tube, allowing large prey to be swallowed, is responsible for the displacement of the bony nasal cavity towards the left and thus for the associated cranial asymmetry. The flaw in this suggestion is that some odontocete species, for instance the killer whale, are able to swallow large, whole prey in spite of the fact that the epiglottic beak is centrally positioned. It may well be that Lillie transposed cause and effect and that the asymmetrically located beak is a consequence not a cause of an asymmetrically located nasal cavity.

Another early explanation of cetacean cranial asymmetry was that it results from the mechanism of propulsion through the water. According to Thompson (1942), for example, the asymmetry can be regarded as a 'counter-spirality' produced directly by the effect on the growing skull of unequal water pressures resulting from a spiral component introduced into cetacean locomotion by a rotatory action of the tail. However, recent studies have shown that during swimming in a straight line the tail and flukes of the dolphin (the skull of which is markedly asymmetrical) move in a vertical plane without any lateral or rotatory movement (see, for example, Slijper, 1962; Gray, 1968).

A third possibility is that the asymmetry is related to a distinction between the two sides in the production of echonavigating sounds. Support for this suggestion is forthcoming from the study by Evans & Prescott (1962) of the sound pressure levels close to the head of *Tursiops truncatus* while swimming. They found these to be asymmetrical about the median plane with a stronger signal output on the right. In *Tursiops truncatus*, as in other odontocetes, the facial asymmetry is associated with a considerable enlargement of the nasal passage on the right side at the expense of that on the left. It could well be, therefore, that the inequality of sound pressure on the two sides is a direct consequence of the difference in volume between the right and left nasal channels.

Alternatively, the asymmetry in the sound beam may be related to the shape of the forehead region of the skull. This region in the odontocete is smoothly concave and, since it lies directly behind the source of the echonavigating sounds, may act as a sound reflector. Norris, Prescott, Asa-Dorian & Perkins (1961) observed that the echolocating sound produced by the bottlenose dolphin is concentrated into a beam at forehead level, above and ahead of the melon (the large mass of fatty tissue located above the rostrum and in front of the distal nasal passage). Evans & Prescott (1962) suggested that this effect is produced by the concave surface of the cranium behind the distal nasal passage acting as a parabolic acoustic mirror capable of beaming high-frequency sound (further control over the range and directionality of the beam may be provided by the melon acting as an acoustic lens – in some dolphin species the shape of the melon can be modified thereby changing its acoustical focus). The discrepancy in the strength of the sound beam between the two sides could, therefore, be a consequence of the asymmetry of the skull affecting the sound reflecting properties of the forehead region.

Whether, however, the development of an asymmetrical sound beam is related to some as yet unknown improvement that it confers upon the echonavigating system, and might, therefore, be regarded as a cause of the facial asymmetry, or whether it is merely a secondary effect of that asymmetry, whose cause lies somewhere else, is not known. The concurrence within the odontocetes, by contrast with the mysticetes, of facial asymmetry and an extremely well-developed echonavigation system certainly makes it tempting to suppose that the two are in some way causally related. Support for this view would be forthcoming if it were found that the evolution of asymmetry and of echonavigation (as indicated by the structure of the auditory apparatus) were closely related in time.

PARANASAL SINUSES

Paranasal air sinuses are characteristic features of the skull of placental mammals. Paulli (1900*c*) found no evidence of pneumatisation in the skull of the monotremes *Ornithorhynchus* and *Tachyglossus* nor in the marsupials *Didelphis*, *Dasyurus*, *Petrogale*, *Macropus* and *Trichosurus*. The studies by this author of the nasal region of modern placental mammals (Paulli, 1900*b*, *c*), which, as still the best and most extensive accounts of the pneumatisation of the mammalian skull, form the basis of the following sections, have indicated that the maxillary sinus is a primitive eutherian possession. It is the only paranasal sinus present in insectivores and bats and its occurrence is constant in the majority of other placental orders. It may already have been present in some cynodonts (Kemp, 1979). Corroborative evidence for the primitive nature of the maxillary sinus would require study of the phylogenetic history of cranial pneumatisation but this has been largely prevented by the lack of suitably fragmented or sectioned skulls of fossil mammals.

MAXILLARY SINUS

1. *Insectivores*

A maxillary sinus is present in the majority of insectivores. It is usually a small chamber extending into the maxilla and lacrimal bone, although in some forms (e.g. *Tenrec*) it extends also into the frontal. No other paranasal sinus is present in the insectivore cranium. The maxillary sinus communicates with the nasal cavity through an opening marginated posteriorly by a curved notch in the anterior border of the lateral ethmoidal plate and anteriorly by the uncinate process (Fig. 93). The maxillary sinus and its opening into the nasal cavity follow the insectivore pattern closely in the majority of mammalian orders. The following is a brief summary of the principal departures from this pattern.

2. *Chiropterans*

The sinus is missing in some very small bats (e.g. *Vesperugo*).

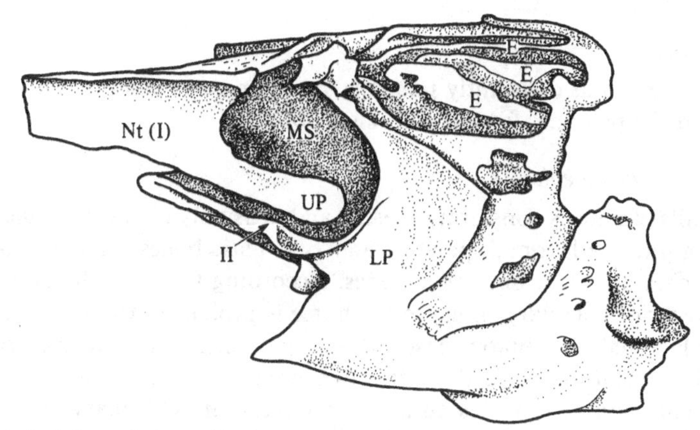

Fig. 93. Left lateral view of isolated ethmoid of *Erinaceus*. Abbreviations: E = ectoturbinals; LP = lateral plate; MS = opening of maxillary sinus; Nt(I) = nasoturbinal (endoturbinal I); UP = uncinate process; II = endoturbinal II. (After Paulli, 1900*c*.)

3. *Hyracoids*

The sinus in *Procavia* extends into the nasal and frontal bones (including the orbital plate of the latter) as well as pneumatising the nasoturbinal and first ectoturbinal.

4. *Carnivores*

According to Paulli, the sinus, although of modest size, is usually present, pneumatising the posterior part of the maxilla and the lacrimal and frontal bones. Negus (1958), however, described the sinus as being totally lacking in many carnivore species, the discrepancy between this author and Paulli probably being due, in part, to the differing techniques they used to examine skulls and also to differences of opinion as to what constitutes a fully independent maxillary sinus as distinct from a mere extension of the nasal cavity. Only in bears is the sinus extensive, extending far into the frontal and also into the nasal bones. The relative size of the sinus is not closely related to total body size in those species, such as the dog, where the latter is widely variable.

5. *Rodents and lagomorphs*

In this group the size of the maxillary sinus appears to be closely related to total body size, the chamber being small or even completely lacking in smaller species (e.g. in *Arvicola terrestris*, although here the large size of the molar dentition may also be a factor). In larger forms, such as the squirrel and the porcupine and especially the capybara (the biggest of all modern rodents), the sinus tends to be extensive, reaching into the nasoturbinal, frontal and nasal bones.

In lagomorphs, the sinus is large and extends into the nasoturbinal.

6. *Edentates*

The maxillary sinus constantly pneumatises the nasoturbinal and may also reach into the palatine, frontal and nasal bones.

7. *Ungulates and elephants*

The maxillary sinus is constantly present and is usually of considerable size, extending into the lacrimal, palatine and zygomatic bones. In some species it encroaches on numerous other bones. According to Paulli the maxillary sinus (posterior maxillary sinus) in the horse is prolonged from the maxilla into the lacrimal, zygomatic, nasoturbinal, palatine, sphenoid and frontal bones. The extensions into the latter two bones are more usually termed 'sphenoidal' and 'frontal' sinuses, respectively, in veterinary texts. An additional sinus (the malar or anterior maxillary sinus) is present in perissodactyls in the anterior part of the maxilla and extends into the maxilloturbinal and zygomatic bones. It opens into the nasal cavity just in front of the posterior maxillary sinus.

Despite the generally large overall size of the ungulate maxillary sinus, it may pneumatise only a very small part of the maxilla itself, being located principally in other bones (e.g. in the hippopotamus).

The maxillary sinus in the elephant is a very large and irregular chamber located in the maxilla and premaxilla and partitioned into numerous subcompartments (see section on frontal sinus, p. 271).

8. *Prosimians*

The size of the maxillary sinus is closely related to overall size. In the larger species of lemurs, for example, an extensive sinus is present which involves the frontal bone, as well as the maxilla and lacrimal, while in smaller species, such as the slender loris, the sinus is reduced to a narrow cavity occupying just the posterior part of the maxilla.

9. *Old and New World monkeys*

A maxillary sinus was found in all the monkeys examined by Paulli, with the exception of *Presbytis*. The sinus varies greatly in size between species, apparently with some relationship to body size, but does not extend outside the maxilla and lacrimal bones.

The sinuses of the hominoids are dealt with collectively in a later section.

10. *Aquatic mammals*

The maxillary sinus is completely lacking, or is present in rudimentary form only, in pinnipeds, sirenians and cetaceans.

To summarise, therefore, it appears that the maxillary sinus is a typical feature of living placental mammals, being found in all species where cranial pneumatisation extending from the nasal cavity is present at all. It develops from the part of the nasal cavity situated immediately anterior to the ethmoid so that its opening is marginated posteriorly by the anterior edge of the ethmoidal lateral plate and anteriorly by the uncinate process. The relative size of the sinus in many mammalian groups exhibits a general relationship to body size and the sinus may accordingly be lacking in some very small species. The sinus frequently extends into other bones besides the maxilla, including the lacrimal, palatine, nasal, frontal, nasoturbinal, zygomatic and sphenoid. In some ungulates only the smaller part of the sinus lies actually within the maxilla and in a few forms it does not extend into this bone at all.

'FRONTAL' AND 'SPHENOIDAL' SINUSES

Most mammals possess, in addition to the maxillary sinus, a system of pneumatic chambers communicating with the olfactory region of the nasal cavity by openings located between the lamellae of the ethmoturbinals. Since this system occupies principally the frontal bone, or, less frequently, the sphenoid, its constituent sinuses are commonly referred to as 'frontal' or 'sphenoidal', as the case may be, but in many species it extends widely to involve other cranial bones. The sinuses develop as evaginations from the intervals between the basal parts of the lamellae, their points of origin being indicated in the adult by their openings into the nasal cavity. The number and extent of the sinuses vary greatly from one species to another but in almost all instances the number is less than that of the intervals between the lamellae of the ethmoturbinals. There is also much individual variation within species and between the two sides within a single individual, much of this variation appearing during the continued enlargement of the sinuses which takes place during the later stages of growth and in adult life.

In spite of the tendency for the system of sinuses to vary in size between the two sides of the cranium, the openings into the nasal cavity are usually bilaterally symmetrical. Believing that the position of these openings is the key to determining homologies between the sinuses of different species, Paulli (1900*b*) devised a scheme of designating the sinuses according to the relationships between their openings and the ethmoturbinals instead of adopting the more usual custom of naming the sinuses according to the bones they occupy. Thus each sinus is given the same number as the ethmoturbinal lying immediately above its opening (e.g. sinus I′ opens below endoturbinal I, and sinus II′ below endoturbinal II, while sinuses 1′ and 2′ open below ectoturbinals 1 and 2, respectively – Figs. 85, 86, 87). Whether or not such designations have the significance in determining homologies that Paulli supposed, they do possess the advantages of being clear and precise and will, for these reasons, be adopted in the following account.

When the sinuses extend widely into the cranial vault or base the corresponding chambers of the two sides may be separated by no more than

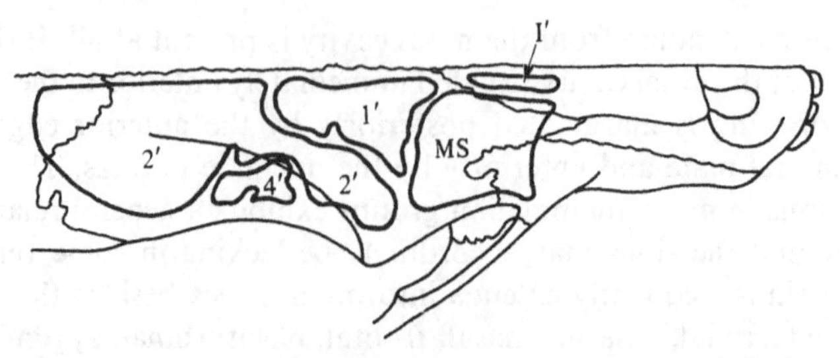

Fig. 94. Dorsal view of cranium of *Ursus arctos* to show paranasal air sinuses in cranial vault. Sinuses numbered according to scheme described in text. For other labelling see list of abbreviations. (After Paulli, 1900*c*.)

a thin bony partition covered by mucous membrane. Because of the presence of marked bilateral asymmetry this septum frequently departs from the median plane. Although multiple communications between the sinuses are frequently present in the bony skull, these were found by Paulli to be always closed by mucous membrane in the recent state. Similarly, the openings of the sinuses into the nasal cavity are frequently marginated by folds of mucous membrane lacking a bony support.

As already noted, the maxillary sinus provides the sole pneumatisation of the cranium in insectivores and bats while in aquatic mammals, cranial pneumatisation is totally lacking, almost certainly having been secondarily lost. In the remaining placental orders the following additional sinuses have been described as present by Paulli (1900*b*, *c*) (again the hominoids are dealt with separately in the subsequent section).

1. *Hyrocoids*

There is one sinus, additional to the maxillary sinus, on each side of the cranium in *Procavia*. It is a high but rather narrow cavity in the presphenoid and opens into the nasal cavity posterior to endoturbinal IV (i.e. the fifth olfactory plate).

2. *Carnivores*

Amongst carnivores the system of sinuses opening into the olfactory region of the nasal cavity is most extensively developed in the bears. In *Ursus arctos*, for example, the following are present (Figs. 86*b*, 94): sinus I′ (i.e. opening between endoturbinal I and ectoturbinal 1) in the nasal bone; sinus 1′ in the frontal; sinus 2′ lying posterolateral to sinus 1′ in the frontal and extending into the parietal; sinus 4′ located in the frontal, lateral to sinus 2′; sinus IV′ in the body of the presphenoid. Paulii found that ectoturbinals 1, 3 and 4 extended partly into sinuses I′, 2′ and 4′, respectively, while the seventh olfactory fold extended likewise into sinus IV′.

In the dog (Fig. 86*a*) a ‘frontal’ sinus is present, opening between the lamellae of ectoturbinals 2 and 3. The size of this sinus appears to be closely related to body size occupying virtually the whole of the frontal bone in large dogs. Ectoturbinal 3 reaches a short distance into the sinus. In very large dogs an additional sinus may be present in the anterior part of the frontal which communicates with the nasal cavity between ectoturbinals 1 and 2.

The domestic cat (Fig. 84) possesses a large sinus in the frontal bone which opens into the nasal cavity at the posterior end of the interval between ectoturbinals 2 and 3. Ectoturbinal 2 reaches a short distance into the sinus. There is also a large sinus in the body of the presphenoid which opens behind endoturbinal IV. The arrangement in the lion is essentially similar, although here sinus 2′ extends beyond the frontal into the parietal bone.

In the hyaena the posterior part of the frontal is occupied by a large sinus which opens between ectoturbinals 2 and 3. Ectoturbinal 2 extends a short distance into the sinus. This sinus becomes enormously enlarged with age, extending into the parietal and pneumatising the sagittal crest (Buckland-Wright, 1969). It exhibits much asymmetry between the two sides. An additional frontal sinus, opening between ectoturbinals 2 and 3, may be found in old animals.

Paulli found that only the maxillary sinus was present in the mustelids that he examined, but other authors (e.g. Allen, 1882; Anthony & Iliesco, 1926) have reported pneumatic sinuses within the frontal bone of these carnivores. This discrepancy is probably due to no more than the mode of development of the 'frontal' sinuses by progressive evagination of the mucous membrane from the intervals between the ethmoturbinals. Thus what appears as part of the nasal cavity at a young stage may well appear as a separate sinus in an older animal.

Paulli found no frontal pneumatisation in the raccoon (although Anthony & Iliesco (1926) reported that the procyonid skull generally possesses frontal and sphenoidal sinuses) and only a small 'frontal' sinus, opening between ectoturbinals 2 and 3 in *Genetta genetta*.

3. *Rodents and lagomorphs*

The majority of members of these two orders possesses only a maxillary sinus although, as already described, in larger forms this may extend beyond the maxilla into the frontal bone. In the porcupine and capybara, however, additional sinuses are present. The skull of the porcupine (*Hystrix cristata*) is characterised by an extensive and complex pneumatisation which includes the following chambers: sinus I′ situated in the frontal and parietal bones; sinus 1′ also situated in the frontal and parietal but extending inferiorly into the temporal; sinuses 2′, 3′ and 4′ occupying the anterior part of the cranial vault and orbital plate of the frontal; sinus IV′, a deep and irregular pocket, extending into the posterior part of the maxilla and palatine; sinus V′, also large, occupying the bodies of the presphenoid and basisphenoid. There are also three sinuses opening from the nasopharynx. Two of these empty behind the posterior edge of the transverse lamina and extend into the medial orbital wall, cranial vault and majority of ethmoturbinals. The third opens immediately behind the posterior nares and is a very large chamber occupying a large part of the maxilla and hard palate (the maxillary sinus being found principally in the nasal bone and nasoturbinal).

The capybara possesses a large 'frontal' sinus which communicates with the nasal cavity between endoturbinal I and ectoturbinal 1 and is prolonged into the anterior part of the parietal and the first ectoturbinal, as well as into the frontal.

4. *Edentates*

In the armadilloes (*Chaetophractus villosus*, *Euphractus sexcinctus* and *Dasypus novemcinctus*) and ant-eaters (*Myrmecophaga tridactyla* and *Tamandua tetradactyla*) 'frontal' sinuses are lacking. In the sloths, as represented by *Choloepus didactylus*, three sinuses are present opening from the olfactory region: sinus I′ extends widely in the cranial vault; sinus 3′ is very small and is located in the anterior part of the vault; sinus V′ extends into the body of the presphenoid behind the posterior prolongation of the nasal cavity into that bone.

Tamandua and *Choloepus* possess pneumatic sinuses in the presphenoid and pterygoid regions of the cranial base which open into the nasopharynx.

5. *Ungulates*

The system of 'frontal' sinuses reaches its maximum extent in the big ungulates and in the elephants. The variation in the system between the different ungulate species makes any attempt at a comprehensive description impossible in the space available here. Instead, an account will be given of the system in a few well-known species as illustrations of the variety of ungulate arrangements.

Amongst the artiodactyls the domesticated pig possesses up to 13 sinuses opening off the olfactory region, the number of sinuses increasing with age. Sinus I′ is present shortly after birth, sinuses 4′ and V′ appear during the first year of life, sinuses 5′, 11′, 12′, II′, 18′, 19′ and III′ develop during the second year, while sinuses 2′, 9′ and 17′ are not present until four years or later. Not all of these sinuses necessarily appear in each animal, however old. The size of the chambers also increases progressively with age and a considerable degree of bilateral asymmetry is usually present in older animals. The position of the sinuses in the cranial vault of an adult male pig studied by Paulli is shown in Fig. 95 (sinus I′ also pneumatises the nasoturbinal). Sinuses 12′, II′, 18′, 19′ and III′, all relatively small, lie in a superoinferior sequence in the medial orbital wall (i.e. in the orbital plate of the frontal and lesser wing of the sphenoid). The later developing sinuses 2′, 9′ and 17′ insinuate themselves between the pre-existing sinuses of the orbital region.

In cattle and camels the system of sinuses opening from the olfactory region is generally even more extensive than that of the pig, pneumatising the

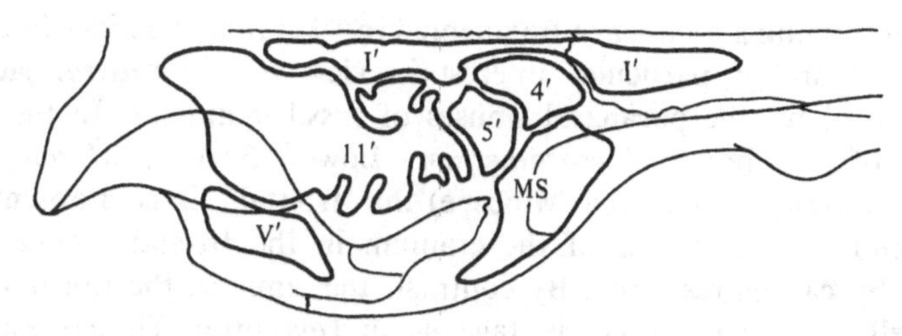

Fig. 95. Dorsal view of cranium of domesticated pig to show paranasal air sinuses in cranial vault. Sinuses numbered according to scheme described in text. For other labelling see list of abbreviations. (After Paulli, 1900*b*.)

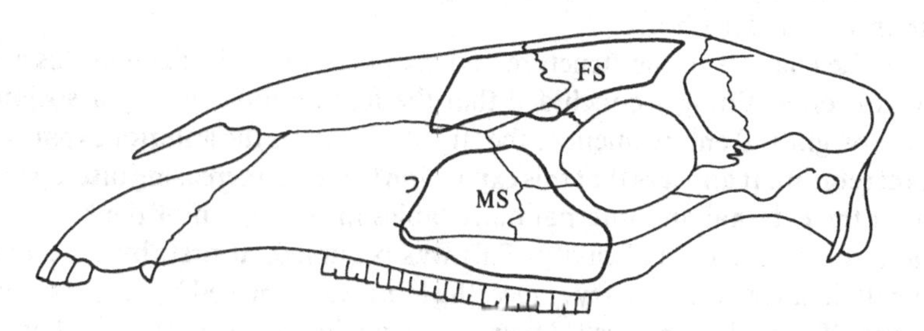

Fig. 96. Left lateral view of skull of horse to show 'frontal' and maxillary sinuses. For labelling see list of abbreviations. (After Edinger, 1950.)

ethmoidal region on all sides (including the ethmoturbinals – Fig. 87*a*) as well as much of the vault and base of the cranium. There is a great extension of one sinus (6′) in cattle which pneumatises the horn peg. In smaller artiodactyls, such as sheep, goats and some deer, the degree of pneumatisation is considerably less, especially that of the cranial base. Pneumatisation is still more drastically reduced in the hippopotamus.

The system of sinuses of perissodactyls is also well developed. In the rhinoceros (*Rhinoceros sondaicus* – Fig. 87*b*), for example, only the vomer, zygomatic and premaxilla lack pneumatisation. There are up to 11 sinuses in the system (sinuses I′, 1′, 3′, 6′, 7′, 8′, 10′, 11′, 12′, 13′ and V′). Sinus V′ is very large and pneumatises the bones of the cranial base. The remainder are situated principally in the bones of the vault, sinuses 10′ and 12′ being the most extensive.

Pneumatisation of the horse skull is almost as extensive (Fig. 96). It is usually described as consisting of a maxillary, frontal and sphenoidal sinus. However, Paulli, as already noted, described both the so-called 'frontal' and 'sphenoidal' chambers as being, in reality, extensions of the maxillary sinus. There are sometimes two independent sinuses present in the presphenoid and basisphenoid, one opening behind ectoturbinal 31, the other behind endoturbinal VI.

Edinger (1950) has investigated the evolution of the 'frontal' sinus in the horse by examining a number of fragmented fossil crania of extinct Equidae. The findings are of particular interest in view of the general lack of information about the paranasal sinuses of fossil mammals. In the early ancestors of the horse (*Hyracotherium* – Lower Eocene, *Mesohippus* – Oligocene, *Parahippus* – Lower Miocene) the 'frontal' sinus is absent, the external and internal tables of the cranium in the frontal region being separated by cancellous bone. By contrast, the sinus in the upper jaw is already well developed at this early stage of equid evolution. The earliest form in which Edinger found a 'frontal' sinus (or frontal extension of the maxillary sinus, if one follows Paulli) is *Merychippus* from the Middle Miocene where a small sinus is present in the region equivalent to that of maximum sinus width in the modern horse.

From the changes in the structure and proportions of the skull in this evolutionary series. Edinger concluded that the feature most closely associated with the origin and enlargement of the 'frontal' sinus is the anterior expansion of the cerebrum. It appears that this expansion led to an increasing discrepancy between the external and internal bony tables in the region of the braincase just anterior to the cranial cavity. This was occupied, at first, by cancellous bone, but, in later evolutionary stages, the bone was replaced by an expanding air sinus. Thus, the one well-documented evolutionary sequence that we possess indicates that sinus formation, as exemplified by the equid 'frontal' sinus, is directly related to evolutionary changes in cranial proportions.

6. *Elephants*

Pneumatisation of the elephant skull presents a comparable form to that seen in the larger ungulates but is even more extensively developed, involving every cranial bone except the zygomatic (Fig. 97). As already described, the maxillary sinus is very large and irregular and occupies much of the maxilla and premaxilla. Opening from the olfactory region are the following sinuses (African elephant): sinuses I′, 1′, 2′, 3′ and 4′ are small and occupy the anterior part of the frontal bone (sinus I′ also extends into the nasoturbinal); sinus 5′, by contrast, is an extremely large chamber that is prolonged over the roof of the cranial cavity through the frontal and parietal bones, into the occipital from where it extends anteroinferiorly into the temporal, this sinus also possessing extensions into the basisphenoid, posterior part of maxilla and palatine; sinuses 9′ and 12′ form a pair of small chambers in the orbital plate of the frontal; sinuses V′ and VI′ are large and irregular and occupy the bodies of the presphenoid, basisphenoid and occipital bones.

The maxillary and basal (V′ and VI′) sinuses have a highly complex, labyrinthine form, consisting of numerous narrow canals radiating from a relatively small central chamber. These canals follow tortuous paths winding between one another in an irregular manner. At intervals along the canals are small dilatations from which further canals are given off.

The sinuses in the cranial vault (principally 5′) are arranged in a different

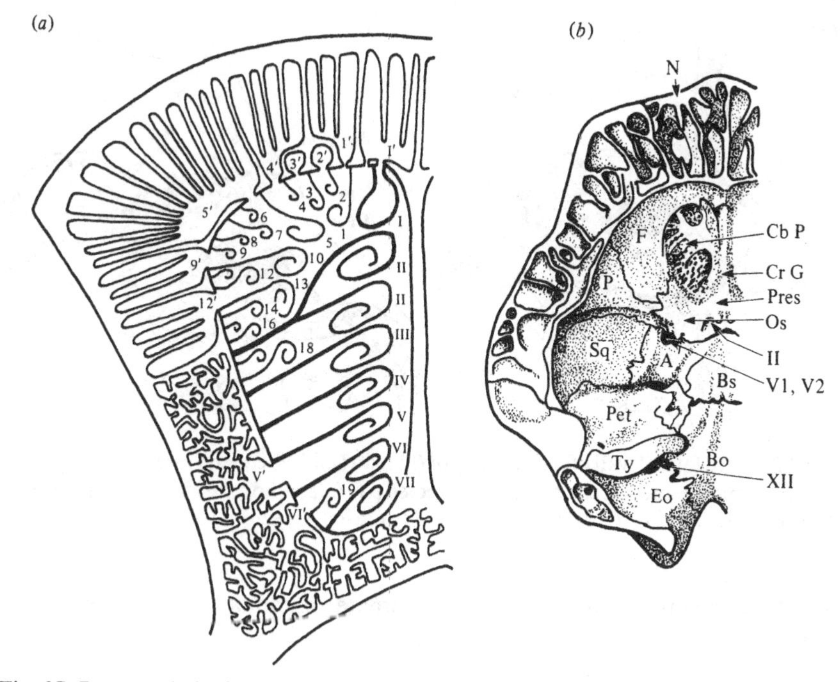

Fig. 97. Paranasal air sinuses in the African elephant. (*a*) Schematic section through nasal cavity; turbinals and sinuses numbered according to scheme described in text. (*b*) Coronally sectioned skull looking anteriorly; for labelling see list of abbreviations. Not to scale. (*a* after Paulli, 1900*b*; *b* after Starck, 1967.)

manner. Here, there are numerous, high (9 cm or more) but thin bony partitions attached to the roof of the sinuses and radiating downwards in a fan-shaped manner. The partitions are close together and thus divide the interior of the sinuses into a multitude of deep but narrow subcompartments.

7. *Prosimians*

In the smaller species (e.g. the slender loris) the only paranasal sinus present is the maxillary. In lemurs there is, in addition, a large sinus (I′) in the anterior part of the frontal bone.

8. *New and Old World monkeys*

In some groups (*Callithrix*, *Cercopithecus* and baboons), paranasal pneumatisation is limited to the maxillary sinus (even this is lacking in *Presbytis*). In *Cebus* and *Aotus* there is, in addition to the maxillary sinus, a small air chamber in the anterior part of the frontal bone (i.e. above the supraorbital ridges) which opens into the nasal cavity between the roof and septum. *Cebus* also possesses a small presphenoidal sinus which, since it opens behind the ethmoturbinals, may represent the posterior sphenoidal extension of the

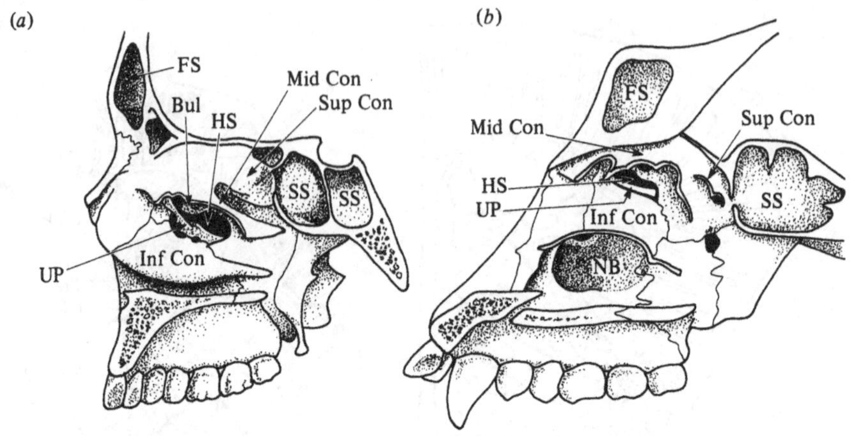

Fig. 98. Right lateral wall of nasal cavity of (*a*) man (middle concha resected); (*b*) gorilla (inferior, middle and superior conchae resected). For labelling see list of abbreviations. Not to scale. (*a* after *Gray's Anatomy*, 1973; *b* after Cave & Haines, 1940.)

olfactory region found in many non-primate mammals rather than a true sinus.

PARANASAL SINUSES IN THE HOMINOIDEA

Detailed information about the paranasal pneumatisation of the hominoid skull has been provided by Cave & Haines (1940). The gibbon and the orang-utan possess the primitive primate complement of sinuses but in man and the African apes two new groups of sinuses have appeared, namely the ethmoidal (here again traditional terminology is confusing, for the systems of so-called 'frontal' sinuses found in most non-primate mammals develop from the ethmoidal region of the nasal cavity; the term 'ethmoidal' is used here with the same connotation it carries in texts of human anatomy, i.e. meaning the small air cells that pneumatise the hominoid ethmoidal labyrinths) and the 'true' frontal sinuses. A 'true' frontal sinus can be distinguished from the 'frontal' sinuses found in many non-primates and in some monkeys and prosimians by the fact that it always opens into the anterosuperior part of the middle meatus and represents developmentally a greatly enlarged ethmoidal air cell.

In the human skull the maxillary sinus is of moderate size and is limited to the maxilla. It tends to enlarge throughout life and may eventually occupy virtually the whole of the interior of the maxillary body. The principal opening of the sinus is into the posterior part of the semilunar groove, although not infrequently one or more secondary ostia may be present posterior to the uncinate process and between the process and the upper border of the inferior concha (Fig. 98). In the recent state these are usually closed by soft tissue.

The human frontal sinus, like that of the African apes, is an ethmoidal derivative. It shows considerable variation in its size and its communication with the nasal cavity although always opening into the anterior or superior part of the middle meatus. It is generally larger in men than in women and does not reach maximum size until early adult life. The septum separating the right and left sinuses is frequently displaced widely from the median plane. The sinus usually opens through the frontonasal canal and infundibulum into the anterior terminal recess of the semilunar groove but occasionally it opens into a frontal recess of the middle meatus or into the suprabullar groove located between the ethmoidal bulla and middle concha. These variations may reflect individual differences in which one of the original ethmoidal air cells becomes enlarged to form the frontal sinus.

The sphenoidal sinus of the human skull is a relatively large and irregular cavity within the body of the sphenoid, separated by a septum from its fellow of the opposite side. The chamber may be partly subdivided by bony plates and may extend into the greater wings. It opens into the sphenethmoidal recess. From study of its development it appears that the sphenoidal sinus represents the prolongation of the olfactory region into the sphenoid found in non-primate mammals, which has lost its complement of olfactory folds, secondary to reduction of the ethmoturbinals, and has become converted into an 'empty' chamber communicating with the nasal cavity by means of a narrowed opening. The sphenoidal sinus of apes and monkeys is probably similarly derived.

The ethmoidal air sinuses reach their maximum elaboration in man where they develop not only from the lateral walls of the middle and superior meatus but frequently from the supreme meatus as well. They appear to represent parts of the original nasal cavity which have become enclosed by the formation of a new lateral wall in the ethmoidal region of the nasal fossa as a result of fusion of ectoturbinals. Ectoturbinals are present in the fetal stage but disappear as discrete structures by a process of fusion as growth proceeds (Negus, 1958). The ethmoidal air cells consist of thin-walled cavities within the labyrinths, the walls of many of them being completed by the contiguous frontal, maxillary, lacrimal, sphenoidal and palatine bones. The cavities are separated from the orbits laterally by the delicate orbital plates. On the basis of their sites of communication with the nasal cavity, the ethmoidal sinuses are divided into anterior, middle and posterior groups. The anterior group, which may comprise as many as 11 individual cells, opens into the infundibulum or frontonasal canal by one or more openings. The middle sinuses, usually three in number, occupy the bulla (probably representing ectoturbinal 1) and open into the middle meatus above the bulla either by a common orifice or by separate orifices. The posterior group, consisting of up to seven sinuses, communicates with the superior meatus usually by a single opening. One of the posterior sinuses may open into the supreme meatus.

In both the gorilla and the chimpanzee, the frontal sinus opens in the suprabullar position. This sinus often becomes very large, especially in the adult male gorilla where it may have a complex multilocular form and may extensively invade the orbital roof. The sphenoidal sinus is also usually well developed, occupying the body of the sphenoid and possessing, characteristically, extensions into the greater wing and pterygoid processes. It opens into the sphenethmoidal recess. The ethmoidal sinuses are limited to no more than three in number – two posterior cells opening into the superior meatus and an anterior cell opening into the semilunar groove. In the gorilla the ethmoidal sinuses become reduced and may disappear in the adult or, alternatively, may never fully develop.

The opening of the maxillary sinus into the middle meatus in the two African apes corresponds exactly in its morphological relationships to the primary ostium of the sinus in man. In the chimpanzee the sinus extends into the hard palate, including its premaxillary part. The maxillary sinus is very large in the gorilla but does not have a palatine extension. It occupies the whole of the body of the maxilla posterolateral to the nasolacrimal bulla and may invade the zygomatic bone laterally. With advancing age the sinus may encroach upon the ethmoidal labyrinth, with subsequent reduction in the size of the contained sinuses, and thence invade the frontal and sphenoid bones, remaining separated from their contained sinuses by thin bony laminae.

The arrangement of the sinuses in the orang-utan departs widely from that seen in the African apes in that ethmoidal and frontal sinuses are lacking and their place is taken by an enormously dilated maxillary sinus. The latter occupies not only the entire maxilla but secondarily invades the ethmoid, frontal, sphenoid, palatine, lacrimal and zygomatic bones. Its sphenoidal extension passes dorsal to the small sphenoidal sinus to extend into the pterygoid plate and zygomatic process of the temporal bone. The maxillary sinus opens directly into the middle meatus by an opening bounded anterosuperiorly by the ethmoid, inferiorly by the vertical plate of the maxilloturbinal and posteriorly by the vertical plate of the palatine. There is no uncinate process.

Frontal and ethmoidal sinuses are also lacking in the gibbon. The maxillary sinus is of moderate size, confined to the maxilla, and usually communicates indirectly with the middle meatus by means of the semilunar groove (i.e. as it does in man and the African apes). The sphenoidal sinus is relatively large and invades the frontal bone, ethmoidal labyrinth and vertical plate of the palatine.

On the basis of the number and structure of their sinuses, the Hominoidea can be divided into three groups. The first consists of the gibbons which possess the primitive primate complement of sinuses and in which pneumatisation of the skull is achieved principally by enlargement of the sphenoidal sinus. The orang-utan forms the second group. Here again the primitive primate patterns of sinuses is present but cranial pneumatisation is due principally to the enormous enlargement of the maxillary sinus. The third group includes man and the African apes in which the neomorphic ethmoidal and 'true' frontal sinuses are present. It seems likely, therefore, that the structure of the nasal cavity and sinuses of the gibbon and orang-utan derive from a relatively primitive primate stage in which the ethmoidal and 'true' frontal sinuses had not yet developed.

FUNCTIONS OF THE PARANASAL SINUSES

The function of air spaces within the cranial bones has always been a matter of considerable uncertainty, largely because the advantages that have been suggested as accruing from their presence are difficult to judge and seem, in many instances, to be so negligible. Of the numerous reasons that have been advanced for the development of pneumatisation, the following have received at one time or another, a measure of support.

1. Probably the most widely held view is that the paranasal sinuses serve to reduce the dead weight of the head. Most of the arguments advanced against this view relate to the human skull where, it is frequently pointed out, the weight saving (as compared with having the sinuses replaced by an equal volume of cancellous bone) would be of the order of a few per cent at most. However, the very much more extensive sinuses found, for example, in ungulates and proboscideans must have a far greater weight-saving effect.

2. Another popular view has been that the sinuses are really functionless, having arisen incidental to the evolutionary changes that have taken place in the shape of the mammalian cranium. Thus Weidenreich (1941) described the sinuses as having no active function, being just the 'regions left vacant between the mechanically essential pillars of the constructing parts of the face and cranium'. The principal merit of this suggestion is that it appears to be the only one capable, on its own, of explaining the very wide variation in the number and arrangement of the pneumatic chambers found between different mammals. The counter argument that the sinuses appear to develop in ontogeny by an active process of bone resorption rather than passively as a result of moving apart of their walls as an integral part of the growth of neighbouring regions, is based on an unduly simplistic view of the recapitulation theory.

3. The observation that the ethmoturbinals extend into the 'frontal' and 'sphenoidal' paranasal sinuses in some carnivores has led to the suggestion that pneumatisation of the skull evolved to allow the area of olfactory epithelium to be increased and the olfactory sense thus improved (Negus, 1958). However, both the range of mammals displaying this phenomenon and the increase in the area of olfactory epithelium achieved appear to be too limited to make this an acceptable explanation of the presence of cranial pneumatisation in mammals generally.

4. Yet another possibility is that the paranasal sinuses act as resonators modifying the sounds produced in the larynx. This suggestion has been applied particularly to man and there is no doubt that blockage of the

sinuses with secretions following, for example, a head cold, may greatly change the character of the voice. In a broader view of mammalian species, however, there is no obvious and consistent correlation between the importance of vocalisation and the extent of pneumatisation.

5. Finally, it has been suggested that the sinuses may have a heat-regulating effect, acting to prevent heat loss from the nasal cavity. The great variability in the degree of pneumatisation between and even within closely related species suggests that this effect can scarcely be of crucial survival significance. Alternatively, the sinuses could act as heat exchange devices although the very small size of their communications with the nasal cavity would appear to limit sharply their efficiency in this respect.

One of the difficulties in judging between these various explanations for the development of cranial pneumatisation is the considerable species and individual variability exhibited in both the number and extent of the paranasal sinuses which gives the impression that pneumatisation is an almost haphazard procedure. However, examination of the positions of the openings of the sinuses reveals a more consistent pattern, at least between closely related mammalian groups, and suggests that in its early development, if not in its final extent, cranial pneumatisation is more orderly than might appear to be the case at first sight.

Two conclusions seem almost self-evident from these observations. First, whatever the function of the sinuses may be it is their *approximate* size and distribution which are important rather than their exact extent. Secondly, in view of the very great variation in cranial pneumatisation between the different orders of mammals, it is by no means certain that the sinuses are serving the same functions in each.

At this stage in the discussion it would be invaluable to be able to refer to the phylogenetic history of the paranasal sinuses but unfortunately the relevant fossil evidence is almost totally lacking. The findings in modern mammals, however, do point strongly to the possibility that the maxillary sinus was the first pneumatic chamber to develop and that it may have been a consistent feature at a very early stage in mammalian evolution. In order to explain this early appearance, one must define a function which would have been relevant to what was almost certainly a relatively small chamber developing in the upper jaw of tiny animals, a task not rendered any easier by our lack of knowledge of the general cranial morphology of the early mammals. Of the possibilities listed above, weight saving and extension of the olfactory area seem unlikely explanations in that the reduction in weight must have been totally insignificant while intrasinus extension of the ethmoturbinals in modern mammals does not involve the maxillary sinus and is not a feature of insectivores or other primitive types. Nothing is known, of course, about voice production in extinct forms but again it seems unlikely that the presence of a maxillary sinus could have had a significant effect upon this character in the early mammals. All that remains, therefore, are the possibilities that the maxillary sinus developed to serve a heat-insulating function or that it developed incidental to restructuring of the upper jaw (and was, therefore, 'functionless' at least initially). In order to judge the second possibility, which seems the more probable of the two, information would be needed about the mechanical restructuring of the upper jaw in early mammals and of the changes in the ontogenetic processes by which this restructuring was achieved, information which seems unlikely to be ever forthcoming.

The sinuses opening from the olfactory area of the nasal cavity appear to be a later mammalian acquisition and such evidence as is available, mainly derived from the study of modern mammals, suggests that they may have evolved independently in the different orders. If so, they are structurally speaking analogous rather than homologous structures. In those groups where they are best developed, notably the ungulates and elephants, there seems to be a correlation between the extent of pneumatisation and total body size. Such a correlation could be attributed to the need for weight saving in what are generally massive animals or, alternatively, the need for drastic cranial restructuring to accommodate the enlarging brain or superstructures, like horns and tusks, and to provide large areas of attachment for powerful masticatory and nuchal muscles. These two possibilities are not, of course, unrelated, pneumatisation providing a means of modifying the size and shape of the cranium with a minimum of additional skeletal material and yet retaining a mechanically efficient structure. There may be, in addition, ontogenetic reasons why this particular method of modifying cranial proportions has been so widely adopted. It may, for example, require simpler, and therefore more flexible, morphogenetic mechanisms for its control than would be required to bring about cranial restructuring by the redisposition of skeletal material alone.

Paulli (1900*b*, *c*) stresses the close correlation between the extent of pneumatisation and body size in other mammalian orders besides the ungulates and elephants but close examination of his findings, together with personal examination of cranial material, suggests that this correlation is much less perfect that Paulli believed. Nevertheless, weight saving might still be a significant factor in the larger carnivores. For the remaining terrestrial mammals the degree of weight saving achieved by pneumatisation from the olfactory region, even allowing for relative effects, would appear to be of no great significance. In these cases it seems more likely that pneumatisation is present solely as a mechanism for modifying cranial proportions.

There are many features of the paranasal sinuses on which the explanations just offered shed no light. For example, why should the sinuses of the elephant possess such a complex internal structure, or why should some hominoids have evolved additional pneumatisation? There seems little point in speculating on such intriguing questions until we have more comparative data relating to cranial pneumatisation in modern mammals and a better understanding of the development and growth, and above all of the evolution, of the paranasal sinuses.

SECTION IV

SKULL GROWTH

8

SKULL GROWTH

Detailed information about the growth of the skull is available for only a small proportion of the many thousands of known mammalian species. Interest has focussed overwhelmingly on the growth of the human skull, as part of the clinical endeavour to prevent and treat the disorders of facial growth which are so prevalent in children living in developed countries. Fortunately, the attempt to unravel the complexities of skull growth in man has involved recourse to experimental and other investigative procedures in the common laboratory animals, including the rat, rabbit, guinea-pig, dog and macaque monkey, as well as in domesticated animals such as the pig and sheep, so that a body of information has been built up indicating that the mechanisms of skull growth have much in common in a variety of species which, though limited in number, covers a considerable range of mammalian orders.

Two main patterns of growth appear to occur typically in the mammalian body. The nervous system and its adnexa grow rapidly during prenatal and early postnatal life, completing most of their growth well before the rest of the body. The human brain, for example, attains some 90 per cent of its adult size by the age of six years. The main bulk of the body follows a more protracted growth course. Again, to take man as an example, growth in overall body size continues until about 18 to 20 years and, although growth is more rapid in the prenatal and infant periods than in later childhood, the contrast between these stages is much less marked than in the neural pattern of growth. In man, and possibly in the great apes, the somatic growth pattern is complicated by a spurt of growth in adolescence.

Comparative studies of skull growth in a number of different mammalian species have demonstrated that the braincase, orbital cavities and otic capsules follow, as would be expected, the neural pattern of growth while the facial skeleton follows the somatic pattern (e.g. rat – Asling & Frank, 1963; Moore, 1966; rabbit – Erickson & Ogilvie, 1958; pig – Todd & Cooke, 1934; hyena – Todd & Schweiter, 1933; sheep – Todd & Wharton, 1934; dog – Scott, 1951; primates – Schultz, e.g. 1926, 1962; Hellman, 1927, 1929; Krogman, 1930*a*, *b*, 1931*a*, *b*, *c*; Moore & Lavelle, 1975). These studies have also indicated that, as a rule, the braincase and the facial skeleton tend to grow more in depth (anteroposterior length) than in height or width, the relationships between cranial dimensions often being of an allometric nature (see, for example, Reeve, 1940; Wood, 1976). The general truth of these statements is a matter of common observation, young mammals, compared to their adults, possessing heads that are short and wide and jaws that are small relative to the braincase and orbits. As we shall see, the differing growth patterns of the facial skeleton and braincase are accommodated by a series of changes in the linear and angular dimensions of the cranial base which forms an intermediate zone between the two major subdivisions of the skull.

The mechanisms of growth by which these changes in skull proportions are achieved are the same, in essence, as those in the skeleton generally although, as might be expected from the complex shape of the skull, being combined in a rather more involved fashion. Because of its rigidly mineralised matrix, bone can grow by surface accretion only. The osteocytes, trapped within their lacunae, cannot divide and form additional matrix. This inability to grow interstitially sets bone clearly apart from cartilage and other connective tissues. Indeed, it has been suggested (see Chapter 2) that this limitation is the reason why cartilage, with its greater growth flexibility, has become the dominant skeletal tissue in immature vertebrates.

A corollary of the fact that bone can grow only by surface addition is that there must also be surface resorption. Without such a mechanism the complex shapes of bones would become progressively distorted as growth proceeds. The combined processes of surface addition and resorption are frequently termed remodelling but, as we shall see, the term is somewhat inappropriate in that remodelling is often a principal growth mechanism not just a remoulding of the fine detail of a bone's shape.

Accretion and resorption are, of course, always complementary. When, for example, a cylindrical bone grows in width, its external surfaces undergo deposition while its internal surfaces (facing into the marrow cavity) undergo resorption at a slightly lower rate. The net result is an increase in the total diameter of the bone, a slightly smaller increase in the diameter of the marrow cavity and a small increase in the thickness of the bony layer. Generally speaking, when the external surface of a region of bone is depository, its internal surface will be resorptive and vice versa (see Enlow (1963) for a full discussion of the principles of bone growth). The manner in which bone tissue is formed leaves its imprint on the internal skeletal architecture, bone formed periosteally being composed of compact circumferential lamellae while that formed endosteally having a whorled structure due to the infilling of the spaces of the cancellous bone which lies deep to the endosteal surfaces. It is thus frequently possible to determine the pattern of remodelling that obtained during the growth period from the adult internal structure of the bone.

In addition to the remodelling changes at the surface, the cartilage replacing bones of the endoskeletal postcranial skeleton receive major growth contributions from the activity of the epiphysial cartilages located at the ends of the bone. In a typical epiphysial cartilage, growth proceeds interstitially within the cartilage but on its diaphysial side the cartilage is resorbed and replaced by bone. The two processes being in balance, the overall result is an increase in the length of the diaphysis rather than in the thickness of the epiphysial cartilage. Nevertheless, the principal determinant of the degree of lengthening of the bone is the rate of proliferation within the epiphysial cartilage not the rate of bone formation. In many advanced tetrapods, including the mammals, secondary centres of ossification appear in the epiphyses and the growth cartilages are then reduced to narrow plates, each sandwiched between the bone of the diaphysis and the bone of the epiphysis, but the processes of endochondral ossification remain essentially similar to those occurring in the epiphysial cartilages of more primitive forms.

The bones that ossify within the endoskeletal chondrocranium resemble, in their mode of development, the cartilage replacing bones of the postcranial skeleton. As we have seen, the ossifying parts of the chondrocranium in the mammalian skull are typically limited to the cranial base, otic and nasal capsules. Of these regions, the cranial base forms a logical starting point from which to begin an account of the growth of the skull, not only because oı its developmental resemblance to the postcranial skeleton, but also because changes in its proportions exert a profound effect upon the shape of the remainder of the skull. The growth of the otic capsule can conveniently be included in the description of the growth of the cranial base but the nasal capsule forms an integral part of the facial skeleton and its growth is better considered with that of the latter part of the skull.

GROWTH OF THE CRANIAL BASE

The cranial base, as traditionally defined in growth studies of the mammalian skull, consists (in the median plane) of a segment, termed the basicranial axis, stretching from the pituitary region to the ventral (anterior) margin of the foramen magnum, an anterior extension from the pituitary region to the junction of the frontal and nasal bones, and a posterior extension from the ventral to the dorsal margin of the foramen magnum. The basicranial axis is broadly equivalent to the parachordal region and the anterior extension to the prechordal region of the developing skull. The growth changes in the linear dimensions of these segments of the cranial base and in the angles between them have been studied in several mammalian species but comparisons are complicated by the variety of anatomical landmarks which have been adopted to define precisely the basicranial axis and its extensions. As will become evident, apparently quite small differences in the landmarks used, or in the way they are defined, can introduce considerable variation into the observed growth changes. Most of the earlier studies were carried out in man and other higher primates and the linear and angular dimensions were initially defined for convenient use in those species (Fig. 99*a*). Perhaps the most commonly used definitions are based on those provided by Huxley (1867) in his classical study of the cranial base of the human skull. He defined the basicranial axis as extending from the posterior extremity of the basioccipital bone (endobasion) to the joint between the presphenoid and

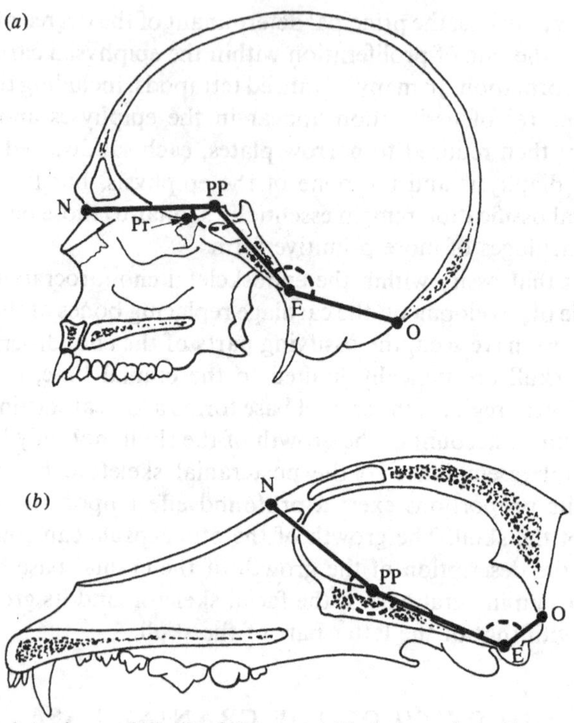

Fig. 99. Sagittal sections of skull of (*a*) man; (*b*) dog to show linear and angular dimensions of cranial base. E = endobasion; N = nasion; O = opisthion; PP = pituitary point; Pr = prosphenion. The anterior and posterior basicranial angles are shown by dashed lines. Not to scale.

ethmoid (prosphenion), choosing the latter landmark because he believed it to mark the site of cranial flexure. Most later authors have used a landmark rather more posteriorly located than the prosphenion to define the anterior limit of the basicranial axis, believing the true point of cranial flexure to be closer to the pituitary fossa. Cameron (1924), for example, chose the pituitary point (anterior edge of the optic groove) in which he has been widely followed. The posterior limit of the anterior extension will obviously coincide with the landmark chosen as the anterior limit of the basicranial axis. The anterior limit of the anterior extension is most usually taken, for convenience in measurement, as the nasion (intersection of internasal and frontonasal sutures on the external cranial surface) but some point near the front edge of the cribriform plate would probably be better (see below). The posterior extension is usually defined as the median diameter of foramen magnum (i.e. from endobasion to opisthion). The angle between the basicranial axis and is anterior extension and that between the axis and its posterior extension were termed by Huxley respectively 'sphenoethmoidal' and 'foraminobasal'. Later authors, using different landmarks for defining the basicranial axis and its extensions, have frequently adopted alternative names for the angles of the cranial base. In order to avoid the confusion that this multiplicity of names can engender, the terms *anterior basicranial* (for the angle between the basicranial axis and anterior extension) and *posterior basicranial* (for the angle between the axis and posterior extension) angles will be used throughout the following account, whatever the precise definitions of the landmarks employed in the study being discussed. Although providing somewhat unwieldy figures for non-primate mammals, it is least confusing to adhere to Huxley's method of locating these angles, viz. the anterior basicranial angle is located between *and below* and the posterior basicranial angle between *and above* the basicranial axis and its respective extensions (Fig. 99).

In the mammalian species that have been studied (man – Ford, 1955, 1956; rat – Dorenbos, 1973; mouse – Hoyte, 1975) an angle of less than 180° is already established between the pre- and parachordal parts of the cranial base at the cartilaginous stage of its development. With ossification, this anterior basicranial angle increases, leading to a straightening of the cranial base. In the human fetus, for example, the angle increases from about 130° to 150° between 10 and 40 weeks, while in the rat the increase during prenatal life is from about 130° to a little in excess of 180°. The fact that the anterior basicranial angle is still below, or only marginally above, 180° at birth reflects the tendency of the facial skeleton to be situated beneath as well as in front of the anterior part of the braincase in immature mammals. In non-primates the angle tends to increase during postnatal growth to a value well in excess of 180°, so that the facial skeleton comes to be located directly anterior to the braincase. In the rat, for example, the anterior basicranial angle at maturity is of the order of 240°.

In the higher primates (Ceboidea, Cercopithecoidea and the apes amongst the Hominoidea) the anterior basicranial angle averages about 150° at the stage of the milk dentition and, as in non-primates, continues to increase throughout the growth period but to a much lesser degree (Zuckerman, 1926; Ashton, 1957; Ashton, Flinn & Moore, 1977). The result is that the anterior basicranial angle remains below 180° throughout life. This difference is attributable to the much enhanced growth of the primate brain which is accommodated, anteriorly, by enlargement of the braincase forwards above the facial skeleton. The primate pattern can be regarded as a tendency to retain the immature shape and correlates with the observation, already noted, that the brain is large relative to the remainder of the head early in the growth period.

In prosimians the degree of cranial flexion is generally less than in anthropoid primates although the anterior basicranial angle is still usually somewhat below 180°. It is of interest that in *Daubentonia* cranial flexion is much greater than in other prosimians. Cartmill (1977) has attributed this to the need to strengthen the skull, by augmenting facial height, against the bending forces produced during use of the incisor teeth (*Daubentonia* possessing rodent-like incisors which it uses in a powerful type of gnawing action). A similar possibility exists in the case of the lagomorphs where cranial flexion, superficially similar to that seen in primates, is present (in fact, this flexion is due to increasing flexure during postnatal growth of the facial skeleton itself rather than to a reduction in the anterior basicranial angle – Moore & Spence, 1969) but DuBrul (1950) relates it to the semi-erect sitting posture habitually adopted by these creatures.

The relationship between the basicranial axis and its posterior extension also undergoes considerable growth changes, the posterior basicranial angle tending generally to decrease with age. In the rat, for instance, this angle is about 110° at birth and decreases postnatally by a few degrees. Similarly, in the higher primates the angle typically decreases. In the great apes, for example, the postnatal decrease is of the order of 5° to 10° from a value of about 130° at the stage of the milk dentition (Ashton, 1957; Ashton *et al.*, 1977).

The primate pattern of angular growth changes is modified in man in association with his still further enlarged brain and adoption of an upright posture (Zuckerman, 1955; Ford, 1958). Here, the anterior basicranial angle, after increasing before birth to about 150°, decreases again postnatally, especially in the period up to six years, reaching a value of around 130° at maturity. The posterior angle increases prenatally from 120° to between 130° and 140° and postnatally by a further 10° or so in the period before the permanent teeth begin to erupt; only then does the angle decrease – by an amount approximately equal to the earlier postnatal increase. The decrease in the anterior basicranial angle and the initial increase in the posterior angle during the postnatal growth period appear to be entirely human characteristics, both being related to the great growth of the brain. The later decrease in the posterior basicranial angle is less readily explained but may be due to the displacement of the opisthion as a result of the upward extension of the nuchal area which is associated with the enlargement of the nuchal muscles in adolescence.

The pattern of growth increases in the linear dimensions of the cranial base appears to be more consistent, from species to species, than that just described for the angular dimensions. In both primates (Zuckerman, 1955; Ford, 1956, 1958; Ashton, 1957; Scott, 1958; Bambha, 1961; Lewis & Roche, 1972) and non-primates (Baer, 1954; Moore, 1966) the basicranial axis has been found to continue to lengthen throughout the growth period and to exhibit, in man at least, an adolescent growth spurt, adhering in these respects to the somatic rather than to the neural growth pattern. The posterior extension of the basicranial axis, as would be expected from a dimension which is really just the diameter of the foramen magnum in the median plane, follows the neural growth pattern closely in all species which have been investigated.

The pattern of growth changes in the anterior extension depends upon the precise landmarks used to delimit this region. When defined as extending from the pituitary region to the nasion, as in the classical anthropometric description, this part of the cranial base may receive contributions from four potential growth sites: midsphenoidal, sphenethmoidal and frontomesethmoid joints and the bone surface at the nasion. Using this definition the anterior extension in man (Zuckerman, 1955) and monkeys and apes (Ashton, 1957) is found to follow the somatic rather than the neural pattern. However, when the anterior limit of the extension is taken as the front edge of the cribriform plate, thus excluding growth at the nasion, the anterior extension has been found to follow the neural pattern of growth closely in both man (Brodie, 1941; Bjork, 1955; Ford, 1958; Scott, 1958) and a number of non-primate species (e.g. rat – Ford & Horn, 1959; Moore, 1966; rabbit – Moore & Spence, 1969). The discrepancy between these findings is due presumably to the thickness of the frontal bone (which lies between the front edge of the cribriform plate and the nasion) increasing, like the enlargement of the facial skeleton, in accordance with the somatic pattern.

The principal sites of growth by which the increase in the linear dimensions of the basicranial axis and its anterior extension is achieved are the cartilaginous joints or synchondroses of the cranial base. The number of these joints depends, of course, upon the manner and timing in which the ossification centres in the cartilaginous cranial base fuse to form definitive bones – the synchondroses representing unossified parts of the chondrocranium remaining between the bones. As well as varying between species, the number of synchondroses decreases during ontogeny as originally separate ossification centres fuse with each other.

In a typical therian mammal the synchondroses remaining in the median plane once the definitive bones have formed include the sphenoccipital (between basioccipital and basisphenoid bones) and midsphenoidal (or presphenoidal between basisphenoid and presphenoid). In those mammals with a mesethmoid ossification (Chapter 2) a sphenethmoidal synchondrosis may persist between this bone and the presphenoid. In man the midsphenoidal synchondrosis disappears at about the time of birth due to the fusion of pre- and basisphenoid, while the cartilage in the sphenethmoidal synchondrosis is replaced, also at about the time of birth, by fibrous tissue (Baume, 1968), so that for most of the postnatal growth period the sphenoccipital synchondrosis is the sole cartilaginous growth site in the median plane. However, the sphenethmoidal joint, despite the replacement of cartilage by fibrous tissue, probably remains a growth site of some significance until about six to eight years. The great apes similarly show early closure of the midsphenoidal synchondrosis, at or shortly after birth, while the sphenethmoidal joint appears to fuse somewhat earlier than in man (Scott, 1958). In *Macaca mulatta* the presphenoid and ethmoid have fused by birth but the sphenoccipital and midsphenoidal synchondroses persist throughout much of the growth period (Michejda, 1971, 1972).

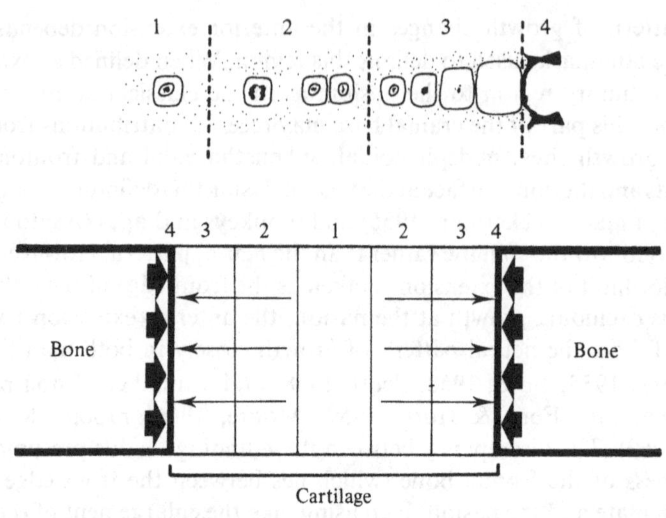

Fig. 100. The zones of a typical synchondrosis. Below: schematic representation of a section through synchondrosis. Above: the cellular changes in the zones of proliferation and hypertrophy; 1 = reserve zone, 2 = proliferative zone, 3 = hypertrophic and degenerative zones, 4 = ossification front. Thin arrows indicate direction of cartilage growth; thick arrows indicate direction of movement of ossification front.

The structure and growth mechanism of synchondroses have been studied in detail in man (Baume, 1968), in *Macaca mulatta* (Michejda, 1972) and in the rabbit, rat and guinea-pig (DuBrul & Laskin, 1961; Baume, 1968; Cleall, Wilson & Garnett, 1968; Koski & Rönning, 1969, 1970; Hoyte, 1971, 1975). The typical synchondrosis (Fig. 100) consists of a central reserve zone flanked on either side by zones in which the hyaline cartilage proliferates interstitially, undergoes cellular hypertrophy and then degenerates to be replaced by bone. In each of the latter, the sequence of events appears to be identical to that found in epiphysial cartilage of a postcranial bone – viz. the chondrocytes undergo cell division, the daughter cells form into regular columns, hypertrophy and, as the remnants of cartilage matrix are mineralised, undergo necrosis; osteogenic tissue then invades the areas between the remnants of mineralised cartilage matrix and forms early trabeculated bone. The two events of cartilage proliferation and cartilage removal and replacement by bone are more or less in balance so that the net effect is an increase in the dimensions of the bones abutting upon the synchondrosis. As the bone enlarges, the early trabeculae laid down at the endochondral ossification front are replaced by more mature and functionally better organised bone.

In the majority of mammalian species, growth at the sphenoccipital synchondrosis appears to be the principal contributor to enlargement of the basicranial axis, although substantial additions may also be made at the midsphenoidal synchondrosis, while the sphenethmoidal synchondrosis, when present, is the main site at which the anterior extension increases in length (Baer, 1954; Hoyte, 1973*a*). The basicranial axis also receives increments from bone deposition along the anterior border of the foramen magnum (Enlow, 1968; Hoyte, 1973*a*). Further increments may derive from local remodelling resulting in repositioning of the pituitary fossa, which provides the landmarks traditionally employed to delimit the anterior extremity of the basicranial axis, although there is uncertainty about the exact direction of this growth movement (in man, the fossa is described by Enlow (1968) as moving upwards and forwards (see below) but by Latham (1972) as moving upwards and backwards). The posterior extension enlarges presumably as a result of the rate of resorption on the posterior border of foramen magnum exceeding that of bone deposition on its anterior margin.

Hoyte's (1975) analysis of previously published data relating to growth in the human cranial base indicates the possibility that growth increments at the basal synchondroses may not be symmetrical. As Hoyte points out, there is no fundamental reason why growth on either side of the central zone should be equal in rate or total amount. Asymmetrical growth has been reported in the sphenethmoidal synchondrosis in the human fetus (greater deposition at the ethmoid – Gordon, 1955, cited by Hoyte, 1975), and in the midsphenoidal synchondrosis of *Macaca mulatta* (greater deposition at presphenoid – Michejda, 1972). Other authors, however, have found growth to be generally symmetrical (e.g. rat – Baer, 1954; mouse – Servoss, 1973; pig and rabbit – Hoyte, 1973*b*). The possibility also exists that growth may be asymmetric along (as opposed to across) a synchondrosis. Vilmann's (1971) observations suggest that asymmetry of this nature, with a dorsoventral gradient of growth, occurs in the sphenoccipital synchondrosis of the rat.

Such differential growth may play a part in producing angular changes in the cranial base. This possibility gains support from Michejda's (1972) finding that there is a close correlation between the timing of changes in the anterior basicranial angle and the period of growth activity at the midsphenoidal synchondrosis in *Macaca mulatta*, suggesting that differential growth at the synchondrosis is involved in producing the angular changes. The early closure of the midsphenoidal synchondrosis in the human cranial base appears to preclude the possibility that this joint is similarly involved in producing the characteristic postnatal changes observed in the human anterior basicranial angle. It seems more likely that the angular changes in man are the result of local remodelling, especially in the region of the sella turcica (see below). There are insufficient data available to judge whether differential synchondrotic growth or local remodelling is the principal growth mechanism involved in producing angular growth changes in the cranial base in mammals generally.

Fusion of the basal synchondroses appears to take place in the majority of non-primates much later than the cessation of growth (Hoyte, 1973*a*; Sawin, Ranlett & Crary, 1959). The synchondroses in the primate skull by contrast may fuse well before growth of the cranial base has ceased. In man,

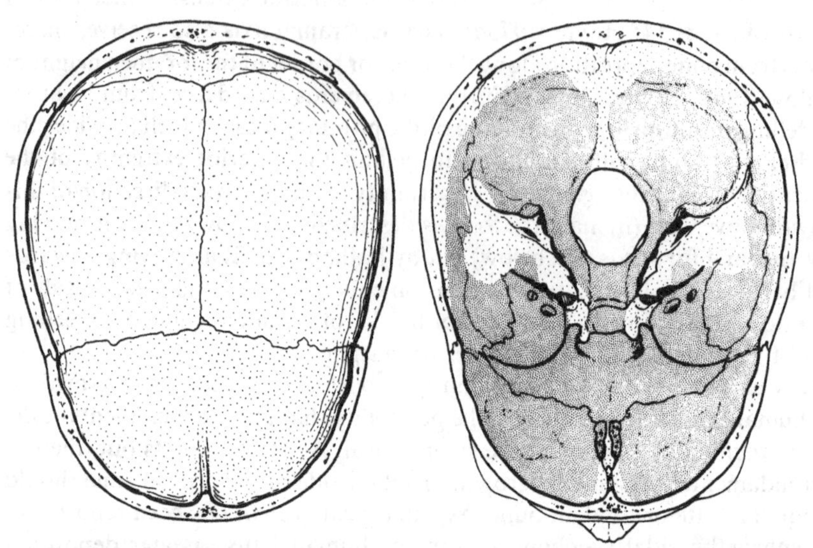

Fig. 101. The distribution of resorptive (dark stipple) and depository (light stipple) endocranial surfaces in the human skull. (Reproduced with permission from Enlow, 1968.)

for example, as already noted, the midsphenoidal synchondrosis has usually closed by birth. There is now evidence that the sphenoccipital synchondrosis also fuses before growth has ceased, fusion possibly beginning in the early teens in both boys and girls (Irwin, 1960; Powell & Brodie, 1963; Ingervall & Thilander, 1972) with active growth ceasing a little earlier (Latham, 1966, 1972). Subsequent to the fusion of this synchondrosis, further growth of the basicranial axis must presumably occur by increments at the endobasion or remodelling of the pituitary region. In the skulls of apes and monkeys closure of the sphenoccipital synchondrosis occurs somewhat later than in the human cranium (Krogman, 1930*b*; Dolan, 1971; Michejda, 1971, 1972) but probably still before the growth of the basicranial axis has ceased.

A further mechanism involved in producing overall cranial enlargement in most, if not all, mammalian species is surface remodelling. This phenomenon has been described in detail for the cranial base of man (Enlow, 1968) and *Macaca mulatta* (Duterloo & Enlow, 1970) on the basis of study of the bony internal architecture, but corresponding information does not appear to be available for other mammalian groups. At its simplest, the cranial floor consists of two cortical plates separated from each other by cancellous bone, although in some regions other structures intervene (e.g. the air sinuses in the body of the sphenoid and the structures of the inner ear and tympanic cavity in the region of the otic capsule). In the human cranial base the predominant remodelling pattern consists of bone deposition on the periosteal surface of the ectocranial cortical plate and resorption from the corresponding surface of the endocranial plate (Fig. 101), the endosteal surfaces of the plates undergoing remodelling activities the converse of those taking place periosteally. In this way the whole cranial floor is moved relatively downwards. This type of growth, often termed cortical drift, helps to produce the requisite amount of enlargement of the braincase with less disruption of the spatial relationships of the structures associated with the cranial base than could be achieved by growth solely at the joints.

Encircling the endocranial surface of the human braincase is a complete circumcranial reversal line of bone growth (Fig. 101). Below this line the endocranial surface as just described is resorptive, but above the line the surface becomes depository. Extending medially from the reversal line to the midline of the cranial base are reciprocal gradients between the amount of growth taking place due to cortical drift and that due to deposition at the joints. In general, as the midline is approached, the remodelling processes producing cortical drift progressively increase while growth at the joints diminishes.

Superimposed upon the overall pattern of cortical drift is a number of regional variations (Fig. 101). The endocranial surfaces of the petrous parts of the temporal bones, for example, form isolated regions of deposition in the generally resorptive cranial floor so that their relative degree of prominence is maintained. Similarly, the endocranial walls of the sella turcica of the sphenoid bone are areas of bone deposition (the floor of the pituitary fossa itself, however, is usually resorptive). Since the endocranial surface of the remainder of the body and of the wings of the sphenoid are resorptive, the region of the sella turcica maintains, and even increases, its relative prominence within the cranial cavity. Such local remodelling may well be the principal factor in the postnatal decrease of the anterior basicranial angle observed in the human cranial base.

The surface remodelling of the cranial base of the macaque monkey is less regular than that in the human skull. Instead of resorption being widespread over the endocranial surface, it is restricted to a number of discrete areas in the frontal region, anteroventral part of the middle cranial fossa and in the occipital region. These contrasts are probably related to the great difference between man and the monkey in the extent of cerebral expansion – in order to accommodate the greatly enlarged cerebral hemispheres of man, a marked degree of cortical drift is necessary to enhance the limited expansion of the braincase brought about by growth of the joints, while in the monkey the more modest growth of the brain can be accommodated by growth at the synchondroses with the addition of only localised areas of resorption.

If this argument is correct it might be expected that cortical drift due to surface remodelling in the cranial base of non-primate species, with their relatively small brains, will be found to be even more restricted than in the macaque monkey, although direct evidence for this is lacking.

The growth of the otic region of the human cranial base has been the subject of several detailed studies. Ford (1955) has described this region as being composed entirely of cartilage up to 18 weeks of prenatal life and as being the only part of the original chondrocranium to reach the lateral edge of the definitive cranial base (this part of the skull being, in man, considerably widened beyond its original chondrocranial limits as part of the general expansion of the braincase). During this time the cartilage is undergoing rapid interstitial growth so that by 18 weeks the otic capsule is virtually adult size (as are the contained structures of the internal ear). The capsule then rapidly ossifies. The otic capsule thus displays an extreme form of the neural growth pattern. The capsule has at first a medial continuity with the cartilage of the parachordal plate but this connection is later lost and replaced by a bridge of connective tissue. The small amounts of growth occurring subsequent to the ossification of the capsule are achieved medially by intramembranous ossification and laterally by endochondral ossification. The growth of the otic capsule (as well as associated structures such as the tympanic membrane and ear ossicles) appears to follow a similar pattern to that seen in man in a wide variety of other mammalian species (Bast & Anson, 1949; Hoyte, 1961).

GROWTH OF THE CRANIAL VAULT

As would be expected, the growth of the cranial vault corresponds closely to that of the contained brain. Both share in the rapid phase of enlargement in prenatal and early postnatal life and in the relatively precocious termination of growth which are typical of the neural growth pattern. The bones of the vault and side walls of the braincase (and also the lateral parts of the cranial base in big-brained species) are predominantly of the dermal variety. It follows that the joints, or sutures, between them will be composed of fibrous connective tissue representing the vestiges of the membrane in which these bones ossify, just as the synchondroses of the cranial base represent the vestiges of the cartilage in which the latter part of the skull is preformed. Again, like the synchondroses, the sutures are important growth sites and their number and arrangement change as the originally separate ossification centres fuse to form the definitive bones.

The development and the microstructure of the fully formed suture have been investigated in detail by Pritchard, Scott & Girgis (1956) in man, sheep, pig, cat, rabbit and rat, and were found to be generally similar in each of these species and also from one skull region to another. During the early stage of development (which Pritchard and his colleagues termed the stage of approaching bone territories) the margins of the bones of the cranial vault grow towards one another, through the condensed mesenchyme (ectomeninx) covering the brain. Two zones can be distinguished in each approaching bone territory: (1) a cambial zone composed of fine radially arranged collagen bundles and numerous osteoprogenitor cells, and with a layer of osteoblasts

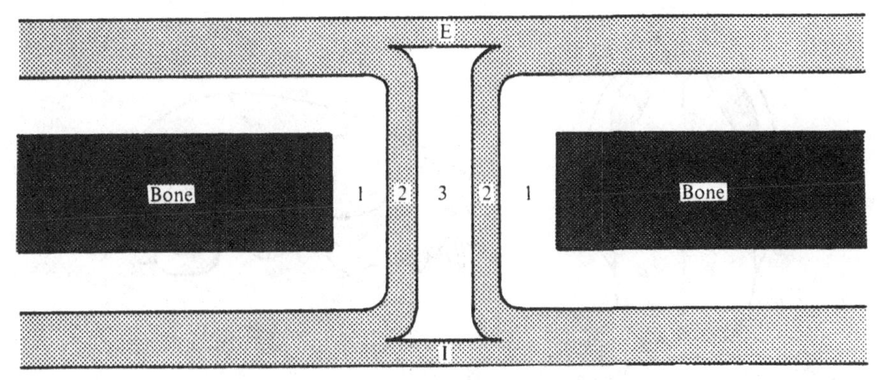

Fig. 102. Schematic representation of the structure of a suture: 1 = cambial zone; 2 = capsular zone; 3 = cellular middle zone. E and I = external and internal uniting layers.

next to (2) a zone of woven bone produced by the activity of the osteoblastic layer. Between the approaching bone territories the fibres of the ectomeninx run in a plane parallel to the principal plane of the bone (i.e. at right angles to the advancing plates of woven bone). As the bone margins advance they split the ectomeninx into outer pericranial and inner dural layers. Just before the cambial layers of two approaching territories meet, a third zone, a fibrous capsular layer, becomes defined at the leading edge of each territory. The adjacent territories then unite with each other to form the definitive sutures, each of which has five layers intervening between the two contiguous bones (Fig. 102):

(1) the cambial zone of the first bone territory;
(2) the capsular zone of the first bone territory;
(3) a loose cellular middle zone;
(4) the capsular zone of the second bone territory;
(5) the cambial zone of the second bone territory.

The cambial and capsular zones are continuous with similar layers of the periosteum on the non-sutural bone surfaces. The capsular layers on the latter surfaces unite across the suture to produce external and internal uniting layers.

These layers persist for some time after the bone territories have met during which time the cells of the cambial layer undergo frequent division and trabeculae of woven bone are rapidly formed at the bone edges, the rate of bone formation greatly exceeding that on non-sutural surfaces. Later in the growth period the cambial zones are much reduced in thickness and the cells of the osteoblastic layer become less active. The capsular layers become denser and united to the bone by radially arranged collagen fibres passing between the elements of the cambial layer. The trabeculae of woven bone are replaced by more mature bone laid down in parallel increments. In the adult stage the cambial layers are reduced to a single layer of flattened cells and collagen fibres

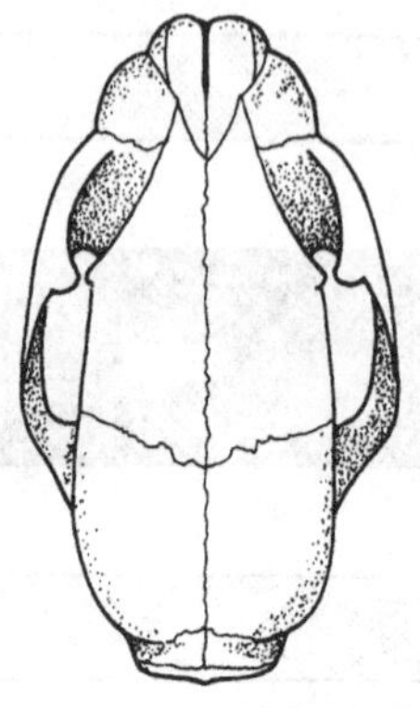

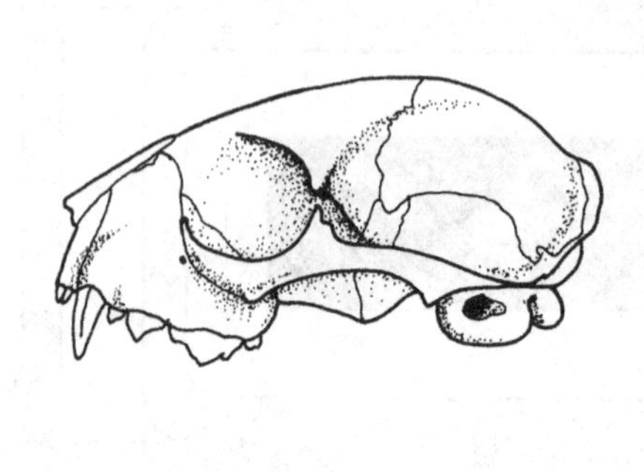

Fig. 103. Dorsal and left lateral view of felid cranium to show arrangement of sutures of braincase.

may be seen passing right across the suture from one bone to the other. During postnatal growth, islands of secondary cartilage may appear in the cranial sutures, either at the bone margins or within the soft tissues. The significance of this tissue is discussed below.

The description of Pritchard and his colleagues makes it clear that the essential growth sites in a suture are the two cambial layers and not the middle layer of loose cellular tissue as had previously been supposed (e.g. Weinmann & Sicher, 1955). The presence of two such growth zones, separated by relatively inactive tissue, allows growth at each bony margin to be independent in direction and amount.

The sutural system of the mammalian braincase, when fully developed, is arranged in the three planes of space (Fig. 103) with the frontoparietal and occipitoparietal sutures lying in the coronal plane, the midline sutures (interfrontal, interparietal, etc.) occupying the sagittal plane and the sutures in the side wall of the braincase (squamoparietal, squamoccipital and the joints between the wings of the sphenoid and the parietal and frontal bones) being disposed in the transverse plane. Thus, increments at the sutures will produce enlargement in the length, breadth and height of the braincase. The greatest growth increments are generally in the sutures in the coronal plane.

Sutural growth alone, however, would be incapable of maintaining the proportions of the braincase for, as growth proceeds, the radius of curvature of the vault and walls must, in general, increase (i.e. the curvature of the constituent bones must decrease). This change in shape is brought about by surface remodelling in which the outer surface of each bone undergoes deposition, which occurs more rapidly towards the periphery of the bone than in its central region, while the inner surface undergoes resorption, again with the peripheral areas being the sites of greatest activity (Fig. 104). In fact, the remodelling changes are generally rather more complex than this simple model indicates, their precise nature depending upon the shape of the braincase, which varies not only from region to region but also from species

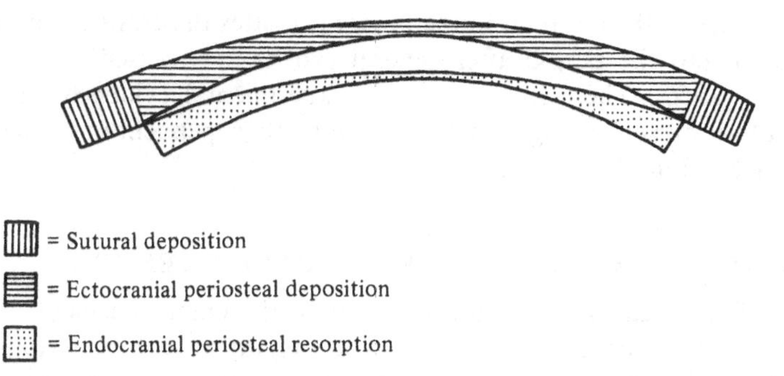

Fig. 104. Schematic representation of remodelling of bones of cranial vault.

to species and may, moreover, change as growth proceeds. In man it appears that the endocranial surface of the cranial vault is principally depository and such resorption as occurs is limited to areas immediately adjacent to the sutures (Enlow, 1968). Remodelling is not limited to the ecto- and endocranial surfaces but also occurs on internal surfaces such as those facing into marrow cavities or air sinuses where, typically, the activities of deposition and resorption are the converse of those taking place on the corresponding external surfaces.

The sutures of the braincase in some mammalian species tend to close after the growth period has ended. This is particularly marked in the higher primates, where the sequence of suture closure is sufficiently constant to provide an approximate guide for ageing specimens. Suture closure is also seen in aged domesticated animals, but information about this phenomenon in feral non-primate mammals is sparse.

The braincase of those mammals with strongly developed jaws and dentition frequently bears a superstructure of crests, whose functions are to increase the area available for the attachment of the big temporal and nuchal muscles and to strengthen the skull against the powerful forces produced when the jaw muscles and dentition are in use. As described in Chapter 5, such crests are particularly well developed in carnivores, but are also a feature of the skulls of great apes and many monkeys, especially those of the male sex. The temporal crests may meet and fuse in the midline on the dorsal surface of the skull where they form a prominent sagittal crest. Posteriorly, the temporal crests (or single sagittal crest) may meet and fuse with the nuchal crest to form a strong transverse bar of bone marking off the nuchal from the temporal area. In modern man, in association with his reduced jaws and jaw musculature, the temporal and nuchal muscle markings are little more than faint ridges of bone, but in some archaic forms of man, for example *Homo erectus*, cresting is more prominent.

In the newly born mammal the jaw muscles are relatively feebly developed, even in carnivores and higher primates, and cresting or even ridging is barely apparent on the surface of the braincase. As the jaw muscles develop, so does the prominence of the cresting. In the higher primates the crests continue to enlarge well into the period after general growth has ceased (Ashton & Zuckerman, 1956). There is considerable experimental evidence that the presence of the muscular ridges and crests is directly dependent upon the pull of the attached muscles (see below).

GROWTH OF THE UPPER FACIAL SKELETON

The part of the facial skeleton associated with the commencement of the respiratory and alimentary tracts follows the somatic growth pattern, while the orbital region, like the eyeballs themselves, completes much of its enlargement at a relatively early stage in accordance with the neural growth pattern. This contrast is especially apparent in the anthropoid skull where the orbit forms an almost complete bony capsule for the eyeball (Schultz, 1962).

Most of the bones of the facial skeleton are of the dermal variety and the joints between them are, in consequence, of the sutural type. In the nasal region, however, there is an extensive area of cartilage, representing the original nasal capsule, which plays an important part in facial growth.

SUTURAL GROWTH

The mode of development of the facial sutures differs from that of the sutures in the cranial vault in that the approaching bone territories in the former case advance through loose mesenchymatous tissue rather than within a preformed fibrous membrane, as occurs in the developing braincase, and a capsular zone, in addition to the cambial and woven bone zone, is present from an early stage. Once adjacent bone territories have united, however, the resulting sutures appear to be identical in structure and in mode of growth in both subdivisions of the skull. As in the cranial vault, islands of secondary cartilage may appear in the facial sutures, either at the margins of the bones or within the sutural soft tissue.

The facial sutures provide a group of principal growth sites within the upper facial skeleton and, as would be expected, their arrangement is such that they can provide increments in the three planes of space. In the human skull at the time of birth, for example, there are three major systems (Scott, 1956 – Fig. 105): (*a*) the circummaxillary group (between the maxillae, on the one hand, and the frontal, nasal, lacrimal, ethmoid, palatine, vomer and zygomatic bones and the pterygoid processes of the sphenoid, on the other); (*b*) the craniofacial group (separating the nasal, lacrimal, facial part of ethmoid, palatine, vomer and zygomatic bones from the frontal, perpendicular plate of ethmoid, temporal and sphenoid bones); (*c*) the sagittal group (comprising internasal, intermaxillary and median palatal sutures and mandibular symphysis). The arrangement of the first two of these groups is such that their growth will

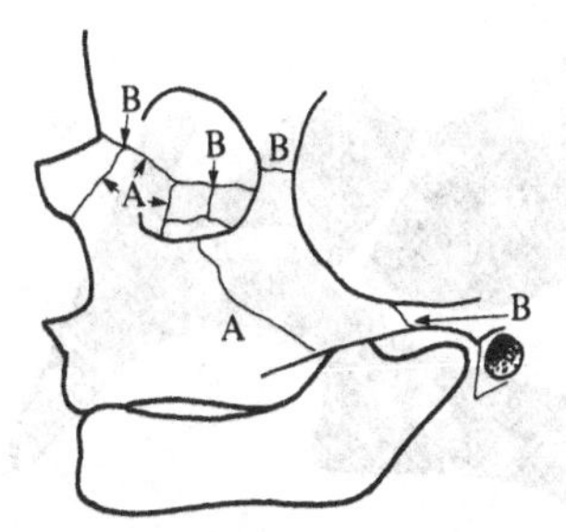

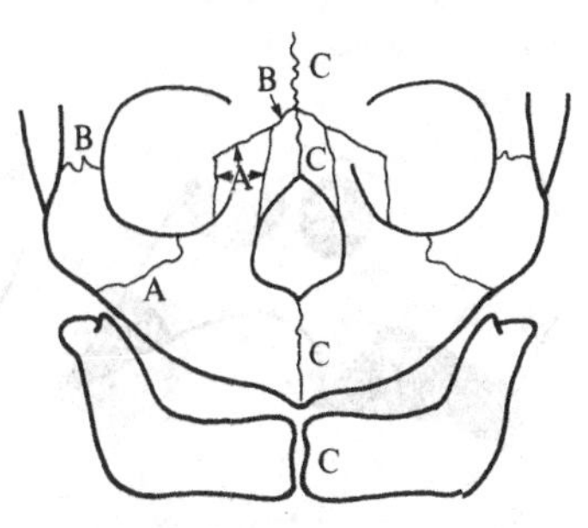

Fig. 105. Left lateral and anterior views of facial skeleton of human fetus at term to show suture systems. A = circummaxillary group; B = craniofacial group; C = sagittal group.

contribute to anteroposterior and vertical enlargement of the facial skeleton, while the sagittal system is so disposed that its growth will lead to an increase in facial width. A corresponding arrangement, allowing for the variations in facial shape, is found in other mammalian species

Although detailed data are lacking, a general comparison of the patterns of facial growth indicates that, in the majority of mammals, the component of sutural growth contributing to anteroposterior enlargement is greatest in amount but that, in primates, with their far less protrusive jaws, this differential effect is less marked (but still present, for example, in *Macaca* – McNamara, Riolo & Enlow, 1976). In man, in association with his characteristic vertical facial profile, sutural growth contributing to the anteroposterior enlargement is still further reduced, being exceeded by the component of sutural growth contributing to facial height. The diminution of anteroposterior growth in primates is enhanced by the tendency for the anterior basicranial angle to remain below 180°. Again, this is further accentuated in man where the angle actually decreases during postnatal growth.

NASAL SEPTUM

The cartilaginous nasal septum is formed partly from the original nasal capsule and partly from the interorbital septum. Its ossification has been studied in some detail in man by Scott (1953) and Latham (1970) (Fig. 106). The vomer ossifies in the mucoperichondrium close to the lower edge of the septal cartilage from two centres which develop at about the eighth week of intrauterine life. The centres extend and unite beneath the lower margin of the cartilage to form the vomerine groove into which the inferior margin of the cartilage is received. Extension also takes place anteriorly, so that the vomer interposes itself between the inferior border of the cartilage and the whole length (both intermaxillary and interpremaxillary portions) of the median palatal suture. The ossification centre for the mesethmoid appears at

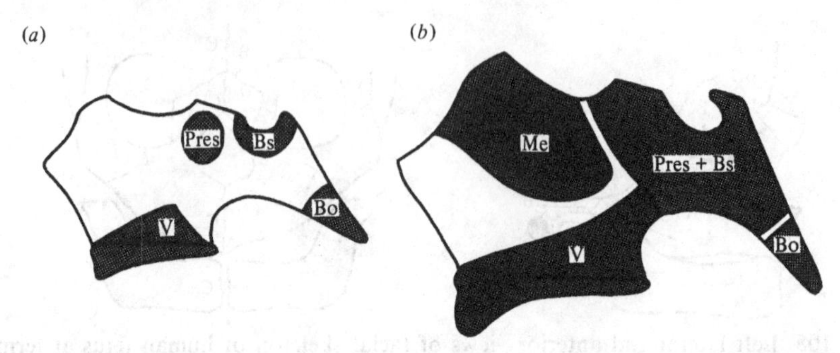

Fig. 106. Diagrammatic representation of human nasal septum to show ossifications: (*a*) at about the time of birth; (*b*) during the second postnatal year. Shading = bone; unshaded = cartilage. Not to scale. For labelling see list of abbreviations.

about the time of birth in the midline of the cartilage of the prechordal part of the cranial base and extends into the septal cartilage to form the perpendicular plate of the ethmoid. For the first few years of life, the mesethmoid remains separated from the vomer by the 'sphenoidal' tail of the septal cartilage. During the third year of life, ossification extends into the sphenoidal tail and the definitive septovomerine joint is formed.

An active endochondral ossification centre was first noted at the junction between the septal cartilage and the perpendicular ethmoidal plate by Baume (1961) in *Macaca mulatta.* He observed that cartilage replacement by bone takes place at this site until at least the fiftieth month of life, while at the septovomerine junction new cartilage is formed within the perichondrium of the septal cartilage and also by interstitial growth. An essentially similar arrangement has since been reported in man and the rat (Baume, 1968). Its presence has led to the suggestion that the growth of the septal cartilage is a principal determinator of the growth in height and depth of the upper face, the septal cartilage having in this respect a role rather like that of the epiphysial cartilage in the growth in length of a postcranial long bone. Recently, Hoyte (1978) has drawn attention to the possibility that the whole of the cartilaginous nasal capsule, not just the septum, acts in this way. In an examination of histological sections of the facial region of the growing rabbit, he found that many of the bones surrounding the nasal capsule have large areas of resorption on their deep surfaces. Hoyte suggests, on the basis of these findings, that the growth of the cartilaginous nasal capsule determines the increase in volume of the nasal cavity, with the growth and remodelling of the surrounding facial bones being an obligatory response.

SURFACE REMODELLING

The growth increments at the sutures and nasal septum are accompanied by extensive remodelling of the external and internal surfaces of the bones of the facial skeleton. Knowledge of these changes is far more detailed for the

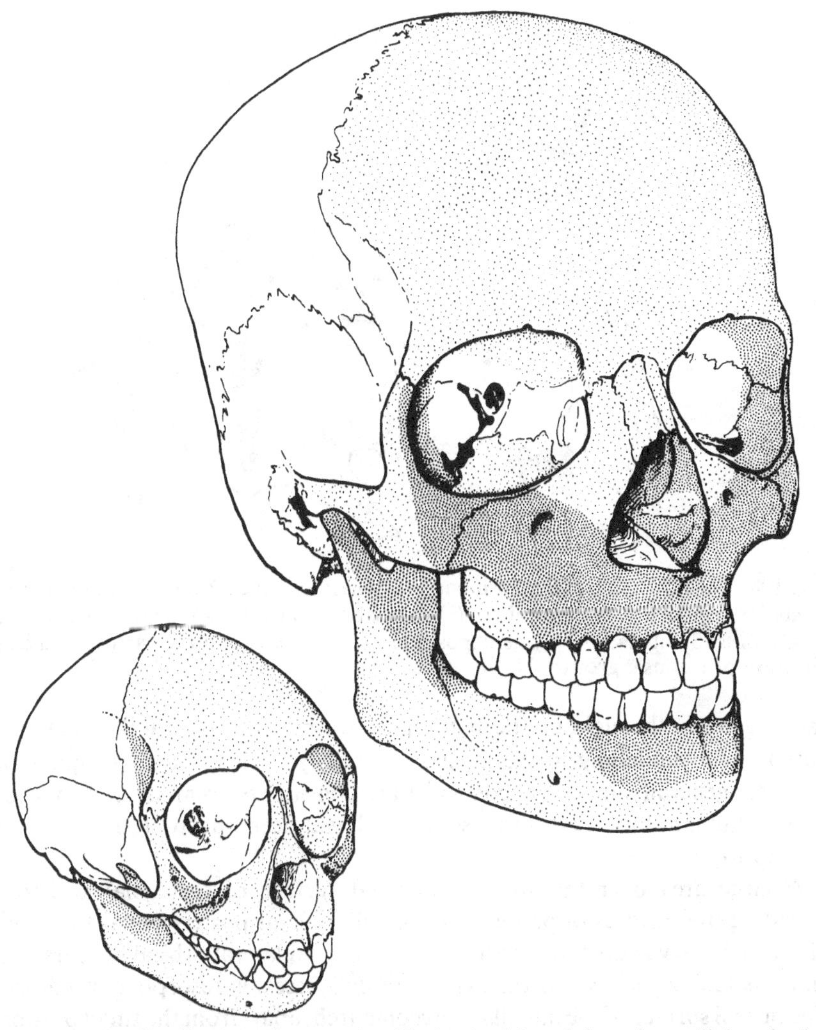

Fig. 107). The distribution of resorptive (dark stipple) and depository (light stipple) periosteal surfaces in the facial skeleton of man and *Macaca*. Reproduced with permission from Enlow, 1966.)

primates than for any other mammalian group, information having been provided for man by Enlow & Bang (1965 – see also Enlow, 1968, 1975) and for *Macaca mulatta* by Enlow (1966) (Fig. 107). In both cases, the method used was based on examination of the internal architecture of the facial bones. Although, as will become apparent, the pattern of remodelling is closely related to the detailed shape of the facial skeleton and may, therefore, differ considerably from one species to another, it is worth recounting in full the remodelling changes that take place in these two higher primates because they illustrate the general principles which almost certainly govern this aspect of

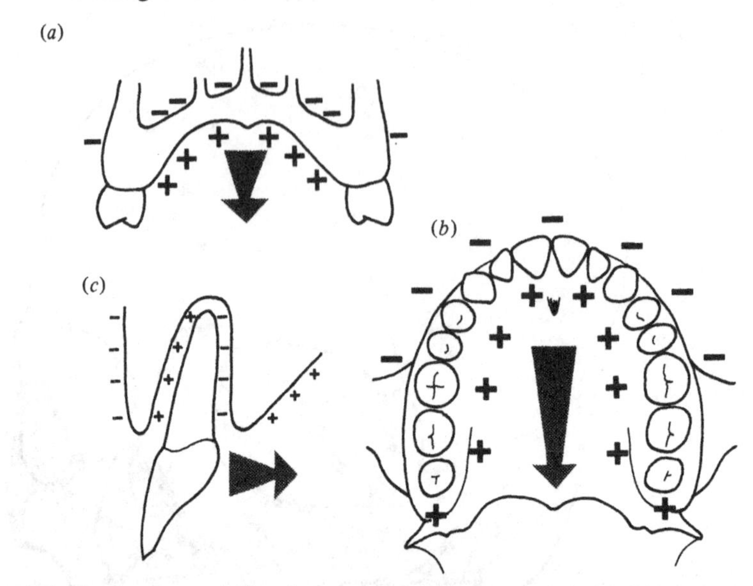

Fig. 108. To show remodelling in human maxillary complex. (*a*) Coronal section in premolar region; (*b*) occlusal view; (*c*) sagittal section through central incisor and its alveolus. Symbols: + = bone deposition; − = bone resorption; arrows indicate direction of intrinsic growth. Not to scale.

facial growth throughout the mammals. Despite the probability that the pattern of remodelling in the macaque monkey is closer to that typical of primates generally, it is more convenient to consider the remodelling changes in the human face first because these have been given a much fuller description.

A large area over the anterior external surface of the facial skeleton, including the surfaces of premaxillae, maxillae as far posteriorly as the roots of the maxillary zygomatic processes, inferior and lateral orbital borders and anterior facing surfaces of the zygomatic processes, is resorptive in nature. The buccal surface of the maxillary alveolar arch, apart from the small portion posterior to the root of the maxillary zygomatic processes, is thus resorptive (Fig. 108*a*, *b*). The lingual periosteal surface of the arch is entirely depository while the region of the maxillary tuberosity, on the posterior aspect of the arch, is a site of particularly rapid deposition. When viewed from below the alveolar arch has a rounded V-shape and the remodelling changes just described are typical of those described by Enlow (1963, 1968) in many other V-shaped bone regions (of which perhaps the most numerous are the metaphyses of the postcranial long bones). The combination of bone deposition on the inner aspect and free ends of the V and resorption from its outer aspect results in the whole region growing posteriorly in the direction of its open end and thus increasing in both length and width (this growth movement is *intrinsic* to the maxillary complex, the whole complex at the same time being displaced anteroinferiorly by the increments at the sutures and septal cartilage). Accompanying these changes on the periosteal surfaces are complementary remodelling processes on the internal endosteal surfaces and the surfaces lining the alveoli for the teeth (Fig. 108*c*).

In coronal section the palate and alveolar arch can be seen to enter into a second V-shaped system, but in this case, the V is inverted and truncated (Fig. 108*a*). Once again, the outer aspect of the V (i.e. the floor of the nasal cavity and maxillary paranasal sinuses and the buccal periosteal surfaces of the alveolar arch) is resorptive, while the inner aspect (i.e. the downward facing periosteal surface of the palate) is depository. The intrinsic growth movement of this second V-shaped system is, therefore, directed inferiorly.

Hence, two overlapping V-shaped systems exist in the maxillary complex. One, the alveolar arch, lies in the horizontal plane; the second, the palate, occupies the coronal plane. Both follow the V-principles of growth enunciated by Enlow (1963, 1968) and together contribute to the growth in length, width and height of the alveolar arches and to the increase in height of the nasal cavity. The growth of the vertically orientated V adds a further component to the downward facial growth contributed by increments at the sutures and nasal septum while growth in the horizontally orientated V, by contrast, reduces the forward component of growth produced at these sites. As a result the downward growth of the human facial skeleton is accentuated but its forward growth reduced, a finding consistent with the orthognathous nature of the adult face.

Within the maxillary complex the maxillary paranasal air sinuses are enlarged by resorption from the internal surfaces of their floors and walls, except medially where deposition takes place, complementing the resorption on the opposing inner facing surfaces of the lateral walls of the nasal cavity.

The maxillary zygomatic process has a cutaneous, anterolaterally facing surface which, as already noted, is an area of bone resorption (Fig. 107). The opposed temporal surface undergoes deposition. The result is that the zygomatic process is moved posteriorly relative to the body of the maxilla. The rate at which this repositioning occurs is considerably lower than that at which the maxillary complex is moved forward by bone deposition at the pterygomaxillary suture. In consequence the zygomatic arch, which is anchored to the maxilla anteriorly and to the postpterygoid region of the cranial base posteriorly, increases in length, despite the remodelling of the maxillary zygomatic process. Further back along the arch the remodelling processes change direction, the lateral periosteal surface becoming depository and the medial periosteal surface resorptive, so producing an outward movement of the arch. The arch increases in height by deposition along both its superior and inferior margins.

The anterior walls of the nasal cavity are formed by the frontal processes of the maxillae. The cutaneous (i.e. external) periosteal surfaces of these processes are depository while their mucosal (internal) surfaces are resorptive.

The nasal bones undergo similar remodelling changes. As a result there is a widening of the anterior nasal aperture and an increasing protrusion of the nasal bridge. This increasing prominence is accentuated by the resorptive nature of the cutaneous periosteal surfaces adjacent to the frontal processes. Within the nasal cavity the mucosal surfaces of the lateral walls are generally resorptive with deposition taking place on the opposing periosteal surfaces of the medial orbital walls and medial walls of the maxillary sinuses. The height of the nasal cavity is increased, as already recounted, by resorption from the upper surface of the palate.

The major part of the bony surfaces facing into the orbital cavity is depository (Fig. 107). Following the V-principle, this will result in the orbit growing forwards (towards its open end) and increasing in overall size. However, there are several regions in which the remodelling departs from this general pattern. The periosteal surface of the zygomatic bone forming the anterior part of the lateral orbital wall undergoes resorption causing the lateral orbital rim to move outwards and backwards. This area of resorption extends on to the lateral part of the orbital floor and inferior orbital rim so that the orbital floor comes to slope downwards and outwards. The orientation of the orbital rim is thus changed during the growth period. At birth, the inferior and superior orbital borders lie in approximately the same coronal plane and the medial and lateral borders are at about the same horizontal level. During postnatal growth, as a result of the differential remodelling at the orbital rim, the supraorbital border comes to lie well forward relative to the infraorbital border and the lateral margin is displaced downwards relative to the medial margin.

The principal difference between the pattern of upper facial remodelling in man and the macaque monkey is that the extensive resorptive area seen on the anterior periosteal surface of the human face is replaced in the macaque by general disposition (Fig. 107). Otherwise the growth of the maxillary arch in the macaque is similar to that in man, the intrinsic direction of growth being principally hindwards and being brought about by rapid deposition at the maxillary tuberosities. Again, as in man, the teeth are repositioned by local remodelling of their alveoli. The depository nature of the anterior facial surface in the monkey contributes to the general forward growth of the face, with the result that the muzzle region becomes increasingly protrusive as growth proceeds.

It thus emerges that the vertical orientation of the human face involves (1) a retention into adult life of a small anterior basicranial angle, (2) resorption from the anterior surface of the facial skeleton and probably (3) a reduced component of forward growth at the sutures and nasal septum as compared with other mammals. These all tend to reduce the forward growth of the facial skeleton with the result that, as compared with the macaque monkey, there is a relatively greater increase in facial height. It is widely assumed that the reduced anterior facial growth in man is related to the general hominid evolutionary trend towards a better balance of the head on the vertebral column which has been part of the adoption of upright posture and bipedal locomotion (Moore, Adams & Lavelle, 1973).

Equivalent information about surface remodelling in the upper facial skeleton of other mammalian groups does not appear to be available. It seems likely, however, that principles similar to those just described for two higher primates may well apply throughout the mammals generally although the details will, of course, vary with the precise circumstances of growth from one species to another. As will become apparent, this expectation has been confirmed in the lower jaw, where information is available for a few non-primate species.

It should be emphasised that the method of growth analysis developed by Enlow is time-consuming and, in consequence, the number of individuals examined in each species is necessarily limited. In man, in particular, and in the macaque monkey to a lesser extent, the size and shape of the face differ considerably between individuals and it is likely that much of this variation is produced by a corresponding variation in the remodelling growth processes. There is as yet, however, little or no information about such variations in the remodelling pattern.

GROWTH OF THE MANDIBLE

In order to achieve and preserve the correct occlusal relationships between the upper and lower dentitions there must obviously be a high degree of coordination between the rates of growth of the mandible and maxillary complex. Both adhere to the somatic growth pattern with a relatively protracted period of enlargement and, in man and possibly the great apes, an adolescent growth spurt.

Despite the close coordination between the mandibular and maxillary growth patterns, there are marked differences in the way these patterns are achieved. The lower jaw is connected to the cranium only at the temporomandibular synovial joints and lacks, therefore, the complex system of sutures and endochondral growth sites which plays such an important part in the growth of the maxillary complex. In consequence, growth of the mandible is dependent upon surface remodelling to an even greater extent than is that of the upper facial skeleton. In the condylar cartilage, however, the mandible does possess one site of growth which appears to be of special importance and which is frequently regarded as having a role analogous to that of the septal cartilage in upper facial growth.

CONDYLAR CARTILAGE

During the development of the mandible several secondary cartilages appear and, by their endochondral ossification, contribute to the growth of the bone. The principal mandibular secondary cartilages, in the majority of mammals, are found in the coronoid, angular and condylar processes. The coronoid and angular cartilages may be active growth centres during the early stages of development, but have usually disappeared by the time of birth or shortly after. The condylar cartilage, by contrast, persists throughout the growth period, and even beyond.

Secondary cartilages may occur at many other sites within the developing mammalian skull besides the lower jaw, including the maxilla, palatine, pterygoid, squamous, jugal, vomer and frontal bones and the cranial and facial sutures. As already emphasised, these secondary cartilages develop quite separately from the primary cartilaginous skeleton of the head. The structure and function of secondary cartilage have been the subject of numerous studies in both mammalian (e.g. Symons, 1952; Girgis & Pritchard, 1958; Moss, 1958; Durkin, 1972) and non-mammalian (principally avian) species (Murray, 1963; Hall, 1975*a*) and the general evolutionary and developmental significance of this type of cartilage has been discussed by Hall (1970, 1975*b*).

Secondary cartilage frequently exhibits a histological structure reminiscent of that seen in cartilage formed in the embryo, possessing large haphazardly arranged chondrocytes with prominent lacunae and a scant matrix. As will shortly be described, it may also respond to certain functional and pathological stimuli in a manner quite different from that of primary cartilage. The fact that secondary cartilage develops in such close proximity to dermal bones suggests that the chondrogenic and osteogenic cells arise from the same pool of undifferentiated cells and there is now much experimental evidence that this is indeed so (see Hall (1970) for a detailed discussion of this evidence). The interchangeability of cell types in developing bone is further emphasised by studies of secondary cartilage formation in sutures (Pritchard *et al.*, 1956; Girgis & Pritchard, 1958; Moss, 1958; Young, 1959). These have shown that two types of tissue may be formed: (1) cartilage with an irregular arrangement of large cells and little matrix and (2) cartilage with a more regular arrangement of smaller cells. Moss described the first type of cartilage as being transformed directly into bone, while the second is replaced by endochondral ossification. Young also noted that secondary cartilage may be transformed directly into bone. When this happens the chondrocytes decrease in size and take on the appearance of osteocytes, while the matrix assumes the staining properties of bone matrix. It seems likely, therefore, that the distinction, in developmental terms, between dermal bone and secondary cartilage is not clear cut, both being produced by cells differentiating from a common pool in response to very locally acting factors. This conclusion correlates with the observation that the number and distribution of secondary cartilages vary widely between closely related species and even within a single species. The factors which determine whether or not secondary cartilage will form are poorly understood, but Hall (1970) suggests that mechanical agents may have an important role and has shown (1975*a*) that secondary cartilage is

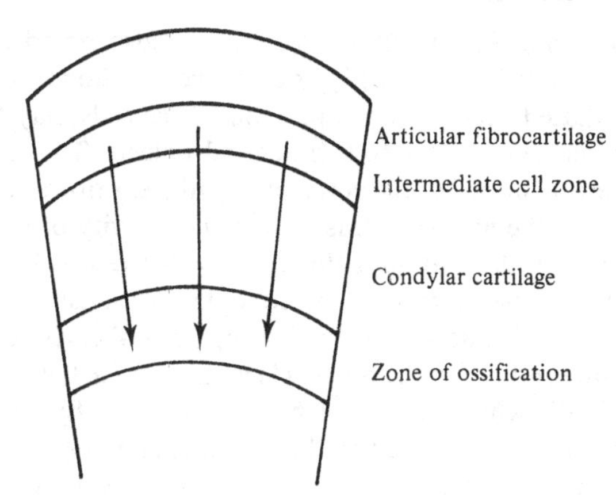

Fig. 109. Diagrammatic representation of zones of condylar cartilage. The arrows indicate the direction of *relative* movement of the chondrocytes formed in the intermediate cell zone.

particularly likely to occur, at least in avian dermal bones, at sites where mobile articulations exist. It is possible, therefore, that the secondary cartilages of the mammalian skull may arise to meet functional, sometimes temporary, needs.

Interest was focussed on the secondary cartilage (the condylar cartilage) of the condylar process of the human mandible by the frequent clinical observation that damage to the condyle in childhood leads to a severe reduction in mandibular growth, while correspondingly in condylar hyperplasia the mandible is much enlarged (e.g. Rushton, 1944; Engel & Brodie, 1947; Sarnat & Engel, 1951). Unlike an epiphysial cartilage, with which it is sometimes inaccurately equated, the condylar cartilage is contiguous with bone on its inferior surface only. Its upper surface is subjacent to the articular layer of the temporomandibuar joint which is of a fibrous or fibrocartilaginous nature. In the human mandible Wilson Charles (1925), Rushton (1944) and Symons (1952) have described an intermediate layer situated between the articular covering and the condylar cartilage and composed of closely packed connective tissue cells (Fig. 109). Both the articular covering and the intermediate layer are of periosteal origin. In a more recent study Blackwood (1966) studied the growth of the condylar cartilage in the rat by using tritiated thymidine. He noted the same three layers previously described in the human mandible. The intermediate layer was found to be the site of rapid mitosis. The cells thus produced pass, as chondrocytes, into the condylar cartilage and are released into the marrow cavity beneath the cartilage some five to six days later. During their passage through the condylar cartilage the chondrocytes hypertrophy but undergo no further division and consequently do not form the columns characteristic of endochondral ossification in primary cartilaginous sites. Similar findings have subsequently been made in the mouse by

Silbermann & Frommer (1972*a*, *b*), who also noted that after their release from the matrix of the condylar cartilage the chondrocytes may revert to less differentiated cells. It is possible that these cells may then redifferentiate to osteogenic cells or fuse to form chondroclasts. The removal of matrix from the inferior surface of the condylar cartilage, prior to ossification, appears to depend, in the mouse at least, upon the activity of these chondroclasts.

There are, then, at least three major differences between the processes of growth and ossification at the condylar cartilage, on the one hand, and at primary cartilaginous sites, such as epiphysial cartilages, synchondroses and the nasal septum, on the other: (1) a primary cartilaginous growth site grows interstitially while the condylar cartilage grows by apposition to its superior surface of cells produced in the intermediate layer; (2) the chondrocytes in a primary cartilaginous growth site, as they divide, form into columns, hypertrophy and then degenerate, whereas the chondrocytes in the condylar cartilage do not divide or form into columns, and emerge, still living, at the ossification face; (3) the removal of cartilage at the ossification face of a primary cartilaginous growth site is by dissolution of the matrix brought about by the medullary tissue which invades the regions formerly occupied by the chondrocytes but in condylar cartilage this removal may be due to chondroclastic activities.

There are, in addition, several findings which suggest that condylar cartilage reacts in a fundamentally different manner from primary cartilage to a wide variety of circumstances and stimuli. When transplanted subcutaneously or intracerebrally or grown in organ culture, for example, the condylar cartilage exhibits, by contrast to epiphysial cartilage, little independent growth potential (Koski & Rönning, 1965; Rönning, 1966; Petrovic, 1972). The two types of cartilage also respond differently to hyper- and hypovitaminosis A (Baume, 1970), and to thyroxine (Becks, Collins, Simpson & Evans, 1946) and papain administration (Irving & Rönning, 1962).

The observations that the chondrocytes of the condylar cartilage are produced in a zone of periosteal origin and that these cells may subsequently become changed into chondroclasts or osteogenic cells are further indications of the developmental plasticity of osteogenic and chondrogenic tissue already commented upon. This point is further emphasised by an interesting study carried out by Meikle (1973) in which the whole temporomandibular joint of seven-day-old rats was transplanted intracerebrally. After an initial cessation of cell division in the intermediate layer, lasting some 24 hours, proliferative activity became re-established but histological and autoradiographic evidence indicated that the cells were now differentiating into osteoblasts, not chondrocytes as previously. On the basis of these findings Meikle suggested that cell division within the intermediate layer operates independently but that the mechanical stress produced by movements at the temporomandibular joint is a necessary stimulus for the differentiation of the cells so produced into chondroblasts. In its absence the cells, being of a periosteal

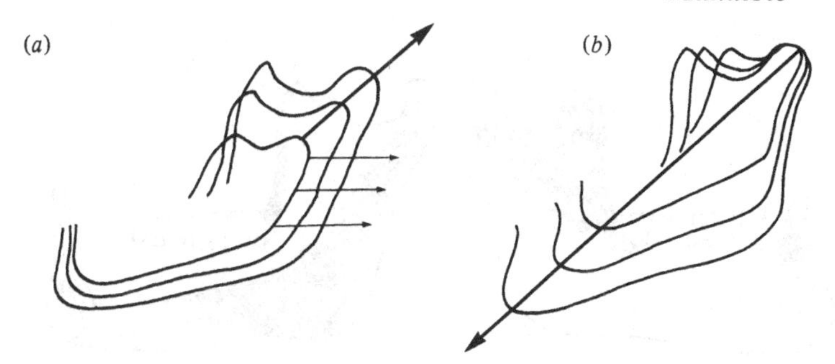

Fig. 110. Outlines of human mandible at different growth stages superimposed to show direction of: (*a*) intrinsic growth (thick arrow = condylar growth; thin arrows = deposition at posterior rameal border); (*b*) actual growth movement.

origin, revert to an osteogenic function. This conclusion fits well with Hall's (1975*a*) observations on avian secondary cartilages.

Although these findings have led to considerable controversy (discussed further below) as to whether or not the condylar cartilage is a primary determinant of the overall rate and amount of mandibular growth, it has been shown repeatedly for a wide variety of mammalian species that much bone formation occurs at the condylar cartilage even if this is secondary or adaptive in nature (e.g. rat – Bhaskar, 1953; rabbit – Bang & Enlow, 1967; pig – Brash, 1934; macaque monkey – McNamara & Graber, 1975; chimpanzee – Johnson, Atkinson & Moore, 1976). When first formed, the condylar cartilage may make up a major part of the condylar region (Fig. 16). The replacement of cartilage by endochondral bone at the inferior surface soon reduces its size but even so the cartilage may remain of appreciable thickness for a considerable part of the early growth period. During this time the cartilage is richly vascular, being penetrated by large blood vessels. Eventually, the cartilage becomes thinner, less vascular and less cellular. It persists in the adult as a very thin layer containing a few inactive cells but can become an active growth site again under the influence of increased secretion of growth hormone (a progressively enlarging mandible being one of the clinical features of acromegaly, the adult version of anterior pituitary hyperactivity).

In general, the condyle projects posterosuperiorly from the mandible so that bone increments at the cartilage increase both the vertical height of the ramus and the anteroposterior depth of the whole bone, the relative contributions to height and depth depending upon whether the condylar process tends more to the vertical or to the horizontal position. Since the condylar process abuts against the articular surface of the squamous, the posterosuperior direction of intrinsic growth produced at the condylar cartilage results in the mandible undergoing an actual growth movement in an anteroinferior direction (Fig. 110). This matches the growth movement produced in the upper facial skeleton by the growth of the sutures and the nasal septum. If the condylar

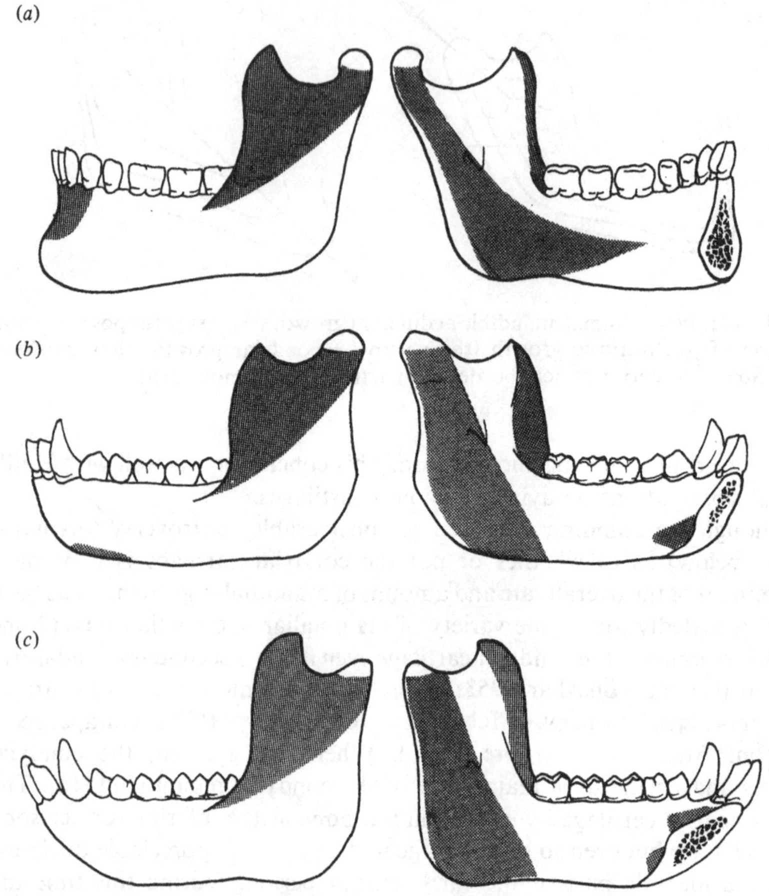

Fig. 111. Lateral (to left) and medial views of mandibles of (*a*) man; (*b*) chimpanzee; (*c*) macaque. Shading indicates bone resorption; no shading indicates bone deposition. Not to scale.

process projects laterally, as well as posterosuperiorly, as it does in the human mandible, increments at the condylar cartilages will also produce an increase in the bilateral width of the lower jaw.

SURFACE REMODELLING

It is well known from intravital staining and implant studies in a variety of mammalian species (e.g. pig – Hunter, 1771; Brash, 1934; Sarnat, 1968; rat – Baer, 1954; macaque monkey – McNamara & Graber, 1975; man – Bjork, 1968) that the posterior rameal border is one of the sites of most rapid bone deposition in the whole mandible and that, in the majority of cases, the anterior border of the ramus (coronoid process) undergoes resorption at an only slightly lower rate. As a result a powerful posteriorly directed component

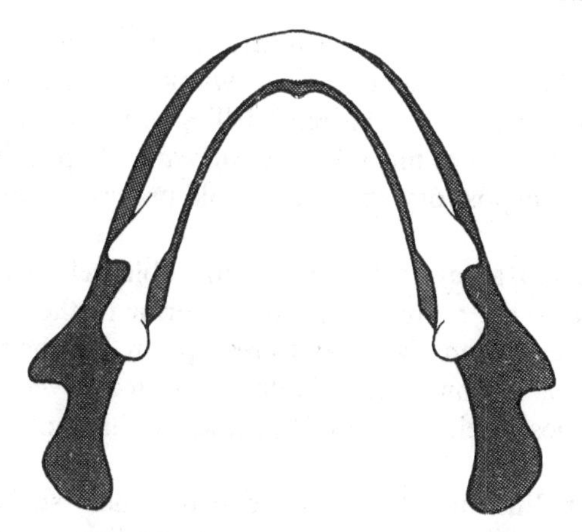

Fig. 112. Immature (unshaded) and mature (shaded) stages of human mandible superimposed in occlusal view to show remodelling changes. (After Enlow, 1968.)

is added to the intrinsic growth of the mandible, matching the posterior component of growth at the condylar cartilage, yet the anteroposterior dimensions of the ramus and body maintain their relative proportions (Fig. 110).

The studies by Enlow & Harris (1964), Enlow (1966) and Johnson *et al.* (1976) have shown that in the human, macaque and chimpanzee mandible the ramus undergoes complex remodelling changes additional to those occurring on its anterior and posterior borders. The patterns of remodelling are broadly similar in all three of these primate species and can be dealt with together (Fig. 111). The buccal periosteal surface of the coronoid process and the adjacent area extending below the sigmoid notch on to the lateral surface of the condylar process is resorptive, while the lingual periosteal surface of the process, posterior to the temporal crest (to which the temporalis muscle is attached), is depository. Since the lingual surface faces upwards as well as inwards and the buccal surfaces downwards as well as outwards, these remodelling changes result in the two coronoid processes growing upwards and their apices moving apart and provide yet another example of the V-principle of bone growth.

The lingual surface of the coronoid process faces somewhat posteriorly as well as upwards and inwards. Hence, bone additions on this surface, together with resorption on the buccal surface and anterior border, result in the coronoid processes also moving posteriorly with a corresponding elongation of the mandibular body. At its base the coronoid process merges with the rather more medially located main part of the ramus and just anterior to this region the ramus itself is continuous with the still more medially located body of the mandible. The deposition on the lingual surface of the coronoid process has the further important effect of bringing about the progressive medial relocation of first the coronoid process into the main part of the ramus and secondly of the ramus into the body as the whole region grows upwards and backwards. It will be noted that this remodelling results in the region in which body and ramus merge being moved posteriorly and, despite the fact that their lingual surfaces are depository, in the coronoid processes moving apart (Fig. 112).

Below and behind a reversal line running obliquely across the lingual periosteal surface of the ramus, from just anterior to the condyle to below the molar teeth, deposition gives way to resorption. On the buccal aspect of the ramus there is a somewhat similarly located reversal line but the remodelling changes on either side of this line are the opposite of those seen lingually.

One important difference between the two rameal surfaces is that the reversal line starts anterior to the condylar process lingually but posterior to the process buccally. As a result both the lingual and buccal periosteal surfaces of the neck of the condylar process are resorptive (Fig. 111). The corresponding endosteal surfaces are, of course, depository. The remodelling processes taking place in this region are thus similar to those seen in the metaphysial reduction of a postcranial long bone and serve the similar purpose of narrowing down a core of bone produced at an endochondral ossification centre.

More inferiorly, the depository zone of the buccal surface of the ramus faces posteriorly as well as laterally while the resorptive zone on the lingual surface faces anteriorly as well as medially, the ramus, in horizontal section, tending to be curved with the convexity facing laterally. As the ramus grows by the addition of bone along its posterior border, regions that previously lay at this border become successively incorporated into the body of the ramus. Because of the curvature just described this relative anterior relocation involves also a lateral relocation of the region previously at the posterior margin which is brought about by the combination of lingual resorption and buccal deposition taking place in the posterior half of the ramus.

The pattern of remodelling in the body of the mandible shows a general resemblance but also certain detailed contrasts between man, the chimpanzee and the macaque monkey. During growth in all three primates the body increases in anteroposterior depth, vertical height and buccolingual width while the separation between right and left halves also increases to accommodate the enlarging tongue. A line drawn through the anteroposterior axes of the molar teeth and continued posteriorly passes lingual to the ramus. As the body increases in depth by the resorption of bone from the anterior border of the ramus, the relocation of regions previously at the base of the coronoid process must, therefore, involve a marked lingual shift. This is brought about by the deposition of bone on the lingual periosteal surface and its resorption from the buccal periosteal surface of the coronoid process as already described.

In the human mandible the molar teeth are supported in alveolar bone which overhangs the sublingual fossa. The zone of resorption on the posterior half of the lingual periosteal surface of the ramus is continued forwards on to the inferior part of the lingual aspect of the mandibular body (Fig. 111*a*). Periosteal resorption and endosteal deposition in this region of the body hollow out the sublingual fossa and produce the overhang of the posterior alveolar arch. In the chimpanzee and macaque monkey the area of resorption does not extend far forwards on to the body and little alveolar overhang is, therefore, produced.

The depository zone extending across the buccal periosteal surface of the ramus continues on to the buccal surface of the body (Fig. 111). Extensive deposition may occur below the molar teeth to produce a marked trihedral eminence. Deposition on the buccal surface appears to be the principal mechanism whereby the buccolingual thickness of the body is increased. Some additional deposition may occur lingually, apart, as we have seen, from the region of the sublingual fossa in the human mandible. The amount of deposition on this aspect of the mandible must be limited, however, in order to avoid encroachment on to the space required by the growing tongue.

Remodelling in the chin region shows contrasts between man, on the one hand, and the chimpanzee and macaque, on the other, which reflect those described for the muzzle region of the upper face. In the human mandible there is an area of resorption on the periosteal surface below the incisor and canine teeth while the mental protuberance is an area of bone deposition. The reversal line separating these two zones varies considerably in its location to produce the wide variations seen in the size and shape of the chin. In the macaque monkey the whole of the chin region is depository, while in the chimpanzee this region is also depository for most of the growth period, although Johnson *et al.* (1976) found in the single mature specimen available to them that deposition had been replaced by resorption over an area similar in extent to the resorptive zone seen in the human mandible. These contrasts presumably relate, as in the upper facial skeleton, to the reduced anterior growth of the face in man, as compared to the ape and monkey, which is part of the hominid evolutionary trend towards a better balance of the head on the vertebral column.

Increase in height of the mandibular body in all three primates is achieved by bone increments along its inferior border and on the alveolar process.

Knowledge of the surface remodelling processes in non-primates is a little more extensive for the mandible than it is for the upper facial skeleton in that Bang & Enlow (1967) have provided information for the growth of the rabbit lower jaw, using methods similar to those they employed in the primate skull. As can be seen from Fig. 113, the general pattern of remodelling is closely comparable to that just recounted for the macaque and human mandible, a finding which gives support to the view that the principles of mandibular growth that obtain in the primate skull may be common to many, if not all,

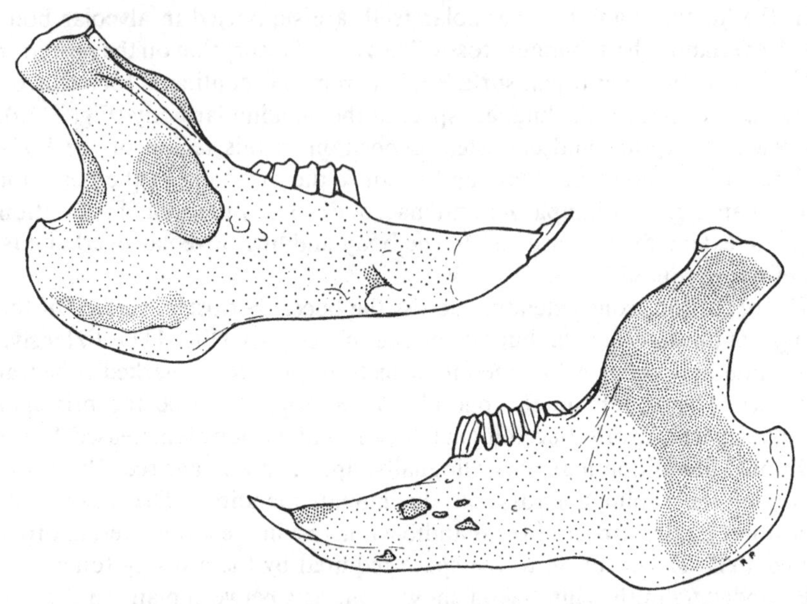

Fig. 113. The distribution of resorptive (dark stipple) and depository (light stipple) periosteal surfaces in the rabbit mandible. Medial view above; lateral view below. (Reproduced with permission from Enlow, 1968.)

mammals. Such differences as exist can be readily related to regional differences in the morphology of the jaw. The coronoid process, for instance, is almost non-existent in the rabbit and so the complex remodelling sequence found in the prominent coronoid process of the anthropoid mandible is also lacking.

The symphysis between the two halves of the mandible fuses early in the growth period in a wide variety of mammalian species. In the remainder the two halves of the lower jaw are usually tied together by ligaments and partial bony fusion may occur with age. It appears that generally the amount of growth taking place at the symphysis is small and limited to the early stages of development, the variations in the structure of this region being related to the mechanics of jaw usage (see Chapter 5) rather than to the requirements of mandibular growth.

CONTROL OF SKULL GROWTH

Probably the most controversial aspect of the investigation of craniomandibular growth has been the attempt to unravel the controlling factors which determine the final size and shape of the skull. Various authors have attributed a special role to certain specific sites of skull growth, notably the primary and secondary cartilaginous proliferative zones and the sutures.

Others have stressed the importance of extrinsic factors, such as the function of the attached musculature or the growth of neighbouring soft structures, in influencing skull growth. Taken in combination, the experimental findings described by the authors on both sides of this controversy as evidence for their own particular point of view leave little doubt that the control of skull growth is multifactorial and that the relative importance of the various factors varies from species to species, from one region of the skull to another, and at various stages of ontogeny.

ROLE OF GROWTH SITES

Synchondroses and nasal septum

Because of their structural similarity to the postcranial epiphysial cartilages, the view has been frequently expressed that the primary cartilaginous growth sites of the skull play a major determining role in the enlargement of the cranial base and upper facial skeleton (e.g. Scott, 1958; Baume, 1961, 1968). There have been numerous investigations of the intrinsic growth potential displayed by these primary cartilages in organ culture or when transplanted to subcutaneous or intracerebral sites, but the results are not consistent. Petrovic and colleagues (e.g. Petrovic, Charlier & Herrmann, 1968) found that both synchondrotic and septal cartilage of the rat exhibits considerable independent growth potential in organ culture, resembling epiphysial cartilage in this respect. Koski & Rönning (1969, 1970), on the other hand, found that, while rat epiphysial cartilage underwent continued growth and differentiation after transplantation, this was not so in the case of cartilage from the sphenoccipital and midsphenoidal synchondroses.

There is a similar inconsistency in the findings following extirpatory procedures. There are, for example, numerous reports of excision of the septal or synchondrotic cartilages in a variety of species being followed by a reduction in the growth of the associated subdivision of the skull (e.g. DuBrul & Laskin, 1961; Wexler & Sarnat, 1961; Sarnat & Wexler, 1967*a*, *b*). But Moss, Bromberg, Song & Eisenman (1968) found that removal of the septal cartilage by careful cauterisation in 20-day-old rats produced no diminution in facial dimensions and suggested that the reduction observed in earlier ablation experiments was probably due to contraction of the scar tissue produced by the surgical trauma. More recently, Stenström & Thilander (1970, 1972) have observed that neither surgical extirpation of the septal cartilage nor the creation of defects at the sphenethmoidal junction in young guinea-pigs produced any marked change in the subsequent growth of the facial skeleton. The clinical evidence, too, suggests that the proliferation of the septal cartilage is not indispensable to facial growth. Latham (1968) and Moss *et al.* (1968) have described cases in man in which the nasal septum is congenitally absent yet in which more or less normal facial growth has proceeded apart from some depression of the bridge of the nose. It is perhaps worth emphasising at this point two particular weaknesses of experimental (or naturally-occurring) extirpation as an investigative procedure which could lead to the importance of a growth site or source of growth stimulus being either over- or underestimated. First, the growth circumstances may be changed so radically that the findings bear little relevance to growth under normal circumstances. Secondly, it seems likely that growth processes possess a high degree of plasticity so that the removal of one growth site (or source of growth stimulus) may produce little apparent change in overall growth because of increased, compensatory growth elsewhere. These criticisms apply with equal force to similar investigative studies, to be described later, of the role of sutures and soft tissue structures.

It is thus impossible, with the evidence presently available, to make a final assessment of the part played by the primary cartilaginous proliferative zones in skull growth. Clearly these cartilages are normally major sites of bone deposition. That the proliferation of the cartilage is a prime determinant of the overall rate and direction of growth of the associated skull subdivision seems less certain – in any event, the absence of these proliferative zones can be compensated for by increased growth elsewhere.

Sutures

Numerous investigators have carried out extirpatory or transplantation procedures on the sutures of experimental animals in an attempt to determine whether or not the proliferation of the cambial layers of these structures provides a separating force between adjacent bones. The results have proved rather more conclusive than those involving extirpation of the synchondroses or nasal septum.

Sarnat and his colleagues, for example, have performed a series of extirpation experiments on the cranial and facial sutures of both primate and non-primate species (see Sarnat (1963, 1971) for reviews of this extensive body of work) and found no grossly apparent disturbance of growth. Essentially similar findings have been made by Watanabe, Laskin & Brodie (1957) in the guinea-pig and by Zoller & Laskin (1969) in the pig. Sutures show little or no independent growth activity in organ culture or when transplanted autogenously (Petrovic *et al.*, 1968).

It seems unlikely, therefore, that suture proliferation is a primary determining force in cranial growth although the caveats entered when discussing synchondrotic and septal cartilage growth apply equally here. There is, of course, no doubt that the sutures are major sites of bone addition in normal growth, but it appears that these additions are in response to a growth stimulus produced elsewhere.

Condylar cartilage

Perhaps more than any other site in the skull, the condylar cartilage has been regarded as possessing an intrinsic growth potential and thus acting as one of the prime determinators of facial growth. This view was engendered by the clinical observation, already noted, that damage to the condyle in young children is frequently followed by a disfiguring diminution of mandibular growth. In order to investigate the role of the condylar cartilage experimentally, recourse has been made once again to organ culture, transplantation and extirpation.

As described above, the condylar cartilage of the young rat, when transplanted subcutaneously or intracerebrally (Koski & Rönning, 1965; Rönning, 1966) or grown in organ culture (Petrovic, 1972), exhibits little independent growth activity, contrasting in this respect with epiphysial cartilage and, to a lesser extent, with synchondrotic and septal cartilage. However, Meikle (1973) has found evidence of independent growth potential in the intermediate zone of the condylar cartilage when the whole temporomandibular joint is transplanted.

One of the principal investigators of the role of the condylar cartilage who has used extirpation procedures is Sarnat (e.g. 1957, 1963). He found that unilateral or bilateral resection of the condyle in growing macaque monkeys led, after a postoperative period of two years or so, to a reduction in the height of the ramus and body of the mandible and concluded that the growth of the condylar cartilage is essential for normal vertical growth of the face. Sarnat later modified this view when similar mandibular changes were found to occur after condylar resection in *adult* squirrel monkeys (Sarnat & Muchnic, 1971). Since there was no histological evidence of growth in the removed condyles, it seems probable that the changes in the facial skeleton were secondary to the structural disturbance of the temporomandibular joint and associated musculature rather than to any interference with the condylar growth mechanism. Bilateral condylectomy in the growing rat (Jarabek & Thompson, 1953; Jolly, 1961; Gianelly & Moorrees, 1965) produces relatively modest disturbances of mandibular growth, while Moss & Rankow (1968) have reported that a similar surgical procedure in a seven-year-old child had little or no effect on facial growth, a finding apparently at variance with the frequently reported cases of severe facial deformity following trauma to the condyle during growth.

Bearing in mind the weaknesses of extirpation as an investigative procedure, it seems on balance that the condylar cartilage plays a less prominent part in determining mandibular growth than has been traditionally believed, although, as with the sutures, there is no doubt that in normal growth it is a major site of bone formation.

ROLE OF FUNCTIONAL COMPONENTS

The general finding that the major sites of proliferation in the facial skeleton have less inherently directed growth potential than has often been assumed has led to a search for other factors which may influence the rate and direction of facial growth. A major advance in this search has been made by Moss and his colleagues by their development of the concept of functional components (for a detailed description and a comprehensive bibliography see, for example, Moss & Salentijn, 1969). According to this concept the head is composed of a number of distinct (but sometimes overlapping) regions, termed functional components, each responsible for a particular function or set of functions. Each component is itself composed of a functional matrix, made up of the structures performing the function, and a skeletal unit which supports and protects the functional matrix. Skeletal units do not coincide with the individual bones of the skeleton. A bone may consist of several skeletal units or, conversely, a skeletal unit may be composed of several bones.

Moss distinguishes two types of functional matrix – the capsular and the periosteal. Capsular matrices are found, for example, in the soft tissues surrounding the brain or eyeball and around functional spaces such as the mouth, nose and pharynx, the skeletal units in these cases being embedded in the capsule surrounding the mass or space. The periosteal matrix is best exemplified by skeletal muscle which is usually attached to its skeletal unit through the periosteum.

An essential part of functional cranial analysis is the belief that the growth and function of the functional matrix play a major determining role in the differentiation, development and growth of the associated skeletal unit. There is now abundant evidence (discussed below) that the presence and function of soft tissues, and especially of skeletal muscle, do indeed influence bone growth and development although there is as yet no general agreement that the response to soft tissue is as dominating as proponents of functional cranial analysis believe. It is widely assumed that it is the mechanical stressing produced by the growth and function of the soft tissues that is the principal factor by which they exert their effect on the skeleton, although the mechanism by which such stressing may influence the differentiation and growth of skeletal tissues is far from completely understood (for comprehensive reviews of the general problem of the control of skeletal differentiation and its significance in vertebrate ontogeny and phylogeny, see Hall, 1970, 1975*b*, and Owen, 1970). The suggestion presently gaining greatest support is probably the bioelectric hypothesis as advocated principally by Bassett and his colleagues (see, for example, Bassett, 1972).

To summarise briefly, it is postulated in this hypothesis that bone growth responds to deforming forces by means of a negative feed-back mechanism. Bone and other connective tissues are capable of acting as transducers converting mechanical energy to electrical energy, probably by virtue of a piezoelectric or similar property of one or more of their components (collagen is certainly and hydroxyapatite possibly piezoelectric). The electric signal is then thought to influence osteogenic cell function, either increasing or decreasing the amount of matrix produced, so that the organisation of the matrix is remodelled to become more resistant to the deforming force and thus to shut off the initiating signal. The means by which this influence is exerted is not known but experimental evidence and theoretical considerations point to several potential ways in which an appropriate cellular response could be provoked, including changes, in response to the electric signal, in cell nutrition, local pH, enzyme activation or suppression, orientation of intra- and extracellular macromolecules such as collagen, cellular migration and proliferation, cellular synthetic activity and the contractility and permeability of cell membranes. In its fully developed form, this hypothesis appears to have the flexibility and sophistication necessary to account for the highly complex response of the skeletal tissues to their functional matrices.

Capsular functional matrices

The principal capsular matrices in the head are those of the neural, orbital, otic, nasal and oral functional components. Of these, only the neural and orbital components have been the subjects of major experimental investigation.

The neural functional component is composed of the neural mass, which includes the meninges and cerebrospinal fluid as well as the brain, surrounded by a neural capsule in which are embedded the bones of the braincase. There is abundant evidence to indicate that the growth of the neural mass is a major determinant of the size of the capsule. Hydrocephaly, for example, whether produced by disease or experimentally (e.g. Young, 1959) is associated with an enlargement of the braincase which may reach extreme proportions. Conversely, a reduction in brain size such as occurs in microcephaly or after experimental removal of brain tissue is accompanied by a corresponding reduction in the size of the braincase (e.g. Johnson, 1967; Singh, Sanyal & Kar, 1974).

As Hoyte (1975) has emphasised, the effects of changes in the neural mass are much greater in the vault than in the cranial base. In Young's (1959) experimental procedures the basal bones were only slightly altered in size while in human microcephaly the severity of the reduction in the cranial base does not appear to be directly correlated with the degree of reduction of brain volume. These findings are not altogether unexpected in that the cranial base is involved in the facial as well as the neural functional components.

Moss & Young (1960) have assembled evidence indicating that, while the magnitude of neural capsular growth is determined by the degree of expansion of the brain and its adnexa, the direction of growth is dependent upon the orientation of fibre tracts within the dura mater. Such fibre systems develop in the embryonic human neural capsule from regions where it is firmly attached to the cranial base (crista galli, orbitosphenoids and otic capsules) and pass dorsally around the brain in approximately the positions which will later be occupied by the sutures of the vault. Moss & Young suggest that the directions of capsular growth, and hence the final proportions of the skull, are determined by the resultants of the growth vectors produced by the constraints imposed on the expanding neural mass by the fibre systems. The location of the initial ossification centres of the calvarial bones may also be determined by the arrangement of the dural fibres, for these centres generally appear approximately at the midpoint of each of the regions delimited by the fibres.

Moss & Young also point out that, while both the inner and outer tables of compact bone in the cranial vault are likely to be greatly influenced by neural growth in animals with thin-walled braincases, the two tables may have a greater measure of independence in those species where the braincase has thicker walls. In animals with powerful jaws, for example, the inner table is responsive to neural growth but the outer table, as described below, appears to be influenced to a much greater degree by the pull of the attached jaw and nuchal muscles.

Although the evidence is more limited, it appears that the growth of the capsule of bones around the orbital cavity is dependent, like that of the braincase, upon the expansion of the enclosed soft tissues. Enucleation in the human infant (Taylor, 1939) and in the growing rabbit, dog, pig and a variety of non-mammalian vertebrates (Washburn & Detwiler, 1943; Sarnat & Shanedling, 1970) results in a diminution in orbital size.

Periosteal functional matrices

The importance of muscular function as a controlling factor in bone development is one of the most debated aspects of skeletal growth. The general stimulatory influence of muscular forces on the growth of the associated bones has been recognised from early times largely from clinical observations such as the frequent diminution in length of limbs paralysed in childhood. There have been similar observations in the skull. Congenital unilateral absence of the muscles of mastication, for example, is associated with marked hypoplasia of the facial skeleton on the affected side (Rogers, 1958). However, experimental investigation of this effect has produced rather inconsistent results.

One of the best of the earlier studies was that of Washburn (1947). He removed the temporalis muscle unilaterally in newborn rats and found at post mortem, some three to five months later, that the temporal crest on the operated side of the cranial vault had failed to develop while the mandibular coronoid process, which was present at the time of operation, had disappeared. On the basis of these findings Washburn classified the skull components, so far as their response to muscular function is concerned, into three categories: (*a*) features such as the temporal crest which depend for their development upon the presence and function of the associated muscles; (*b*) features such as the coronoid process, which, while self-differentiating, require the presence of functioning muscle to attain full development; (*c*) components like the braincase which grow independently of muscle function.

The distinction between categories (*a*) and (*b*) may reflect no more than a difference in the time at which the structures develop. When Horowitz & Shapiro (1951) repeated Washburn's surgical procedure in rats 30 days old, they observed that the coronoid process and temporal crests were well developed at this age, but both completely disappeared within two months of operation. Components in category (*c*) are probably little influenced by muscular forces because their growth is responding predominantly to other morphogenetic influences. To express Washburn's findings in the terms of functional component analysis: the coronoid process is the skeletal unit of the mandible specifically associated with the temporal muscle; the braincase is, as we have seen, powerfully influenced by the growth of the neural capsular matrix; the temporal crest, even in the thin-walled rat skull, can be regarded as a skeletal unit which is part of the outer table of the braincase and which is influenced more by its attached muscle than by the neural capsular matrix.

Subsequent experimental investigations have shown repeatedly and in a wide variety of species that the growth of the skeletal units of the skull specifically associated with muscles is very responsive to the functioning of those muscles. To give just a few examples, Avis (1959) in cats, Schumacher (1968) and Schumacher & Dokladal (1968) in the dog, sheep and rabbit have shown that the coronoid process is dependent, as in the rat, for its full development upon the presence and activity of the temporalis muscle. Similarly, the growth of the angular process in rats (Horowitz & Shapiro, 1955; Avis, 1961) is responsive to the function of the masseter and medial pterygoid muscles while condylar growth in the macaque monkey (Baume & Derichsweiler, 1961) and man (Moffet, Johnson, McCabe & Askew, 1964) responds to the function of the lateral pterygoid muscle.

Before going on to describe several experimental studies which have been designed specifically to examine the validity of functional cranial analysis, it will be helpful to enumerate the skeletal units of the lower and upper jaws. In the mandible of man, for instance (Moss, 1960; Moss & Rankow, 1968; Moss & Simon, 1968), the following six skeletal units are recognised (with functional matrix given in parentheses) (Fig. 114*a*):

(1) basal (inferior alveolar neurovascular bundle);
(2) condylar (temporomandibular joint and lateral pterygoid muscle);
(3) coronoid (temporalis muscle);
(4) angular (masseter and medial pterygoid muscles);
(5) alveolar (mandibular dentition);
(6) symphysial (facial and genial muscles).

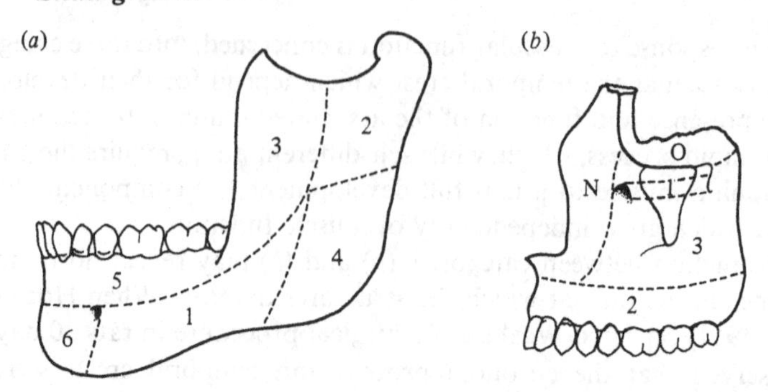

Fig. 114. Left lateral views of (*a*) mandible and (*b*) maxilla to show skeletal units. Mandible: 1 = basal, 2 = condylar, 3 = coronoid, 4 = angular, 5 = alveolar, 6 = symphysial; maxilla: 1 = basal, 2 = alveolar, 3 = pneumatic. Surfaces marked N and O in maxilla partake in the nasal and orbital capsular units respectively.

Moss & Greenberg (1967) have suggested that the following three skeletal units are separable in the maxillary complex (Fig. 114*b*):

(1) basal (infraorbital neurovascular bundle);
(2) alveolar (maxillary dentition);
(3) pneumatic (maxillary paranasal air sinus).

The palatal, orbital and nasal surfaces of the maxillae are consigned to the skeletal units associated with the oral, orbital and nasal capsular functional matrices.

Moss & Meehan (1970) investigated the relationship between the coronoid skeletal unit and its functional matrix by removing, in rats, the middle and posterior fibres of the left temporalis down to the level of the zygomatic arch. During the first few postoperative days the only changes were a slight diminution in size of the coronoid process but from the end of the first week the diminution became increasingly marked. The proportions of the neighbouring skeletal units were unaltered. Moore (1973) investigated the effect of the much severer surgical procedure of bilateral masseter muscle ablation in newly born rats on the mandibular functional components. In this case, statistically significant reductions in growth were observed (at five months) not only in the angular process, the skeletal unit specifically associated with the masseter, which was reduced by 25 per cent, but also in the condylar process (reduced 11 per cent), coronoid process (10 per cent) and basal unit (reduced 5 per cent).

Taken together, the findings in these studies indicate that the angular and coronoid skeletal units, at least, do possess a considerable measure of autonomy in their development and growth and that each is capable of responding to its own periosteal matrix, with limited effects upon contiguous skeletal units. When the change in a functional matrix is particularly severe, the effects may 'spill over' on to neighbouring skeletal units through their functional matrices. Thus, for example, following bilateral removal of the very large masseter muscles of the rat, the temporalis and lateral pterygoid muscles undergo a diminution in their mass of some 10 to 15 per cent, presumably due to an overall reduction of masticatory function. As a result the condylar and coronoid units are diminished, although to a lesser extent, than is the angular unit. The basal, alveolar and symphysial units are much less reduced because their functional matrices are relatively unaffected by the operative procedure.

The shape of the basal unit of the lower jaw is believed to be determined principally by the path of the inferior neurovascular bundle. As demonstrated by Moss & Salentijn (1970) in the human mandible and by Hildyard, Moore & Corbett (1976) in the mandible of the African great apes, the course of the neurovascular bundle is that of a logarithmic spiral (i.e. a spiral in which the whorls cut the radius vector into lengths which increase in geometrical progression and the angle between the tangent to the curve and the radius vector remains constant). In order to provide a conforming pattern of growth the basal unit of the upper jaw must, presumably, follow a spiral of similar form. The functional significance of the basal units possessing this shape is that the space enclosed by them (the oral cavity) is a gnomon (i.e. as it grows, it changes in size but not in shape). Growth in the form of a logarithmic spiral is seen in many organs which grow by accretion, such as the shells of molluscs, horns and continuously growing incisor teeth (Huxley, 1932; Thompson, 1942; Herzberg & Schour, 1941). The essential features of growth in these structures are that it takes place entirely by accretion, that the increments of new material are themselves incapable of interstitial growth, that the site of increments is limited to one end of the growing organ and that there is a constant gradient of incremental accretion occurring across this growth site. The first three of these requirements are certainly met in the growth of the basal units – bone grows by accretion and is incapable of interstitial growth, and the principal site of deposition is at the posterior end of the basal units in both jaws. There is also much indirect evidence (discussed by Hildyard *et al.*, 1976), in the human mandible at least, that a constant gradient of growth exists across the growth site. Whether logarithmic expansion is a feature of the growth of the basal units of the jaws in other mammalian species is not known (although preliminary investigations by Hildyard and Moore indicate that this is so in the macaque and baboon) but it seems likely since the circumstances of growth appear to be similar to those obtaining in hominoids.

REFERENCES

Adams, L. A. (1919). A memoir on the phylogeny of the jaw muscles in recent and fossil vertebrates. *Ann. N.Y. Acad. Sci.* **28**, 51–166.

Adelmann, H. B. (1925). The development of the neural folds and cranial ganglia of the rat. *J. comp. Neurol.* **39**, 19–171.

Ahlgren, J. (1966). Mechanisms of mastication. *Acta odont. scand.* **24**, Suppl. 44, 1–109.

Ahlgren, J. (1976). Masticatory movements in man. In *Mastication* (ed. D. J. Anderson & B. Matthews). Bristol: Wright & Sons.

Ahlgren, J. & Öwell, B. (1970). Muscular activity and chewing force: a polygraphic study of human mandibular movements. *Archs oral Biol.* **15**, 271–80.

Allen, H. (1882). On a revision of the ethmoid bone in the Mammalia, with special reference to the description of this bone and of the sense of smell in the Chiroptera. *Bull. Mus. comp. Zool. Harv.* **10**, 135–64.

Allin, E. F. (1975). Evolution of the mammalian middle ear. *J. Morphol.* **147**, 403–38.

Allis, E. P. (1931). Concerning the homologies of the hypophysial pit and the polar and trabecular cartilages of fishes. *J. Anat.* **65**, 247–65.

Anderson, S. D. (1976). The intratympanic muscles. In *Scientific foundations of otolaryngology* (ed. R. Hinchliffe & D. Harrison). London: Heinemann.

Anthony, R. L. F. & Iliesco, G. M. (1926). Etude sur les cavités nasales des carnassiers. *Proc. zool. Soc. Lond.* 1926, 989–1015.

Ardran, G. M. & Kemp, F. H. A. (1960). Biting and mastication. A cineradiographic study. *Dent. Practnr, Bristol* **11**, 23–34.

Ardran, G. M., Kemp. F. H. A. & Ride, W. D. L. (1958). A radiographic analysis of mastication and swallowing in the domestic rabbit *Oryctolagus cuniculus* (L.) *Proc. zool. Soc. Lond.* **130**, 257–74.

Ariëns Kappers, C. U., Huber, G. C. & Crosby, E. C. (1936). *The comparative anatomy of the nervous system of vertebrates, including man.* New York: Macmillan Co.

Ashton, E. H. (1957). Age changes in the basicranial axis of the Anthropoidea. *Proc. zool. Soc. Lond.* **129**, 61–74.

Ashton, E. H., Flinn, R. M. & Moore, W. J. (1977). The basicranial axis in certain fossil hominoids. *J. Zool.* **176**, 577–91.

Ashton, E. H. & Zuckerman, S. (1954). The anatomy of the articular fossa (fossa mandibularis) in man and apes. *Am. J. phys. Anthrop.* **12**, 29–50.

Ashton, E. H. & Zuckerman, S. (1956). Cranial crests in the Anthropoidea. *Proc. zool. Soc. Lond.* **126**, 581–634.

Asling, C. W. & Frank, H. R. (1963). Roentgen cephalometric studies on skull development in rats. *Am. J. phys. Anthrop.* **21**, 527–43.

Avis, V. (1959). The relation of the temporal muscle to the form of the coronoid process. *Am. J. phys. Anthrop.* **17**, 99–104.

Avis, V. (1961). The significance of the angle of the mandible: an experimental and comparative study. *Am. J. phys. Anthrop.* **19**, 55–61.

Baer, M. J. (1954). Patterns of growth of the skull as revealed by vital staining. *Hum. Biol.* **26**, 80–126.

Balfour, F. M. (1876–78). On the development of elasmobranch fishes. *J. Anat. Physiol., Lond.* **10**, 377–410, 517–70, 672–88; **11**, 128–72, 406–90, 674–706; **12**, 177–216.

Bambha, J. K. (1961). Longitudinal cephalometric roentgenographic study of face and cranium in relation to body height. *J. Am. dent. Assoc.* **63**, 776–99.

Bang, S. & Enlow, D. H. (1967). Postnatal growth of the rabbit mandible *Archs oral Biol.* **12**, 993–8.

Barghusen, H. R. (1968). The lower jaw of cynodonts (Reptilia, Therapsida) and the evolutionary origin of mammalian adductor musculature. *Postilla* **116**, 1–49.

Barghusen, H. R. (1972). The origin of the mammalian jaw apparatus. In *Morphology of the maxillo-mandibular apparatus* (ed. G. H. Schumacher). Leipzig: Georg Thieme.

Barghusen, H. R. & Hopson, J. A. (1970). Dentary–squamosal joint and the origin of mammals. *Science* **168**, 573–5.

Barnett, C. H., Davies, D. V. & MacConaill, M. A. (1961). *Synovial joints, their structure and mechanics.* London: Longmans.

Bartelmez, G. W. (1922). The origin of the otic and optic primordia in man. *J. comp. Neurol.* **34**, 201–32.

Bartelmez, G. W. (1923). The subdivisions of the neural folds in man. *J. comp. Neurol.* **35**, 231–47.

Bartelmez, G. W. (1960). Neural crest from the forebrain in mammals. *Anat. Rec.* **138**, 269–81.

Bartelmez, G. W. (1962). The proliferation of neural crest from forebrain levels in the rat. *Contr. Embryol.* **37**, 1–12.

Bartelmez, G. W. (1922). The origin of the otic and optic primordia in man. *J. comp.* primary optic vesicle in man. *Contr. Embryol.* **35**, 55–71.

Bassett, C. A. L. (1972). A biophysical approach to craniofacial morphogenesis. In *Mechanisms and regulation of craniofacial morphogenesis* (ed. B. C. Moffett). Amsterdam: Swets & Zeitlinger.

Bast, T. H. & Anson, B. J. (1949). *The temporal bone and the ear.* Springfield: Charles C. Thomas.

Baume, L. J. (1961). The postnatal growth activity of the nasal cartilage septum. *Helv. odont. Acta* **5**, 9–13.

Baume, L. J. (1968). Patterns of cephalofacial growth and development. A comparative study of the basicranial growth centers in rat and man. *Int. dent. J., Lond.* **18**, 489–513.

Baume, L. J. (1970). Differential response of condylar, epiphyseal, synchondrotic, and articular cartilages of the rat to varying levels of vitamin A. *Am. J. Orthod.* **58**, 537–51.

Baume, L. J. & Derichsweiler, H. (1961). Response of condylar growth cartilage to induced stresses. *Science, N.Y.* **134**, 53–4.

Becht, G. (1953). Comparative biologic–anatomical researches on mastication in some mammals. *Proc. K. ned. Akad. Wet.* Ser. C **56**, 508–26.

Becks, H., Collins, D. A., Simpson, M. E. & Evans, H. M. (1946). Growth and transformation of the mandibular joint in the rat. III. The effect of growth hormone and thyroxin injections in hypophysectomised female rats. *Am. J. Orthod.* **32**, 447–51.

Beecher, R. M. (1977*a*). *Functional significance of the mandibular symphysis.* Ann Arbor: University of Michigan Microfilms.

Beecher, R. M. (1977*b*). Function and fusion at the mandibular symphysis. *Am. J. phys. Anthrop.* **47**, 325–36.

Bellairs, A. d'A. (1969). *The life of reptiles*, Vol. I. London: Weidenfeld & Nicolson.

Berry, H. M. & Hoffmann, F. A. (1956). Cinefluorography with image intensification for observation of temporomandibular joint movements. *J. Am. dent. Assoc.* **53**, 517–27.

Bhaskar, S. N. (1953). Growth pattern of the rat mandible from 13 days insemination age to 30 days after birth. *Am. J. Anat.* **92**, 1–53.

Bjork, A. (1955). Cranial base development. *Am. J. Orthod.* **41**, 198–225.
Bjork, A. (1968). The use of metallic implants in the study of facial growth in children: method and application. *Am. J. phys. Anthrop.* **29**, 243–54.
Blackwood, H. J. J. (1966). Growth of the mandibular condyle of the rat studied with tritiated thymidine. *Archs oral Biol.* **11**, 493–500.
Brash, J. C. (1934). Some problems in the growth and developmental mechanics of bone. *Edinb. med. J.* **41**, 363–87.
Brink, A. S. (1960). A new type of primitive cynodont. *Palaeont. afr.* **7**, 115.
Brodie, A. G. (1934). The significance of tooth form. *Angle Orthod.* **4**, 335–50.
Brodie, A. G. (1941). On the growth pattern of the human head. *Am. J. Anat.* **68**, 209–62.
Brooks, D. N. (1976). Acoustic impedance. In *Scientific foundations of otolaryngology* (ed. R. Hinchliffe & D. Harrison). London: Heinemann.
Broom, R. (1895). On the homology of the palatine process of the mammalian premaxillary. *Proc. Linn. Soc. N.S.W.* **10**, 477–85.
Broom, R. (1911). On the structure of the skull in cynodont reptiles. *Proc. zool. Soc. Lond.* 1911, 893–925.
Broom, R. (1926). On the mammalian presphenoid and mesethmoid bones. *Proc. zool. Soc. Lond.* 1926, 257–64.
Broom, R. (1927). Some further points on the structure of the mammalian basicranial axis. *Proc. zool. Soc. Lond.* 1927, 233–44.
Broom, R. (1936). On the structure of the skull in the mammal-like reptiles of the suborder Therocephalia. *Phil. Trans. R. Soc.* B **226**, 1–42.
Broom, R. (1938). The origin of cynodonts. *Ann. Transv. Mus.* **19**, 279.
Buckland-Wright, J. C. (1969). Craniological observations on *Hyaena* and *Crocuta* (Mammalia). *J. Zool.* **159**, 17–29.
Bugge, J. (1974). The cephalic arterial system in insectivores, primates, rodents and lagomorphs, with special reference to the systematic classification. *Acta anat.* Suppl. **62**, 1–159.
Butler, P. M. (1948). On the evolution of the skull and teeth of the Erinaceidae with special reference to fossil material in the British Museum. *Proc. zool. Soc. Lond.* **118**, 446–500.
Butler, P. M. (1952). The milk molars of Perissodactyla, with remarks on molar occlusion. *Proc. zool. Soc. Lond.* **121**, 777–817.
Butler, P. M. (1973). Molar wear facets in Early Tertiary North American primates. In *Craniofacial biology of primates* (*Symposia of the fourth international congress of primatology*, Vol. 3) (ed. M. R. Zingeser). Basel: Karger.
Cachel, S. M. (1979). A functional analysis of the primate masticatory system and the origin of the anthropoid post-orbital system. *Am. J. phys. Anthrop.* **50**, 1–18.
Cameron, J. (1924). The prosphenion. *Am. J. phys. Anthrop.* **7**, 281–2.
Campion, G. G. (1905). Some graphic records of movements of the mandible in the living subject and their bearing on the mechanism of the joint and the construction of articulators. *Dent. Cosmos* **47**, 39–42.
Cartmill, M. (1972). Arboreal adaptations and the origin of the order Primates. In *The functional and evolutionary biology of primates* (ed. R. H. Tuttle). Chicago: University of Chicago Press.
Cartmill, M. (1974). Strepsirhine basicranial structures and the affinities of the Cheirogaleidae. In *Phylogeny of the primates* (ed. W. P. Luckett & F. S. Szalay). New York & London: Plenum Press.
Cartmill, M. (1977). *Daubentonia, Dactylopsila*, woodpeckers and klinorhynchy. In *Prosimian anatomy, biochemistry and evolution* (ed. R. D. Martin, G. A. Doyle & A. C. Walker). London: Duckworth.

Case, E. C. (1910). Description of a skeleton of *Dimetrodon incisivus* Cope. *Bull. Am. Mus. nat. Hist.* **28**, 189–96.
Cave, A. J. E. (1948). The nasal fossa in the primates. In *BMA proceedings of the annual meeting 1948*. London: Butterworth.
Cave, A. J. E. (1949). Notes on the nasal fossa of a young chimpanzee. *Proc. zool. Soc. Lond.* **119**, 61–3.
Cave, A. J. E. (1973). The primate nasal fossa. *Biol. J. Linn. Soc. Lond.* **5**, 377–87.
Cave, A. J. E. & Haines, R. W. (1940). The paranasal sinuses of the anthropoid apes. *J. Anat.* **74**, 493–523.
Celestino da Costa, A. (1920). Note sur la crête ganglionnaire cranienne chez le Cobuye. *C. r. Séanc. Soc. Biol.* **83**, 1651–7.
Chibon, P. (1967). Marquage nucléaire par la thymidine tritiée des dérivés de la crête neurale chez l'amphiban Urodèle *Pleurodeles waltlii* Michah. *J. Embryol. exp. Morphol.* **18**, 343–58.
Cleall, J. F., Wilson, G. N. & Garnett, D. S. (1968). Normal craniofacial skeletal growth of the rat. *Am. J. phys. Anthrop.* **29**, 225–42.
Crompton, A. W. (1958). The cranial morphology of a new genus and species of ictidosaurian. *Proc. zool. Soc. Lond.* **130**, 183–216.
Crompton, A. W. (1963*a*). On the lower jaw of *Diarthrognathus* and the origin of the mammalian lower jaw. *Proc. zool. Soc. Lond.* **140**, 697–753.
Crompton, A. W. (1963*b*). The evolution of the mammalian lower jaw. *Evolution, Lancaster, Pa.* **17**, 431–9.
Crompton, A. W. (1972). The evolution of the jaw articulation in cynodonts. In *Studies in vertebrate evolution* (ed. K. A. Joysey & T. S. Kemp). Edinburgh: Oliver & Boyd.
Crompton, A. W. & Hiiemae, K. M. (1970). Molar occlusion and mandibular movements during occlusion in the American opossum, *Didelphis marsupialis L. J. Linn. Soc. (Zool.)* **49**, 21–47.
Crompton, A. W. & Hotton, N. H. (1967). Functional morphology of the masticatory apparatus of two dicynodonts (Reptilia, Therapsida). *Postilla* **109**, 1–51.
Crompton, A. W., Thexton, A. J., Parker, P. & Hiiemae, K. M. (1977). The activity of the jaw and hyoid musculature in the Virginian opossum, *Didelphis virginiana*. In *The biology of marsupials* (ed. B. Stonehouse & D. Gilmore). London: Macmillan Co.
Daget, J. (1965). Le crâne des téléostéens. *Mém. Mus. natn. Hist. nat., Paris* NS(A) **31**, 163–341.
Dallos, P. (1973). *The auditory periphery biophysics and physiology*. New York & London: Academic Press.
Davis, D. D. (1955). Masticatory apparatus in the spectacled bear *Tremarctos ornatus. Fieldiana, Zool.* **37**, 25–46.
Dawes, B. (1930). The development of the vertebral column in mammals, as illustrated by its development in *Mus musculus. Phil. Trans R. Soc.* B. **218**, 115–70.
de Beer, G. R. (1922). The segmentation of the head in *Squalus acanthias. Q. J. microsc. Sci.* **66**, 457–74.
de Beer, G. R. (1924). The prootic somites of *Heterodontus* and of *Amia. Q. J. microsc. Sci.* **68**, 17–38.
de Beer, G. R. (1928). The early development of the chondrocranium of *Salmo fario. Q. J. microsc. Sci.* **71**, 259–312.
de Beer, G. R. (1937). *The development of the vertebrate skull*. London: Oxford University Press.
de Beer, G. R. (1971). *Homology, an unsolved problem*. London: Oxford University Press.

de Beer, G. R. & Barrington, E. J. W. (1934). The segmentation and chondrification of the skull of the duck. *Phil. Trans. R. Soc.* B **223**, 411–68.
de Beer, G. R. & Fell, W. A. (1937). The development of the Monotremata – part III. The development of the skull of *Ornithorhynchus*. *Trans. zool. Soc. Lond.* **23**, 1–42.
de Vree, F. & Gans, C. (1973). Masticatory responses of pygmy goats (*Capra hircus*) to different foods. *Am. Zool.* **13**, 1342–3.
de Vree, F. & Gans, C. (1975). Mastication in pygmy goats, *Capra hircus*. *Annls Soc. r. zool. Belg.* **105**, 255–306.
Dieulafe, L. (1906). Morphology and embryology of the nasal fossae of vertebrates (translated by H .W. Loeb). *Ann. Otol. Rhinol. Lar.* **15**, 1–60, 267–349, 513–84.
Dolan, K. (1971). Cranial suture closure in two species of South American monkeys. *Am. J. phys. Anthrop.* **35**, 109–18.
Dorenbos, J. (1973). Morphogenesis of the spheno-occipital and the presphenoidal synchondrosis in the cranial base of the fetal Wistar rat. *Acta morph. neerl-scand.* **11**, 63–74.
DuBrul, E. L. (1950). Posture, locomotion and the skull in Lagomorpha. *Am. J. Anat.* **87**, 277–313.
DuBrul, E. L. & Laskin, D. M. (1961). Preadaptive potentialities of the mammalian skull: an experiment in growth and form. *Am. J. Anat.* **109**, 117–32.
DuBrul, E. L. & Sicher, H. (1954). *The adaptive chin.* Springfield, Illinois: Charles C. Thomas.
Durkin, J. F. (1972). Secondary cartilage: a misnomer? *Am. J. Orthod.* **62**, 15–41.
Duterloo, H. S. & Enlow, D. H. (1970). A comparative study of cranial growth in *Homo* and *Macaca*. *Am. J. Anat.* **127**, 357–68.
Edgeworth, F. H. (1935). *The cranial muscles of vertebrates.* Cambridge: Cambridge University Press.
Edinger, T. (1950). Frontal sinus evolution (particularly in the Equidae). *Bull. Mus. comp. Zool. Harv.* **103**, 412–96.
Engel, M. B. & Brodie, A. G. (1947). Condylar growth and mandibular deformities. *Surgery, St Louis* **22**, 976–92.
Enlow, D. H. (1963). *Principles of bone remodelling.* Springfield, Illinois: Charles C. Thomas.
Enlow, D. H. (1966). A comparative study of facial growth in *Homo* and *Macaca*. *Am. J. phys. Anthrop.* **24**, 293–308.
Enlow, D. H. (1968). *The human face.* New York: Harper & Row.
Enlow, D. H. (1975). *Handbook of facial growth.* Philadelphia: W. B. Saunders Co.
Enlow, D. H. & Bang, S. (1965). Growth and remodelling of the human maxilla. *Am. J. Orthod.* **51**, 446–64.
Enlow, D. H. & Harris, D. B. (1964). A study of the postnatal growth of the human mandible. *Am. J. Orthod.* **50**, 25–50.
Erickson, L. C. & Ogilvie, A. L. (1958). Aspects of growth in the cranium, mandible and teeth of the rabbit as revealed through the use of alizarin and metallic implants. *Angle Orthod.* **28**, 47–56.
Estes, R. (1961). Cranial anatomy of the cynodont reptile *Thrinaxodon liorhinus*. *Bull. Mus. comp. Zool. Harv.* **125**, 165–80.
Evans, W. E. & Prescott, J. H. (1962). Observations of the sound production capabilities of the bottlenose porpoise: a study of whistles and clicks. *Zoologica, N.Y.* **47**, 121–8.
Ewer, R. F. (1958). Adaptive features in the skulls of the African Suidae. *Proc. zool. Soc. Lond.* **131**, 135–55.
Flower, W. H. (1869). On the value of the characters of the base of the cranium in

the classification of the order Carnivora, and on the systematic position of *Bassaris* and other disputed forms. *Proc. zool. Soc. Lond.* 1869, 4–37.
Flower, W. H. (1871). On the connexion of the hyoid arch with the cranium. *Rep. Br. Assoc. Advmt Sci.* 1871, 136–7.
Ford, E. H. R. (1955). The growth of the foetal skull. MD thesis, Cambridge University.
Ford, E. H. R. (1956). The growth of the foetal skull. *J. Anat.* **90**, 63–72.
Ford, E. H. R. (1958). Growth of the human cranial base. *Am. J. Orthod.* **44**, 498–506.
Ford, E. H. R. & Horn, G. (1959). Some problems in the evaluation of differential growth in the rat's skull. *Growth* **23**, 191–204.
Fraser, F. C. & Purves, P. E. (1960*a*). Anatomy and function of the cetacean ear. *Proc. R. Soc. B* **152**, 62–77.
Fraser, F. C. & Purves, P. E. (1960*b*). Hearing in cetaceans: evolution of the accessory air sacs and the structure and function of the outer and middle ear in recent cetaceans. *Bull. Br. Mus. nat. Hist.* **7**, 1–140.
Frazzetta, T. H. (1962). A functional consideration of cranial kinesis in lizards. *J. Morphol.* **111**, 287–320.
Frazzetta, T. H. (1966). Studies on the morphology and function of the skull in the Boidae (Serpentes). Part II. Morphology and function of the jaw apparatus in *Python sebae* and *Python molurus*. *J. Morphol.* **118**, 217–96.
Friant, M. (1960). L'évolution du cartilage de Meckel humain, jusqu'à la fin du sixième mois de la vie foetale. *Acta anat.* **41**, 228–39.
Friant, M. (1966). Vue d'ensemble sur l'évolution du 'cartilage de Meckel' de quelque groupes de mammifères. *Acta zool., Stockh.* **47**, 67–80.
Froriep, V. A. (1887). Bemerkungen zur Frage nach der Wirbeltheorie des Kopfskelettes, *Anat. Anz.* **2**, 815–35.
Fürbringer, M. (1897). Ueber die spino-occipitalen Nerven der Selachier und Holocephalen und ihre vergleichende Morphologie. In *Festschrift für Carl Gegenbaur*, Bd 3, Leipzig: Engelmann.
Gardiner, B. G. & Bartram, A. W. H. (1977). The homologies of ventral cranial fissures in osteichthyans. In *Problems in vertebrate evolution* (ed. S. M. Andrews, R. S. Miles & A. D. Walker). New York & London: Academic Press.
Gaupp, E. (1908). Zur Entwicklungsgeschichte und vergleichenden Morphologie des Schädels von *Echidna aculeata* vor. typica. *Denkschr. med.-naturw. Ges. Jena* **8**, 539–788.
Gaupp, E. (1912). Die Reichertsche Theorie. *Arch. Anat. EntwGesch.* 1912, 1–63.
Gegenbaur, C. (1878). *Elements of comparative anatomy* (translated by F. J. Bell). London: Macmillan Co.
Gegenbaur, C. (1898). *Vergleichende Anatomie der Wirbelthiere mit Beruchsichtigung der Wirbellosen*, Vol. I. Leipzig: Engelmann.
Gianelly, A. A. & Moorrees, C. F. A. (1965). Condylectomy in the rat. *Archs oral Biol.* **10**, 101–6.
Gibbs, C. H., Messerman, T., Reswick, J. B. & Derda, H. J. (1971). Functional movements of the mandible. *J. prosth. Dent.* **26**, 604–20.
Girgis, F. G. & Pritchard, J. J. (1958). Effects of skull damage on the development of sutural patterns in the rat. *J. Anat.* **92**, 39–51.
Goodrich, E. S. (1911). On the segmentation of the occipital region of the head in the Batrachia Urodela. *Proc. zool. Soc. Lond.* 1911, 101–20.
Goodrich, E. S. (1918). On the development of the segments of the head in *Scyllium*. *Q. J. microsc. Sci.* **63**, 1–30.
Goodrich, E. S. (1930). *Studies on the structure and development of vertebrates*. London: Macmillan Co.

Gordon, H. J. (1955). Human cranial base development during the late embryonic and fetal periods. MS (Orthod.) thesis, University of Illinois, Chicago. (Cited by Hoyte, 1975.)
Gould, E., Negus, N. C. & Novick, A. (1964). Evidence for echolocation in shrews. *J. exp. Zool.* **156**, 19–38.
Gray, H. (1973). *Gray's Anatomy*, 35th edition (ed. R. Warwick & P. L. Williams). Edinburgh: Longmans.
Gray, J. (1968). *Animal locomotion*. London: Weidenfeld & Nicholson.
Greaves, W. S. (1978). The jaw lever system in ungulates: a new model. *J. Zool.* **184**, 271–85.
Green, H. L. H. H. (1930). A description of the egg tooth of *Ornithorhynchus*, together with some notes on the development of the palatine processes of the premaxilla. *J. Anat.* **64**, 512–22.
Green, H. L. H. H. & Presley, R. (1978). The dumb-bell bone of *Ornithorhynchus*. *J. Anat.* **127**, 216.
Gregory, W. K. & Noble, G. K. (1924). The origin of the mammalian alisphenoid bone. *J. Morphol.* **39**, 435–63.
Griffin, D. R. (1958). *Listening in the dark*. New Haven: Yale University Press.
Gysi, A. (1921). Studies on the leverage problem of the mandible. *Dent. Dig.* **27**, 74–84, 144–50, 203–8.
Haan, F. W. R. (1957). Hearing in whales. *Acta oto-lar.* Suppl. 134, 1–114.
Haan, F. W. R. (1960). Some aspects of mammalian hearing under water. *Proc. R. Soc. B* **152**, 54–62.
Haines, R. W. (1938). The primitive form of epiphysis in the long bones of tetrapods. *J. Anat.* **72**, 323–43.
Hall, B. K. (1970). Cellular differentiation in skeletal tissues. *Biol. Rev.* **45**, 455–84.
Hall, B. K. (1975*a*). Differentiation and maintenance of articular (secondary) cartilage on avian membrane bones. *Ann. rheum. Dis.* Suppl. 34, 145.
Hall, B. K. (1975*b*). Evolutionary consequences of skeletal differentiation. *Am. Zool.* **15**, 329–50.
Hall, B. K. (1978). *Developmental and cellular skeletal biology*. New York & London: Academic Press.
Halstead, L. B. (1969). Calcified tissues in the earliest vertebrates. *Calcif. Tissue Res.* **3**, 107–24.
Harpman, J. A. & Woollard, H. H. (1938). The tendon of the lateral pterygoid. *J. Anat.* **73**, 112–15.
Harrington, R. W. (1955). The osteocranium of the American cyprinid fish, *Notropis bifrenatus*, with an annotated synonymy of teleost skull bones. *Copeia* 1955, 267–90.
Hellman, M. (1927). A preliminary study in development as it affects the human face. *Dent. Cosmos* **69**, 250–69.
Hellman, M. (1929). The face and teeth of man. A study of growth and position. *J. dent. Res.* **9**, 179–201.
Henson, O. W. (1970). The ear and audition. In *Biology of bats*, Vol. II (ed. W. A. Wimsatt). New York & London: Academic Press.
Herrick, J. C. (1899). The cranial and first spinal nerves of *Menidia*. *J. comp. Neurol.* **9**, 153–455.
Herring, S. W. & Scapino, R. P. (1973). Physiology of feeding in miniature pigs. *J. Morphol.* **141**, 427–60.
Herzberg, F. & Schour, I. (1941). The pattern of appositional growth in the incisor of the rat. *Anat. Rec.* **80**, 497–506.
Hiiemae, K. M. (1967). Masticatory function in the mammals. *J. dent. Res.* **46**, 883–93.
Hiiemae, K. (1971*a*). The structure and function of the jaw muscles in the rat (*Rattus norvegicus* L.). II. Their fibre type and composition. *J. Linn. Soc.* (*Zool.*) **50**, 101–9.

Hiiemae, K. (1971*b*). The structure and function of the jaw muscles in the rat (*Rattus norvegicus* L.). III. The mechanics of the muscles. *J. Linn. Soc.* (*Zool.*) **50**, 111–32.
Hiiemae, K. M. (1976). Masticatory movements in primitive mammals. In *Mastication* (ed. D. J. Anderson & B. Matthews). Bristol: Wright & Sons.
Hiiemae, K. M. (1978). Mammalian mastication, a review of the activity of the jaw muscles and the movements they produce in chewing. In *Development, function and evolution of teeth* (ed. P. M. Butler & K. A. Joysey). New York & London: Academic Press.
Hiiemae, K. M. & Ardran, G. M. (1968). A cinefluorographic study of mandibular movement during feeding in the rat (*Rattus norvegicus*). *J. Zool.* **154**, 139–54.
Hiiemae, K. M. & Crompton, A. W. (1971). A cinefluorographic study of feeding in the American opossum, *Didelphis marsupialis*. In *Dental morphology and evolution* (ed. A. A. Dahlberg). Chicago: University of Chicago Press.
Hiiemae, K. & Houston, W. J. B. (1971). The structure and function of the jaw muscles in the rat (*Rattus norvegicus* L.). I. Their anatomy and internal architecture. *J. Linn. Soc.* (*Zool.*) **50**, 75–99.
Hiiemae, K. M. & Jenkins, F. A. (1969). The anatomy and internal architecture of the muscles of mastication in the American opossum, *Didelphis marsupialis*. *Postilla* **140**, 1–49.
Hiiemae, K. M. & Kay, R. F. (1972). Trends in the evolution of primate mastication. *Nature, Lond.* **240**, 486–7.
Hiiemae, K. M. & Kay, R. F. (1973). Evolutionary trends in the dynamics of primate mastication. In *Craniofacial biology of primates* (*Symposia of the fourth international congress of primatology*, Vol. 3) (ed. M. R. Zingeser). Basel: Karger.
Hiiemae, K. M., Thexton, A. J. & Crompton, A. W. (1978). Intra-oral food transport: the fundamental mechanism of feeding. In *Muscle adaptation in the craniofacial region* (ed. D. Carison & T. McNamara). Ann Arbor: University of Michigan Press.
Hildyard, L. T., Moore, W. J. & Corbett, M. E. (1976). Logarithmic growth of the hominoid mandible. *Anat. Rec.* **186**, 405–12.
Hill, J. P. & Watson, K. M. (1958). The early development of the brain in marsupials. *J. Anat.* **92**, 493–7.
Hinchliffe, R. & Pye, A. (1969). Variation in the middle ear of the Mammalia. *J. Zool.* **157**, 227–88.
Hogben, L. T. (1919). The progressive reduction of the jugal in the Mammalia. *Proc. zool. Soc. Lond.* 1919, 71–8.
Holmgren, N. (1940). Studies on the head in fishes. Part I. Development of the skull in sharks and rays. *Acta zool., Stockh.* **21**, 51–267.
Hopson, J. A. (1966). The origin of the mammalian middle ear. *Am. Zool.* **6**, 437–50.
Hopson, J. A. & Crompton, A. W. (1969). Origin of mammals. *Evol. Biol.* **3**, 15–72.
Horowitz, S. L. & Shapiro, H. H. (1951). Modifications of mandibular architecture following removal of temporalis muscle in the rat. *J. dent. Res.* **30**, 276–80.
Horowitz, S. L. & Shapiro, H. H. (1955). Modification of skull and jaw architecture following removal of the masseter muscle in the rat. *Am. J. phys. Anthrop.* **13**, 301–8.
Hörstadius, S. (1950). *The neural crest: its properties and derivatives in the light of experimental research.* London: Oxford University Press.
Howes, G. B. (1896). On the mammalian hyoid, with especial reference to *Lepus, Hyrax* and *Choloepus*. *J. Anat. Physiol., Lond.* **30**, 513–26.
Hoyte, D. A. N. (1961). The postnatal growth of the ear capsule in the rabbit. *Am. J. Anat.* **108**, 1–16.
Hoyte, D. A. N. (1971). Mechanisms of growth in the cranial vault and base. *J. dent. Res.* **50**, 1447–61.
Hoyte, D. A. N. (1973*a*). Basicranial elongation: 3. Differential growth between synchondroses and basion. *Proc. 3rd Europ. Anat. Congr.* 1973, 231–2.

Hoyte, D. A. N. (1973*b*). Basicranial elongation: 2. Is there differential growth within a synchondrosis? *Anat. Rec.* **175**, 347.
Hoyte, D. A. N. (1975). A critical analysis of the growth in length of the cranial base. *Birth Defects* **11**, 255–82.
Hoyte, D. A. N. (1978). The nasal capsule in the rabbit: an expansile force in facial growth. *J. Anat.* **126**, 428.
Humphreys, H. F. (1921). Function in the evolution of man's dentition. *Br. dent. J.* **42**, 939–57.
Humphreys, H. F. (1932). Age changes in the temporomandibular joint and their importance in orthodontics. *Int. J. Orthod. Dent. Child.* **18**, 809–15.
Hunt, R. M. (1974). The auditory bulla in Carnivora: an anatomical basis for reappraisal of carnivore evolution. *J. Morphol.* **143**, 21–76.
Hunter, J. (1771). *Natural history of the human teeth.* London: John Johnson.
Huxley, J. S. (1932). *Problems of relative growth.* London: Methuen.
Huxley, T. H. (1859). On the theory of the vertebrate skull. *Proc. R. Soc.* **9**, 381–457.
Huxley, T. H. (1867). On two widely contrasted forms of the human cranium. *J. Anat. Physiol., Lond.* **1**, 60–77.
Huxley, T. H. (1874). On the structure of the skull and heart of *Menobranchus lateralis*. *Proc. zool. Soc. Lond.* 1874, 186–204.
Hylander, W. L. (1975). The human mandible: lever or link? *Am. J. phys. Anthrop.* **43**, 227–42.
Hylander, W. L. (1977). *In vivo* bone strain in the mandible of *Galago crassicaudatus*. *Am. J. phys. Anthrop.* **46**, 309–26.
Ingervall, B. & Thilander, B. (1972). The human spheno-occipital synchondrosis. *Acta odont. scand.* **30**, 349–56.
Irving, J. T. & Rönning, O. V. (1962). The selective action of papain on calcification sites. *Archs oral Biol.* **7**, 357–63.
Irwin, G. L. (1960). Roentgen determination of the time of closure of the spheno-occipital synchondrosis. *Radiology* **75**, 450–3.
Jarabek, J. R. & Thompson, J. R. (1953). Growth of the mandible of the rat following bilateral resection of the mandibular condyles. *Am. J. Orthod.* **39**, 58.
Jarvik, E. (1954). On the visceral skeleton in *Eusthenopteron*, with a discussion of the palato-quadrate in fishes. *K. svenska VetenskAkad. Handl.* **5**, 1–14.
Jarvik, E. (1959). Dermal fin-rays and Holmgren's principle of delamination. *K. svenska VetenskAkad. Handl.* **6**, 3–49.
Jefferies, R. P. S. (1968). The subphylum *Calcichordata* (Jefferies 1967) primitive fossil chordates with echinoderm affinities. *Bull. Br. Mus. nat. Hist.* **16**, 243–339.
Jenkins, F. A. & Parrington, F. R. (1976). The postcranial skeletons of the triassic mammals *Eozostrodon*, *Megazostrodon* and *Erythrotherium*. *Phil. Trans. R. Soc. B* **273**, 387–431.
Johnson, C. E. (1913). The development of the prootic head somites and eye muscles in *Chelydra serpentina*. *Am. J. Anat.* **14**, 119–86.
Johnson, D. R. (1967). Extra-toes: a new mutant gene causing multiple abnormalities in the mouse. *J. Embryol. exp. Morphol.* **17**, 543–81.
Johnson, P. A., Atkinson, P. J. & Moore, W. J. (1976). The development and structure of the chimpanzee mandible. *J. Anat.* **122**, 467–77.
Johnston, M. C. (1966). A radioautographic study of the migration and fate of cranial neural crest cells in the chick embryo. *Anat. Rec.* **156**, 143–56.
Jollie, M. (1971). Some developmental aspects of the head skeleton of the 35–37 mm *Squalus acanthias* foetus. *J. Morphol.* **133**, 17–40.
Jollie, M. (1975). Development of head skeleton and pectoral girdle in *Esox*. *J. Morphol.* **147**, 61–88.

Jolly, M. (1961). Condylectomy in the rat. An investigation of the ensuing repair processes in the region of the temporomandibular articulation. *Aust. dent. J.* **6**, 243–56.

Jones, F. W. (1938). The so-called maxillary antrum of the gorilla. *J. Anat.* **73**, 116–19.

Kallen, F. G. & Gans, C. (1972). Mastication in the little brown bat, *Moytis lucifugus*. *J. Morphol.* **136**, 385–420.

Kanagasuntheram, R. (1967). A note on the development of the tubotympanic recess in the human embryo. *J. Anat.* **101**, 731–41.

Kay, R. F. & Hiiemae, K. M. (1974). Jaw movement and tooth use in recent and fossil primates. *Am. J. phys. Anthrop.* **40**, 227–56.

Kellogg, R. (1928). The history of whales – their adaptation to life in the water. *Q. Rev. Biol.* **3**, 29–76, 174–208.

Kellogg, W. N. (1961). *Porpoises and sonar*. Chicago: University of Chicago Press.

Kellogg, W. N., Kohler, R. & Morris, H. N. (1953). Porpoise sounds as sonar signals. *Science, N.Y.* **117**, 239–43.

Kemp, T. S. (1972*a*). Whaitsiid Therocephalia and the origin of cynodonts. *Phil. Trans. R. Soc. B* **264**, 1–54.

Kemp, T. S. (1972*b*). The jaw articulation and musculature of the whaitsiid Therocephalia. In *Studies in vertebrate evolution* (ed. K. A. Joysey & T. S. Kemp). Edinburgh: Oliver & Boyd.

Kemp, T. S. (1979). The primitive cynodont *Procynosuchus*: functional anatomy of the skull and relationships. *Phil. Trans. R. Soc. B* **285**, 73–122.

Kermack, D. M., Kermack, K. A. & Mussett, F. (1968). The Welsh pantothere *Kuehneotherium praecursoris*. *J. Linn. Soc.* (*Zool.*) **47**, 407–23.

Kermack, K. A. (1963). The cranial structure of the triconodonts. *Phil. Trans. R. Soc. B* **246**, 83–103.

Kermack, K. A. & Kielan-Jaworowska, Z. (1971). Therian and non-therian mammals. In *Early mammals* (ed. D. M. Kermack & K. A. Kermack). New York & London: Academic Press.

Kermack, K. A. & Mussett, F. (1958). The jaw articulation of the Docodonta and the classification of Mesozoic mammals. *Proc. R. Soc. B* **149**, 204–15.

Kermack, K. A., Mussett, F. & Rigney, H. W. (1973). Lower jaw of *Morganucodon*. *J. Linn. Soc.* (*Zool.*) **53**, 87–175.

Kernan, J. D. (1915). The development of the occipital region of the domestic cat with an interpretation of the paracondyloid process. *Anat. Rec.* **10**, 213–15.

Kingsbury, B. F. (1926). Branchiomerism and the theory of head segmentation. *J. Morphol.* **42**, 83–109.

Klatsky, H. (1939). Cinephotography and cinefluorography of the masticatory apparatus in function. *Am. J. Orthod.* **25**, 205–10.

Klatsky, H. (1940). A cinefluorographic study of the human masticatory apparatus in function. *Am. J. Orthod.* **26**, 664–70.

Koski, K. & Rönning, O. (1965). Growth potential of transplanted components of the mandibular ramus of the rat III. *Suom. HammaslääkSeur. Toim* **61**, 292–7.

Koski, K. & Rönning, O. (1969). Growth potential of subcutaneously transplanted cranial base synchondroses of the rat. *Acta odont. scand.* **27**, 343–57.

Koski, K. & Rönning, O. (1970). Growth potential of intracerebrally transplanted cranial base synchondroses in the rat. *Archs oral Biol.* **15**, 1107–8.

Krebs, B. (1971). Evolution of the mandible and lower dentition in dryolestids. In *Early mammals* (ed. D. M. Kermack & K. A. Kermack). New York & London: Academic Press.

Krogman, W. M. (1930*a*). Studies in growth changes in the skull and face of anthropoids. I. The eruption of the teeth in anthropoids and Old World monkeys. *Am. J. Anat.* **46**, 303–13.

Krogman, W. M. (1930*b*). Studies in growth changes in the skull and face of anthropoids. II. Ectocranial and endocranial suture closure in anthropoids and Old World apes. *Am. J. Anat.* **46**, 315–53.

Krogman, W. M. (1931*a*). Studies in growth changes in the skull and face of anthropoids. III. Growth changes in the skull and face of the gorilla. *Am. J. Anat.* **47**, 89–115.

Krogman, W. M. (1931*b*). Studies in growth changes in the skull and face of anthropoids. IV. Growth changes in the skull and face of the chimpanzee. *Am. J. Anat.* **47**, 325–42.

Krogman, W. M. (1931*c*). Studies in growth changes in the skull and face of anthropoids. V. Growth changes in the skull and face of the orang-utan. *Am. J. Anat.* **47**, 343–65.

Landacre, F. L. (1914). Embryonic cerebral ganglia and the doctrine of nerve components. *Folia neuro-biol.* **8**, 601–15.

Latham, R. A. (1966). Observations on the growth of the cranial base in the human skull. *J. Anat.* **100**, 435.

Latham, R. A. (1968). A new concept of the early maxillary growth mechanism. *Eur. orthod. Soc.* **44**, 53–63.

Latham, R. A. (1970). Maxillary development and growth: the septo-premaxillary ligament. *J. Anat.* **107**, 471–8.

Latham, R. A. (1972). The sella point and postnatal growth of the human cranial base. *Am. J. Orthod.* **61**, 156–62.

Lavelle, C. L. B. & Moore, W. J. (1973). The incidence of agenesis and polygenesis in the primate dentition. *Am. J. phys. Anthrop.* **38**, 671–80.

Lawrence, B. & Schevill, W. E. (1956). The functional anatomy of the delphinid nose. *Bull. Mus. comp. Zool. Harv.* **114**, 103–51.

Lay, D. M. (1972). The anatomy, physiology, functional significance and evolution of specialised hearing organs of gerbilline rodents. *J. Morphol.* **138**, 41–120.

Le Douarin, N. (1973). A biological cell labelling technique and its use in experimental embryology. *Devl Biol.* **30**, 217–22.

Le Douarin, N. & Barq, G. (1969). Sur l'utilisation des cellules de la caille japonaise comme 'marqueurs biologiques' en embryologie expérimental. *C. r. hebd. Séanc. Acad. Sci., Paris* Sér. D **269**, 1543–6.

Le Gros Clark, W. E. (1934). *Early forerunners of man.* London: Baillière, Tindall & Cox.

Le Gros Clark, W. E. (1955). *The fossil evidence for human evolution. An introduction to the study of palaeo-anthropology.* Chicago: University of Chicago Press.

Le Gros Clark, W. E. (1959). *The antecedents of man: an introduction to the evolution of the primates.* Edinburgh: Edinburgh University Press.

Le Lièvre, C. (1974). Rôle des cellules mésectodermiques issués des crêtes neurales céphaliques dans la formation des arcs branchiaux et du squelette viscéral. *J. Embryol. exp. Morphol.* **31**, 453–77.

Le Lièvre, C. S. & Le Douarin, N. M. (1975). Mesenchymal derivatives of the neural crest: analysis of chimaeric quail and chick embryos. *J. Embryol. exp. Morphol.* **34**, 125–54.

Lewis, A. B. & Roche, A. F. (1972). Elongation of the cranial base in girls during pubescence. *Angle Orthod.* **42**, 358–67.

Lillie, D. G. (1910). Observations on the anatomy and general biology of some members of the larger Cetacea. *Proc. zool. Soc. Lond.* 1910, 769–92.

Lindblom, G. (1960). On the anatomy and function of the temporomandibular joint. *Acta odont. scand.* **17**, Suppl. 28, 1–460.

Lubosch, W. (1906). Über Variationen am Tuberculum articulare des Kiefergelenks des Menschen und ihre morphologische Bedeutung. *Gegenbaurs morph, Jb.* **35**, 322–53.

Luschei, E. S. & Goodwin, G. M. (1974). Patterns of mandibular movement and jaw muscle activity during mastication in the monkey. *J. Neurophysiol.* **37**, 954–66.

MacConaill, M. A. (1950). The movements of bones and joints. III. The synovial fluid and its assistants. *J. Bone Jt Surg.* **32*B***, 244–52.

McKenna, M. C. (1966). Paleontology and the origin of primates. *Folia Primatol.* **4**, 1–25.

MacMillan, H. W. (1930). Unilateral vs. bilateral balanced occlusion. *J. Am. dent. Assoc.* **17**, 1207–21.

McNamara, J. A. & Graber, L. W. (1975). Mandibular growth in the rhesus monkey (*Macaca mulatta*). *Am. J. phys. Anthrop.* **42**, 15–24.

McNamara, J. A., Riolo, M. L. & Enlow, D. H. (1976). Growth of the maxillary complex in the rhesus monkey (*Macaca mulatta*). *Am. J. phys. Anthrop.* **44**, 15–26.

MacPhee, R. D. E. (1977). Ontogeny of the ectotympanic–petrosal plate relationship in strepsirhine prosimians. *Folia Primatol.* **27**, 245–83.

Marshall, A. M. (1881). On the head cavities and associated nerves of elasmobranchs. *Q. J. microsc. Sci.* **21**, 72–97.

Meikle, M. C. (1973). *In vivo* transplantation of the mandibular joint of the rat; an autoradiographic investigation of cellular changes at the condyle. *Archs oral Biol.* **18**, 1011–20.

Michejda, M. (1971). Ontogenic changes of the cranial base in *Macaca mulatta. Proc. 3rd Int. Congr. Primatol., Zurich* 1, 215–25.

Michejda, M. (1972). The role of basicranial synchondroses in flexure processes and ontogenic development of the skull base. *Am. J. phys. Anthrop.* **37**, 143–50.

Miller, G. S. (1923). The telescoping of the cetacean skull. *Smithson. misc. Collns* **76**(5), 1–70.

Mills, J. R. E. (1955). Ideal dental occlusion in the primates. *Dent. Practnr, Bristol* **6**, 47–61.

Mills, J. R. E. (1963). Occlusion and malocclusion in the teeth of primates. In *Dental anthropology* (ed. D. R. Brothwell). Oxford: Pergamon Press.

Mills, J. R. E. (1978). The relationship between tooth patterns and jaw movements in the Hominoidea. In *Development, function and evolution of teeth* (ed. P. M. Butler & K. A. Joysey). New York & London: Academic Press.

Moffett, B. C., Johnson, L. C., McCabe, J. B. & Askew, H. C. (1964). Articular remodeling in the adult human temporomandibular joint. *Am. J. Anat.* **115**, 119–42.

Møhl, B. (1968). Hearing in seals. In *The behaviour and physiology of pinnipeds* (ed. R. J. Harrison, R. C. Hubbard, R. S. Peterson, C. E. Rice & R. J. Schusterman). New York: Appleton-Century-Crofts.

Møller, E. (1966). The chewing apparatus: An electromyograph study of the action of the muscles of mastication and its correlation to facial morphology. *Acta physiol. scand.* **69**, Suppl. 280, 1–229.

Moore, W. J. (1966). Skull growth in the albino rat (*Rattus norvegicus*). *J. Zool.* **149**, 137–44.

Moore, W. J. (1973). An experimental study of the functional components of growth in the rat mandible. *Acta anat.* **85**, 378–85.

Moore, W. J., Adams, L. M. & Lavelle, C. L. B. (1973). Head posture in the Hominoidea. *J. Zool.* **169**, 409–16.

Moore, W. J. & Lavelle, C. L. B. (1975). *Growth of the facial skeleton in the Hominoidea.* New York & London: Academic Press.

Moore, W. J. & Spence, T. F. (1969). Age changes in the cranial base of the rabbit (*Oryctolagus cuniculus*). *Anat. Rec.* **165**, 355–62.

Moss, M. L. (1958). Fusion of the frontal suture in the rat. *Am. J. Anat.* **102**, 141–66.
Moss, M. L. (1959). Functional anatomy of the temporomandibular joint. *N.Y. Univ. J. Dent.* **29**, 315–19.
Moss, M. L. (1960). Functional analysis of human mandibular growth. *J. prosth. Dent.* **10**, 1149–59.
Moss, M. L. (1968). The origin of vertebrate calcified tissue. In *Current problems of lower vertebrate phylogeny* (ed. T. Ørvig). Stockholm: Almqvist & Wiksell.
Moss, M. L. (1969). Comparative histology of dermal sclerifications in reptiles. *Acta anat.* **73**, 510–33.
Moss, M. L., Bromberg, B. E., Song, I. C. & Eisenman, G. (1968). The passive role of nasal septal cartilage in mid-facial growth. *Plastic reconstr. Surg.* **41**, 536–42.
Moss, M. L. & Greenberg, S. N. (1967). Functional cranial analysis of the human maxillary bone. I. Basal bone. *Angle Orthod.* **37**, 151–64.
Moss, M. L. & Meehan, M. A. (1970). Functional cranial analysis of the coronoid process in the rat. *Acta anat.* **77**, 11–24.
Moss, M. L. & Rankow, R. M. (1968). The role of the functional matrix in mandibular growth. *Angle Orthod.* **38**, 95–103.
Moss, M. L. & Salentijn, L. (1969). The primary role of functional matrices in facial growth. *Am. J. Orthod.* **55**, 566–77.
Moss, M. L. & Salentijn, L. (1970). The logarithmic growth of the human mandible. *Acta anat.* **77**, 341–60.
Moss, M. L. & Simon, M. R. (1968). Growth of the human mandibular angle process: a functional cranial analysis. *Am. J. phys. Anthrop.* **28**, 127–38.
Moss, M. L. & Young, R. W. (1960). A functional approach to craniology. *Am. J. phys. Anthrop.* **18**, 281–92.
Moy-Thomas, J. A. & Miles, R. S. (1971). *Palaeozoic fishes*, 2nd edition. London: Chapman & Hall.
Moyers, R. E. (1950). An electromyographic analysis of certain muscles involved in temporomandibular movement. *Am. J. Orthod.* **36**, 481–515.
Murray, P. D. F. (1963). Adventitious (secondary) cartilage in the chick embryo and the development of certain bones and articulations in the chick skull. *Aust. J. Zool.* **11**, 368–430.
Negus, V. (1958). *The comparative anatomy and physiology of the nose and paranasal sinuses.* Edinburgh & London: Livingstone.
New, D. A. T. (1966). *The culture of vertebrate embryos.* London: Logos Press.
Noden, D. M. (1973). The migratory behaviour of neural crest cells. In *Fourth symposium on oral sensation and perception. Development in the fetus and infant* (ed. J. F. Bosma). Bethesda: US Department of Health, Education and Welfare.
Norris, H. W. (1925). Observations upon the peripheral distribution of the cranial nerves of certain ganoid fishes (*Amia lepidosteus*, *Polyodon*, *Scaphirhynchus* and *Acipenser*). *J. comp. Neurol.* **39**, 345–432.
Norris, K. S. (1964). Some problems of echolocation in cetaceans. In *Marine bio-acoustics* (ed. W. N. Tavolga). Oxford: Pergamon Press.
Norris, K. S. (1969). The echolocation of marine mammals. In *The biology of marine mammals* (ed. H. T. Andersen). New York & London: Academic Press.
Norris, K. S., Prescott, J. H., Asa-Dorian, P. V. & Perkins, P. (1961). An experimental demonstration of echolocation behaviour in the porpoise, *Tursiops truncatus* (Montagu). *Biol. Bull. mar. biol. Lab., Woods Hole* **120**, 163–76.
Novacek, M. J. (1977). Aspects of the problem of variation, origin and evolution of the eutherian auditory bulla. *Mammal Rev.* **7**, 131–49.
Novick, A. & Griffin, D. R. (1961). Laryngeal mechanism in bats for the production of orientation sounds. *J. exp. Zool.* **148**, 125–46.

Olson, E. C. (1944). Origin of mammals based upon cranial morphology of the therapsid suborders. *Bull. geol. Soc. Am. Spec. Pap.* **55**, 1–136.
Olson, E. C. (1959). The evolution of mammalian characters. *Evolution, Lancaster, Pa.* **13**, 344–53.
Olson, E. C. (1971). *Vertebrate paleozoology.* New York: Wiley-Interscience.
Osborn, H. F. (1936). *Proboscidea: a monograph of the discovery, evolution, migration and extinction of the mastodonts and elephants of the world,* Vols. I and II. New York: American Museum of Natural History.
Owen, M. (1970). The origin of bone cells. *Int. Rev. Cytol.* **28**, 213–38.
Owen, R. (1866). *On the anatomy of vertebrates,* Vols. 1, 2 and 3. London: Longmans Green & Co.
Panchen, A. L. (1972). The interrelationships of the earliest tetrapods. In *Studies in vertebrate evolution* (ed. K. A. Joysey & T. S. Kemp). Edinburgh: Oliver & Boyd.
Parker, W. K. (1874). On the structure and development of the skull in the pig (*Sus scrofa*). *Phil. Trans. R. Soc.* **164**, 289–336.
Parker, W. K. (1885*a*). On the structure and development of the skull in the Mammalia – Part II Edentata. *Phil. Trans. R. Soc.* **176**, 1–120.
Parker, W. K. (1885*b*). On the structure and development of the skull in the Mammalia – Part III Insectivores. *Phil. Trans. R. Soc.* **176**, 121–276.
Parrington, F. R. (1934). On the cynodont genus *Galesaurus*, with a note on the functional significance of the changes in the evolution of the theriodont skull. *Ann. Mag. nat. Hist.* **13**, 38–67.
Parrington, F. R. (1946). On the cranial anatomy of cynodonts. *Proc. zool. Soc. Lond.* **116**, 181 97.
Parrington, F. R. (1955). On the cranial anatomy of some gorgonopsids and the synapsid middle ear. *Proc. zool. Soc. Lond.* **125**, 1–40.
Parrington, F. R. (1967). The origins of mammals. *Advmt Sci., Lond.* **24**, 165–73.
Parrington, F. R. (1971). On the upper triassic mammals. *Phil. Trans. R. Soc. B* **261**, 231–72.
Parrington, F. R. (1974). The problem of the origin of the monotremes. *J. nat. Hist.* **8**, 421–6.
Parrington, F. R. (1978). A further account of the Triassic mammals. *Phil. Trans. R. Soc. B* **282**, 177–204.
Parrington, F. R. & Westoll, T. S. (1940). On the evolution of the mammalian palate. *Phil. Trans. R. Soc. B* **230**, 305–55.
Patterson, C. (1975). The braincase of pholidophorid and leptolepid fishes, with a review of the actinopterygian braincase. *Phil. Trans. R. Soc. B* **269**, 275–579.
Patterson, C. (1977). Cartilage bones, dermal bones and membrane bones, or the exoskeleton versus the endoskeleton. In *Problems in vertebrate evolution* (ed. S. M. Andrews, R. S. Miles & A. D. Walker). New York & London: Academic Press.
Paulli, S. (1900*a*). Über die Pneumaticität des Schädels bei den Säugethieren. I. Über den Bau des Siebbeins. Über die Morphologie des Siebbeins und die der Pneumaticität bei den Monotremen und den Marsupialiern. *Morphol. Jb.* **28**, 147–78.
Paulli, S. (1900*b*). Über die Pneumaticität des Schädels bei den Säugethieren. II. Über die Morphologie des Siebbeins und die Pneumaticität bei den Ungulaten und Probosciden. *Morphol. Jb.* **28**, 179–251.
Paulli, S. (1900*c*). Über die Pneumaticität des Schädels bei den Säugethieren. III. Über die Morphologie des Siebbeins und Pneumaticität bei den Insectivoren, Hyracoideen, Chiropteren, Carnivoren, Pinnipedien, Edentaten, Rodentiern, Prosimien und Primaten. *Morphol. Jb.* **28**, 483–564.

Pearson, A. A. (1938). The spinal accessory nerve in human embryos. *J. comp. Neurol.* **68**, 243–66.

Pehrson, T. (1922). Some points on the development of teleostomian fishes. *Acta zool., Stockh.* **3**, 1–63.

Petrovic, A. G. (1972). Mechanisms and regulation of mandibular condylar growth. In *Mechanisms and regulation of craniofacial morphogenesis* (ed. B. C. Moffett). Amsterdam: Swets & Zeitlinger.

Petrovic, A., Charlier, J. P. & Herrmann, J. (1968). Les mécanismes de croissance du crâne. Recherches sur le cartilage de la cloison nasale et sur les sutures craniennes et faciales de jeune rats en culture d'organe. *Bull. Assoc. Anat., Paris* **143**, 1376–82.

Platt, J. B. (1893). Ectodermic origin of the cartilages of the head. *Anat. Anz.* **8**, 506–9.

Pocock, R. I. (1928). The structure of the auditory bulla in the Procyonidae and the Ursidae, with a note on the bulla of *Hyaena. Proc. zool. Soc. Lond.* 1928, 963–74.

Powell, T. V. & Brodie, A. G. (1963). Closure of the spheno-occipital synchondrosis. *Anat. Rec.* **147**, 15–23.

Presley, R. (1978). Ontogeny of some elements of the auditory bulla in mammals. *J. Anat.* **126**, 428.

Presley, R. (1979). The primitive course of the internal carotid artery in mammals. *Acta anat.* **103**, 238–44.

Presley, R. & Steel, F. L. D. (1976). On the homology of the alisphenoid. *J. Anat.* **121**, 441–59.

Presley, R. & Steel, F. L. D. (1978). The pterygoid and ectopterygoid in mammals. *Anat. Embryol.* **154**, 95–110.

Pritchard, J. J., Scott, J. H. & Girgis, F. G. (1956). The structure and development of cranial and facial sutures. *J. Anat.* **90**, 73–86.

Pye, A. (1966*a*). The structure of the cochlea in Chiroptera I. Microchiroptera: Emballonuroidea and Rhinolophoidea. *J. Morphol.* **118**, 495–510.

Pye, A. (1966*b*). The Megachiroptera and Vespertilionoidea of the Microchiroptera. *J. Morphol.* **119**, 101–20.

Pye, A. (1967). The structure of the cochlea in Chiroptera III. Microchiroptera: Phyllostomatoidea. *J. Morphol.* **121**, 241–54.

Pye, A. (1970). The structure of the cochlea in Chiroptera. A selection of Microchiroptera from Africa. *J. Zool.* **162**, 335–43.

Pye, A. (1977). The structure of the cochlea in some myomorph and caviomorph rodents. *J. Zool.* **182**, 309–21.

Pye, A. (1979). The structure of the cochlea in some mammals. *J. Zool.* **187**, 39–53.

Pye, J. D. (1968). Hearing in bats. In *Hearing mechanisms in vertebrates* (ed. A. V. S. de Reuck & J. Knight). London: Churchill.

Ralls, K. (1967). Auditory sensitivity in mice. *Peromyscus* and *Mus musculus. Anim. Behav.* **15**, 123–8.

Ramprashad, F., Corey, S. & Ronald, K. (1972). Anatomy of the seal's ear (*Pagophilus groenlandicus*). In *Functional anatomy of marine mammals*, Vol. I (ed. R. J. Harrison). New York & London: Academic Press.

Rees, L. A. (1954). The structure and function of the mandibular joint. *Br. dent. J.* **96**, 125–33.

Reeve, E. C. R. (1940). Relative growth in the snout of anteaters. *Proc. zool. Soc. Lond. Ser. A* **110**, 47–80.

Reinbach, W. (1952). Zur Entwicklung des Primordial-craniums von Dasypus novemcinctus Linné. *Z. Morphol. Anthrop.* **44**, 375–444; **45**, 1–72.

Repenning, C. A. (1972). Underwater hearing in seals: functional morphology. In *Functional anatomy of marine mammals*, Vol. I (ed. R. J. Harrison). New York & London: Academic Press.

Riesner, S. E. (1938). Temporomandibular reactions to occlusal anomalies. *J. Am. dent. Assoc.* **25**, 1938–53.

Robinson, J. T. (1954). Nuchal crests in the australopithecines. *Nature, Lond.* **174**, 1197–8.

Robinson, J. T. (1958). Cranial cresting patterns and their significance in the Hominoidea. *Am. J. phys. Anthrop.* **16**, 397–428.

Rogers, W. M. (1958). The influence of asymmetry of the muscles of mastication upon the bones of the face. *Anat. Rec.* **131**, 617–32.

Romer, A. S. (1937). The braincase of the Carboniferous crossopterygian *Megalichthys nitidus. Bull. Mus. comp. Zool., Harv.* **82**, 3–73.

Romer, A. S. (1941). Notes on the crossopterygian hyomandibular and braincase. *J. Morphol.* **69**, 141–60.

Romer, A. S. (1956). *Osteology of the reptiles.* Chicago: University of Chicago Press.

Romer, A. S. (1962). *The vertebrate body*, 3rd edition. Philadelphia: Saunders Co.

Romer, A. S. (1969). Cynodont reptile with incipient mammalian jaw articulation. *Science* **166**, 881–2.

Romer, A. S. & Parsons, T. S. (1977). *The vertebrate body*, 5th edition. Philadelphia: Saunders Co.

Romer, A. S. & Price, L. I. (1940). Review of the Pelycosauria. *Bull. geol. Soc. Am. Spec. Pap.* **28**, 1–538.

Rönning, O. (1966). Observations on the intracerebral transplantation of the mandibular condyle. *Acta odont. scand.* **24**, 443–57.

Roux, G. H. (1947). The cranial development of certain Ethiopian 'insectivores' and its bearing on the mutual affinities of the group. *Acta zool., Stockh.* **28**, 165 307.

Rushton, M. A. (1944). Growth at the mandibular condyle in relation to some deformities. *Br. dent. J.* **76**, 57–68.

Sarnat, B. G. (1957). Facial and neurocranial growth after removal of the mandibular condyle in the macaca rhesus monkey. *Am. J. Surg.* **94**, 19–30.

Sarnat, B. G. (1963). Postnatal growth of the upper face: some experimental considerations. *Angle Orthod.* **33**, 139–61.

Sarnat, B. G. (1968). Growth of bones as revealed by implant markers in animals. *Am. J. Phys. Anthrop.* **29**, 255–86.

Sarnat, B. G. (1971). Surgical experimentation and gross postnatal growth of the face and jaws. *J. dent. Res.* **50**, 1462–76.

Sarnat, B. G. & Engel, M. B. (1951). A serial study of mandibular growth after removal of the condyle in the macaca rhesus monkey. *Plastic reconstr. Surg.* **7**, 364–80.

Sarnat, B. G. & Muchnic, H. (1971). Facial skeletal changes after mandibular condylectomy in the adult monkey. *J. Anat.* **108**, 323–38.

Sarnat, B. G. & Shanedling, P. D. (1970). Orbital volume following evisceration, enucleation, and exenteration in rabbits. *Am. J. Ophthalmol.* **70**, 787–99.

Sarnat, B. G. & Wexler, M. R. (1967*a*). The snout after resection of nasal septum in adult rabbits. *Archs Otolar.* **86**, 463–6.

Sarnat, B. G. & Wexler, M. R. (1967*b*). Rabbit snout growth after resection of central linear segments of nasal septal cartilage. *Acta oto-lar.* **63**, 467–78.

Sawin, P. B., Ranlett, M. & Crary, D. D. (1959). Morphogenetic studies of the rabbit: XXV. The spheno-occipital synchondrosis of the dachs (chondrodystrophy) rabbit. *Am. J. Anat.* **105**, 257–80.

Scapino, R. C. (1972). Adaptive radiation of mammalian jaws. In *Morphology of the maxillo-mandibular apparatus* (ed. G. H. Schumacher). Leipzig: Georg Thieme.

Scapino, R. C. (1976). The jaw symphysis in cats and bears. *Am. J. phys. Anthrop.* **44**, 204.

Schulte, H. von W. & Tilney, F. (1915). Development of the neuraxis in the domestic cat to the stage of twenty-one somites. *Ann. N.Y. Acad. Sci.* **24**, 319–46.
Schultz, A. H. (1926). Fetal growth of man and other primates. *Q. Rev. Biol.* **1**, 465–521.
Schultz, A. H. (1962). Metric age changes and sex differences in primate skulls. *Z. Morphol. Anthrop.* **52**, 239–55.
Schultz, A. H. (1969). *The life of primates.* London: Weidenfeld & Nicolson.
Schumacher, G. H. (1961). *Funktionelle Morphologie der Kaumuskulatur.* Jena: Verlag.
Schumacher, G. H. (1968). Der maxillo-mandibuläre Apparat unter dem Einflussformgestaltender Factoren. *Nova Acta Leopoldina* **33**, 1–186.
Schumacher, G. H. & Dokladal, M. (1968). Über unterschiedliche Sekundärveränderungen am Schädel als Folge von Kaumuskelresektionen. *Acta anat.* **69**, 378–92.
Schweitzer, J. M. (1951). *Oral rehabilitation.* London: Kimpton.
Scott, J. H. (1951). The comparative anatomy of jaw growth and tooth eruption. *Dent. Rec.* **71**, 149–67.
Scott, J. H. (1953). The cartilage of the nasal septum (a contribution to the study of facial growth). *Br. dent. J.* **95**, 37–43.
Scott, J. H. (1956). Growth at facial sutures. *Am. J. Orthod.* **42**, 381–7.
Scott, J. H. (1958). The cranial base. *Am. J. phys. Anthrop.* **16**, 319–48.
Servoss, J. M. (1973). An *in vivo* and *in vitro* autoradiographic investigation of growth in synchondrosal cartilage. *Am. J. Anat.* **136**, 479–85.
Seydel, O. (1891). Über die Nasenhöhle der höheren Säugethiere und des Menschen. *Morphol. Jb.* **17**, 44–99.
Shute, C. C. D. (1956). The evolution of the mammalian eardrum and tympanic cavity. *J. Anat.* **90**, 261–81.
Shute, C. D. D. (1972). The composition of vertebrae and the occipital region of the skull. In *Studies in vertebrate evolution* (ed. K. A. Joysey & T. S. Kemp). Edinburgh: Oliver & Boyd.
Sicher, H. (1944). Masticatory apparatus in the giant panda and bears. *Fld Mus. Nat. Hist., (Zool. Series)* **29**, 61–73.
Silbermann, M. & Frommer, J. (1972*a*). The nature of endochondral ossification in the mandibular condyle of the mouse. *Anat. Rec.* **172**, 659–67.
Silbermann, M. & Frommer, J. (1972*b*). Vitality of chondrocytes in the mandibular condyle as revealed by collagen formation. An autoradiographic study with ^{3}H-proline. *Am. J. Anat.* **135**, 359–69.
Simpson, G. G. (1959). Mesozoic mammals and the polyphyletic origin of mammals. *Evolution, Lancaster, Pa.* **13**, 405–14.
Simpson, G. G. (1961). Evolution of Mesozoic mammals. In *International colloquium on the evolution of lower and nonspecialised mammals*, Part I (ed. G. Vandenbroek). Brussels: Paleis der Academiën.
Simpson, G. G. (1971). Mesozoic mammals revisited. In *Early mammals* (ed. D. M. Kermack & K. A. Kermack). New York & London: Academic Press.
Singh, S., Sanyal, A. K. & Kar, A. K. (1974). The effect of cyclophosphamide on the morphogenesis of the cerebellum in chick embryos. *Anat. Rec.* **178**, 127–37.
Slijper, E. J. (1962). *Whales.* London: Hutchinson.
Smith, J. M. & Savage, R. J. G. (1959). The mechanics of mammalian jaws. *Sch. Sci. Rev.* **40**, 289–301.
Smith, R. J. (1978). Mandibular biomechanics and temporomandibular joint function in primates. *Am. J. phys. Anthrop.* **49**, 341–50.
Sprague, J. M. (1943). The hyoid region of placental mammals with especial reference to the bats. *Am. J. Anat.* **72**, 385–472.
Sprague, J. M. (1944). The hyoid region in the Insectivora. *Am. J. Anat.* **74**, 175–216.

Starck, D. (1967). Le crâne des mammifères. In *Traité de Zoologie*, Vol. XVI (ed. Pierre-P. Grassé). Paris: Masson et Cie.
Starck, D. (1974). The development of the chondrocranium in primates. In *Phylogeny of the primates* (ed. W. P. Luckett & F. S. Szalay). New York & London: Plenum Press.
Starks, E. C. (1926). Bones of the ethmoid region of the fish skull. *Stanford Univ. Publs, Biological Sciences* **4**, 139–338.
Stenström, S. J. & Thilander, B. L. (1970). Effects of nasal septal cartilage resections on young guinea-pigs. *Plastic reconstr. Surg.* **45**, 160–70.
Stenström, S. J. & Thilander, B. L. (1972). Healing of surgically created defects in the septal cartilages of young guinea-pigs. *Plastic reconstr. Surg.* **49**, 194–9.
Straus, W. L. & Howell, A. B. (1936). The spinal accessory nerve and its musculature. *Q. Rev. Biol.* **11**, 387–405.
Strong, O. S. (1895). The cranial nerves of amphibia. *J. Morphol.* **10**, 101–230.
Symons, N. B. B. (1952). The development of the human mandibular joint. *J. Anat.* **86**, 326–32.
Szalay, F. S. (1972). Cranial morphology of the Early Tertiary *Phenacolemur* and its bearing on primate phylogeny. *Am. J. phys. Anthrop.* **36**, 59–76.
Szalay, F. S. (1974). Phylogeny of primate higher taxa: the basicranial evidence. In *Phylogeny of the primates* (ed. W. P. Luckett & F. S. Szalay). New York & London: Plenum Press.
Tattersall, I. (1977). Facial structure and mandibular mechanics in *Archaeolemur*. In *Prosimian anatomy, biochemistry and evolution* (ed. R. D. Martin, G. A. Doyle & A. C. Walker). London: Duckworth.
Tavolga, W. N. (1964). *Review of marine bio-acoustics. State of the art.* New York: American Museum of Natural History.
Taylor, W. O. G. (1939). Effect of enucleation of one eye in childhood upon the subsequent development of the face. *Trans. ophthal. Soc. UK* **59**, 361–71.
Terry, R. J. (1917). The primordial cranium of the cat. *J. Morphol.* **29**, 281–434.
Thompson, D'Arcy W. (1942). *On growth and form. A new edition.* Cambridge: Cambridge University Press.
Thompson, J. R. (1946). The rest position of the mandible and its significance to dental science. *J. Am. dent. Assoc.* **33**, 151–80.
Thomson, K. S. (1965). The endocranium and associated structures in the Middle Devonian rhipidistian fish *Osteolepis*. *Proc. Linn. Soc. Lond.* **176**, 181–95.
Thomson, K. S. (1967). Mechanisms of intracranial kinetics in fossil rhipidistian fishes (Crossopterygii) and their relatives. *J. Linn. Soc.* (*Zool.*) **46**, 223–53.
Todd, T. W. & Cooke, E. W. (1934). The later developmental features in the skull of *Sus barbatus barbatus*. *Proc. zool. Soc. Lond.* 1934, 685–96.
Todd, T. W. & Schweiter, F. P. (1933). The later stages of developmental growth in the hyena skull. *Am. J. Anat.* **52**, 81–123.
Todd, T. W. & Wharton, R. E. (1934). Later postnatal skull growth in the sheep. *Am. J. Anat.* **55**, 79–95.
Toldt, C. von (1905). Der Winkelforsatz des Unterkiefers beim Menschen und bei den Säugetieren und die Beziehungen der Kaumuskeln zu demselben. II Teil. *Sber. Akad. Wiss. Wien* **114**, 315–476.
Tonndorf, J. (1966). Bone conduction studies in experimental animals. *Acta oto-lar.* Suppl. 213, 1–132.
Tonndorf, J. & Khanna, S. M. (1976). Mechanics of the auditory system. In *Scientific foundations of otolaryngology* (ed. R. Hinchliffe & D. Harrison). London: Heinemann.
Tumarkin, A. (1968). Evolution of the auditory conducting apparatus in terrestrial mammals. In *Hearing mechanisms in vertebrates* (ed. A. V. S. De Reuck & J. Knight). London: Churchill.

Turnbull, W. D. (1970). Mammalian masticatory apparatus. *Fieldiana, Geol.* **18**, 153–356.
van der Klaauw, C. J. (1929). On the development of the tympanic region of the skull in the Macroscelididae. *Proc. zool. Soc. Lond.* **37**, 491–560.
van der Klaauw, C. J. (1931). The auditory bulla in some fossil mammals. *Bull. Am. Mus. nat. Hist.* **52**, 1–352.
van Kampen, P. N. (1905). Die Tympanalgegend des Säugetierschädels. *Morphol. Jb.* **34**, 321–722.
van Valen, L. (1965). Tree shrews, primates and fossils. *Evolution, Lancaster, Pa.* **19**, 137–51.
van Wijhe, J. W. (1882). Ueber das Visceralskelett und die Nerven des Kopfes des Ganoiden and von *Ceratodus. Niederländisches Archiv für Zoologie* **5**, 207–320.
Vilmann, H. (1971). The growth of the cranial base in the Wistar albino rat studied by vital staining with alizarin red S. *Acta odont. scand.* **29**, Suppl. 59, 1–44.
Washburn, S. L. (1947). The relation of the temporal muscle to the form of the skull. *Anat. Rec.* **99**, 239–48.
Washburn, S. L. & Detwiler, S. R. (1943). An experiment bearing on the problems of physical anthropology. *Am. J. phys. Anthrop.* **1**, 171–90.
Watanabe, M., Laskin, D. M. & Brodie, A. G. (1957). The effect of autotransplantation on growth of the zygomatico-maxillary suture. *Am. J. Anat.* **100**, 319–36.
Watson, D. M. S. (1914). The Deinocephalia, an order of mammal-like reptiles. *Proc. zool. Soc. Lond.* 1914, 749–86.
Watson, D. M. S. (1921). The basis of classification of the Theriodontia. *Proc. zool. Soc. Lond.* 1921, 35–98.
Watson, D. M. S. (1948). *Dicynodon* and its allies. *Proc. zool. Soc. Lond.* **118**, 823–77.
Watson, D. M. S. (1951). *Palaeontology and modern biology*. New Haven: Yale University Press.
Watson, D. M. S. (1953). The evolution of the middle ear. *Evolution, Lancaster, Pa.* **7**, 159–77.
Webster, D. B. (1975). Auditory systems of the Heteromyidae: postnatal development of the ear of *Dipodomys merriami. J. Morphol.* **146**, 377–94.
Webster, D. B. & Strother, W. F. (1972). Middle ear morphology and auditory sensitivity of heteromyid rodents. *Am. Zool.* **12**, 727.
Webster, D. B. & Webster, M. (1975). Auditory systems of Heteromyidae: functional morphology and evolution of the middle ear. *J. Morphol.* **146**, 343–76.
Webster, D. B. & Webster, M. (1978). Auditory systems of Heteromyidae: cochlear diversity. *J. Morphol.* **152**, 153–70.
Weidenreich, F. (1941). The brain and its role in the phylogenetic transformation of the human skull. *Trans. Am. phil. Soc.* **31**, 321–442.
Weijs, W. A. (1973). Morphology of the muscles of mastication in the albino rat *Rattus norvegicus* (Berkenhout, 1769). *Acta morph. neerl.-scand.* **11**, 321–40.
Weijs, W. A. & Dantuma, R. (1975). Electromyography and mechanics of mastication in the albino rat. *J. Morphol.* **146**, 1–34.
Weinmann, J. P. & Sicher, H. (1955). *Bone and bones*, 2nd edition. London: Kimpton.
Westoll, T. S. (1936). On the structures of the dermal ethmoid shield of *Osteolepis. Geol. Mag.* **73**, 157–71.
Westoll, T. S. (1937). On a specimen of *Eusthenopteron* from the old red sandstone of Scotland. *Geol. Mag.* **74**, 507–24.
Westoll, T. S. (1943*a*). The hyomandibular of *Eusthenopteron* and the tetrapod middle ear. *Proc. R. Soc. B* **131**, 393–414.
Westoll, T. S. (1943*b*). The origin of tetrapods. *Biol. Rev.* **18**, 78–98.
Westoll, T. S. (1944). New light on the mammalian ossicles. *Nature, Lond,* **154**, 770–1.

Westoll, T. S. (1945). The mammalian middle ear. *Nature, Lond.* **155**, 114–15.
Weston, J. A. (1970). The migration and differentiation of neural crest cells. *Adv. Morphogen.* **8**, 41–114.
Wever, E. G. & Lawrence, M. (1954). *Physiological acoustics*. Princeton: Princeton University Press.
Wever, E. G. & Vernon, J. A. (1961). The protective mechanisms of the bat's ear. *Ann. Otol. Rhinol. Lar.* **70**, 1–17.
Wexler, M. R. & Sarnat, B. G. (1961). Rabbit snout growth: effect of injury to septovomeral region. *Archs Otolar.* **74**, 305–13.
Wilson, G. H. (1920). The anatomy and physics of the temporomandibular joint. *J. natn. dent. Assoc.* **7**, 414–20.
Wilson, G. H. (1921). The anatomy and physics of the temporomandibular joint. *J. natn. dent. Assoc.* **8**, 236–41.
Wilson, J. T. (1906). On the fate of the 'taenia clino-orbitalis' (Gaupp) in *Echidna* and in *Ornithorhynchus* respectively; with demonstration of specimens and stereo-photographs. *J. Anat. Physiol., Lond.* **40**, 85–90.
Wilson Charles, S. (1925). The temporo-mandibular joint and its influence on the growth of the mandible. *Br. dent. J.* **46**, 845–55.
Wood, A. E. (1962). The early tertiary rodents of the family Paramyidae. *Trans. Am. phil. Soc.* NS **52**, 1–261.
Wood, A. E. (1965). Grades and clades among rodents. *Evolution, Lancaster, Pa.* **19**, 115–30.
Wood, B. A. (1976). The nature and basis of sexual dimorphism in the primate skeleton. *J. Zool.* **180**, 15–34.
Young, J. Z. (1950). *The life of vertebrates*. Oxford: Clarendon Press.
Young, R. W. (1959). The influence of cranial contents on postnatal growth of the skull in the rat. *Am. J. Anat.* **105**, 383–415.
Zoller, R. M. & Laskin, D. M. (1969). Growth of the zygomaticomaxillary suture in pigs after sectioning the zygomatic arch. *J. dent. Res.* **48**, 573–8.
Zuckerman, S. (1926). Growth-changes in the skull of the baboon, *Papio porcarius*. *Proc. zool. Soc. Lond.* 1926, 843–73.
Zuckerman, S. (1954). Correlation of change in the evolution of higher primates. In *Evolution as a process* (ed. J. S. Huxley, A. C. Hardy & E. B. Ford). London: Allen & Unwin.
Zuckerman, S. (1955). Age changes in the basicranial axis of the human skull. *Am. J. phys. Anthrop.* **13**, 521–39.
Zwislocki, J. (1965). Analysis of some auditory characteristics. In *Handbook of mathematical psychology*, Vol. III (ed. R. Luce, R. Bush & E. Galanter). New York: Wiley & Sons.

[illegible] the middle ear. Nature, Lond. [illegible] 114–15.

[illegible] and differentiation of [illegible] cells. Adv. Morphogen. [illegible]

[illegible] University Press.

[illegible] mechanisms of the [illegible] Ann. [illegible]

[illegible] growth: effect of injury to [illegible]

Watson, [illegible] of the temporomandibular joint [illegible]

Watson, [illegible] of the temporomandibular joint [illegible]

[illegible] with [illegible] stereo [illegible]

[illegible] mandibular joint and its influence on the [illegible] 24–55.

[illegible] of the [illegible] Proc. [illegible]

[illegible] Pt. 13.

[illegible] sexual [illegible] in the primate [illegible]

[illegible] (1950) [illegible]

[illegible] (1959). The influence of [illegible] on postnatal growth of the [illegible]

[illegible] Growth [illegible] after [illegible]

Zuckerman, [illegible] changes in the skull of [illegible] baboon, Papio porcarius. Proc. zool. Soc. [illegible]

Zuckerman, S. (1955) [illegible] of change in the [illegible] of the higher primates. In [illegible] C. Hardy [illegible] Ford. London [illegible]

[illegible] of the human skull. Am. J. phys. Anthrop. [illegible]

[illegible] (196[illegible]) [illegible] In Handbook of [illegible] New York: Wiley & Sons.

INDEX

Italicised page numbers refer to illustrations